Crossing the Boundaries of Life

CONVENING SCIENCE
Discovery at the Marine Biological Laboratory

Crossing the Boundaries of Life

Günter Blobel and the Origins
of Molecular Cell Biology

KARL S. MATLIN

The University of Chicago Press
Chicago and London

The University of Chicago Press, Chicago 60637

The University of Chicago Press, Ltd., London

Published 2022

Printed in the United States of America

31 30 29 28 27 26 25 24 23 22   1 2 3 4 5

ISBN-13: 978-0-226-81923-5 (cloth)

ISBN-13: 978-0-226-81934-1 (paper)

ISBN-13: 978-0-226-81935-8 (e-book)

DOI: https://doi.org/10.7208/chicago/9780226819358.001.0001

Publication of this book is supported in part by the Edwin S. Webster Foundation.

Library of Congress Cataloging-in-Publication Data

Names: Matlin, Karl S., author.
Title: Crossing the boundaries of life : Günter Blobel and the origins
    of molecular cell biology / Karl S. Matlin.
Other titles: Convening science.
Description: Chicago : University of Chicago Press, 2022. |
    Series: Convening science: discovery at the Marine Biological
    Laboratory | Includes bibliographical references and index.
Identifiers: LCCN 2021043711 | ISBN 9780226819235 (cloth) |
    ISBN 9780226819341 (paperback) | ISBN 9780226819358 (ebook)
Subjects: LCSH: Blobel, Günter. | Molecular biology—History. |
    Cytology—History.
Classification: LCC QH506 .M388 2022 | DDC 572/.33—dc23
LC record available at https://lccn.loc.gov/2021043711

Dedicated
To all those who look at cells in microscopes and wonder how they work
and
To Holly, Gabe, and Nick

Life means that the form is retained even though matter is being transformed.

THOMAS MANN, *The Magic Mountain*

Many will prefer to start with a unified whole, develop the parts from it, and then retrace the parts directly to the whole. The nature of the organism provides us with the best reason to do this: the most perfect organism appears before us as a unified whole, discrete from all other beings.

JOHANN WOLFGANG VON GOETHE, *Observation on Morphology in General*

CONTENTS

This book originated many years ago when I was a graduate student in cell biology at the Rockefeller University in New York City. At the time, in the 1970s, my mentor's laboratory was located across the corridor from Günter Blobel's laboratory. Because we shared many pieces of equipment and because there were some overlapping scientific interests, I got to know all the members of the Blobel lab very well, including Blobel himself. Indeed, Blobel, a gregarious man, introduced himself to me soon after I arrived at Rockefeller.

Blobel's main interest was proving what became known as the signal hypothesis. For reasons that are now hard to understand, I was very interested in the work of the Blobel lab, and I closely followed all the experimental machinations that occurred there in the mid-1970s. When Blobel and Bernhard Dobberstein published their classic papers in 1975, I realized that their accomplishments were significant scientific achievements.

Dobberstein, who had become a friend, soon left Rockefeller for a position at the European Molecular Biology Laboratory in Heidelberg, Germany. Through him I learned that a lab that I was considering for postdoctoral training was at the same institution, and about two years later I followed him there. My interest in the signal hypothesis continued, and, while sharing a lab and an office with Dobberstein, I took in the experimental back-and-forth between David Meyer, a postdoc in Dobberstein's group, and Peter Walter, Blobel's graduate student, as they discovered the signal recognition particle and its receptor.

After I established my own laboratory back in the United States and began to focus on my own work, I continued to pay attention to the efforts of Blobel and others to establish the molecular mechanisms of protein targeting and translocation across the membrane of the endoplasmic reticulum.

Eventually I realized that I was in the unique position of knowing a great deal about the experimental (and personal) details underpinning this work while, at the same time, having a degree of distance from it because I had not worked directly with Blobel or anyone else on the topic. I began to contemplate writing a book about the history of the signal hypothesis and, in 1998, started to conduct interviews. In 1999 Blobel won the Nobel Prize, confirming to me that the studies that I had followed for many years were important and motivating me to continue my project. Progress was, however, slow because I had never attempted anything like this before.

Prior to a trip to Europe in 2000, I contacted Hans-Jörg Rheinberger at the Max Planck Institute for the History of Science in Berlin. Rheinberger had written about the history of protein synthesis research and, as both a scientist and a historian, I thought that he might be able to tell me if the topic of the signal hypothesis was of interest. When I met with him, I presented the project as a narrative history, emphasizing my inside knowledge of the many "good stories" connected with the discoveries. He confirmed to me that the project was worthwhile but urged me to go beyond the historical narrative to make it something of more general intellectual interest.

Over the next few years, I continued to collect interviews and other resources. My work on the topic became known within the community of cell biologists, and I began to be invited to write short historical articles on the signal hypothesis for scientific journals. In 2007 I moved to the University of Chicago and got involved with the Committee on Conceptual and Historical Studies of Science through the invitation of William Wimsatt, a philosopher of biology. At the same time, I began to reconsider more broadly my motives for writing the book. Although the importance of the signal hypothesis and the stories connected to the discoveries remained a central motivation, I realized that what was really driving my work was a desire to understand why I believed that cell biology provides more effective explanations of biological phenomena than a gene-centric version of reductionist molecular biology. This book is an attempt to answer that question.

As I tell the story of cell biology and the signal hypothesis, some might question my objectivity because of my close relationship to the Blobel laboratory and the early studies. Blobel was an intense competitor and was not shy about strongly defending his opinions when challenged by other scientists. He considered the inner circle of his laboratory his family, and even though I was never a true member of the family, I was certainly an adopted child. Although this gives me the advantage of knowing details about discoveries that cannot be easily recovered from publications or even interviews, it also creates the possibility that I may at times overemphasize particular points of view. Being sensitive to this possible bias, I have tried to balance the inescapable fact that I am both the historian and a historical artifact by seeking outside perspectives to validate, refute, or (most often) provide nuanced interpretations of my inside knowledge. I hope I have been successful.

The goal of trying to understand how cell biology explains biological phenomena required me to apply ideas from the philosophy of biology. Over the past thirty to forty years, philosophers of biology have realized that many biologists are more interested in discovering biological mechanisms than articulating biological theories. Reductionism, a concept classically considered in physics in terms of theory reduction, has been revised for biology to apply to levels of biological organization. Under these circumstances the practice of scientific research is paramount and strategies of investigation that deal with part-whole relationships are key to achieving mechanistic molecular understanding of biological processes. Among the leaders in this effort are Wimsatt and William Bechtel, and I try to appropriately use their work in the last part of the book to generalize what I refer to as the epistemic strategy of cell biology. I also find that Rheinberger's concept of *epistemic things* helps provide insights into the recursive nature of research in cell biology that begins with a whole cell and eventually progresses to molecular understanding. My use of the philosophical literature is by no means comprehensive or exhaustive, and I apologize to those scholars who I have not mentioned even though their work is also applicable to the problems that I discuss.

This book is not the first to deal with the history of post–World War II cell biology and will, I hope, not be the last. Bechtel's excellent *Discovering Cell Mechanisms: The Creation of Modern Cell Biology* (2006) and Nicolas Rasmussen's equally excellent *Picture Control* (1997) on the history of biological electron microscopy are just two examples. Works by Rheinberger, Carol Moberg, Andrew Reynolds, Hannah Landecker, and Mathias Grote, as well as my recent coedited volume, also deserve mention (Rheinberger 1997; Landecker 2007; Moberg 2012; Matlin, Laubichler, and Maienschein 2018; Reynolds 2018b; Grote 2019). Despite these examples, the history of cell biology has not gotten the attention that it deserves. Other contributions, such as Hanna Worliczek's recent dissertation, "Molecularizing Microscopic Imaging: A History of Immunofluorescence Microscopy as a Visual Epistemic Tool of Modern Cell Biology (1959–1980)" (Worliczek 2020),[1] on the rather unexpected development of immunofluorescence microscopy at the Cold

---

1. The dissertation is in German, titled *Wege zu einer molekularisierten Bildgebung: Eine Geschichte der Immunofluoreszencmikroskopie als visuelles Erkenntnisinstrument der modernen Zellbiologie 1959–1980*.

Spring Harbor Laboratory, an institution not known for a focus on cell biology, should find their way to publication before too long.

I have many people to thank for their direct or indirect help with my project. Early on, my friend the philosopher Harold Kincaid was a sounding board for my dissatisfaction with molecular biology and early attempts to understand why cell biology is different, with our discussions contributing to his publication, "Molecular Biology and the Unity of Science" (Kincaid 1990). Judith Yaross Lee and Joe Slade helped me understand how to go about writing a book. Bill Wimsatt invited me into the community of scholars interested in the history and philosophy of biology at the University of Chicago and has continued to be both a mentor and an unlimited source of ideas. Also at Chicago, Robert Richards, through his own work, provided me with examples of scholarship in the history of science at the highest level of excellence, and his advice is always most helpful. At Wimsatt's urging, I attended a seminar on the history of cell biology at the Marine Biological Laboratory (MBL) in Woods Hole, Massachusetts, where I met the historian of biology Jane Maienschein for the first time. Later, a chance encounter with Jane on the bridge in Woods Hole led to a close and productive collaboration that continues to this day. Her encouragement, mentorship, and friendship have enabled me to complete this project. Others who encouraged me throughout the long development of this book include Holly Harte, who listened to chapter after chapter read aloud; Hanna Worliczek, who was not afraid to challenge my interpretations or to open my eyes to things that I missed; and Keith Mostov, who was always enthusiastic about the project. Bernhard Dobberstein, Maienschein, Worliczek, Wimsatt, Graham Warren, Nica Borgese, and Alan Engelberg also read and commented on the manuscript. Finally, I am most grateful to Karen Merikangas Darling at the University of Chicago Press for her support and help in facilitating publication.

Some of the material in the book is modified from my earlier publications in *Visions of Cell Biology* (Matlin 2018) and *Nature Reviews Molecular Cell Biology* (Matlin 2002, 2011). I am grateful to the publishers for allowing me to reuse parts of these works.

Support for research and production of the manuscript came from many sources. Grants from the National Library of Medicine and from the Edwin S. Webster Foundation (with the assistance of Suzanne Sears) provided critical financial resources. Other grants for the MBL History Project to Jane Maienschein from the National Science Foundation, the James S. McDonnell Foundation, and Arizona State University enabled me to interact with other historians and philosophers of biology at the MBL in Woods Hole and elsewhere. The librarians at the University of Chicago and the MBL–Woods Hole

Oceanographic Institution (WHOI) Library provided me with the books and journal articles I needed to follow more than one hundred years of discoveries in cytology and cell biology. The archivists in the History of Medicine Division at the National Library of Medicine gave me access to the George Palade papers. Hanna Worliczek translated German passages from books and journals and shared her research on the history of immunofluorescence microscopy. Michelle LaBonte brought the Bernard Davis papers at Harvard Medical School to my attention. Anna Guerrero created the diagram of the epistemic cycle. Irina Livezeanu translated a letter by George Palade from Romanian into English. To all these individuals and institutions, I give my sincere thanks.

Finally, I am grateful to all who shared their memories and perspectives on the events described in this book. They include Mark Adelman, Qais Al-Awqati, David Anderson, Jon Beckwith, Carl Blobel, Günter Blobel, Nica Borgese, Bernhard Dobberstein, Nancy Dwyer, Ann Erikson, Eric Fries, Larry Gerace, Reid Gilmore, Bernie Gilula, Ben Glick, Barbara Goldman, Tim Harrison, Katharine Howell, Flora Katz, Gert Kreibich, Vishu Lingappa, Harvey Lodish, Maria Luisa Maccecchini, David Meyer, Cesar Milstein, Keith Mostov, Chris Nicchitta, George Palade, Carl Potter, Tom Rapoport, James Rothman, David Sabatini, Geoff Schatz, Randy Schekman, Dennis Shields, Pam Silver, Sandy Simon, Kai Simons, Ted Steck, Don Steiner, Alan Tartakoff, Peter Walter, Graham Warren, Colin Watts, and Bill Wickner. I am in their debt.

A Very Small Difference . . .

On Christmas Day 1974 Günter Blobel, with characteristic energy and enthusiasm, rushed into his laboratory to see the results of the experiment he had been planning for years. His experiment was designed to find out how the cell directs certain protein molecules during their synthesis into a pathway leading out of the cell, a process called secretion. Oddly enough, even though he was asking a question about how cells work, Blobel did not include any cells at all in his experiment. Instead, he combined factors needed to synthesize proteins with tiny membrane fragments called microsomes that he had purified from secretory cells and incubated them together.

The part of his experiment involving the synthesis of proteins without cells, known as protein synthesis *in vitro* (literally, "in glass"),[1] was not that novel, having been successfully tried decades earlier (Siekevitz 1952; Rheinberger 1997, 150). What *was* new was his addition of microsomes to his mixtures. Microsomes were not necessary for protein synthesis, but Blobel added them because they are tiny pieces of the cell, a kind of cell surrogate in the form of closed membrane vesicles, that enabled him to ask questions about cell functions in the absence of living cells. The results of the experiment were recorded on a sheet of X-ray film that Blobel developed that day in the darkroom. When he held the film to the light, he saw a series of short, horizontal bands arrayed in vertical columns or lanes, like irregular steps on a ladder (fig. 1). The bands corresponded to proteins that Blobel had

---

1. Here and elsewhere in the book, "in vitro" refers to experiments conducted with *cell-free* preparations, including mixtures of enzymes and cell fragments isolated by differential centrifugation. "In vivo," on the other hand, refers to experiments conducted with living cells, tissues, or even whole multicellular organisms. In practice, the terms "in vitro" (systems) and "cell-free" (systems) are frequently used synonymously.

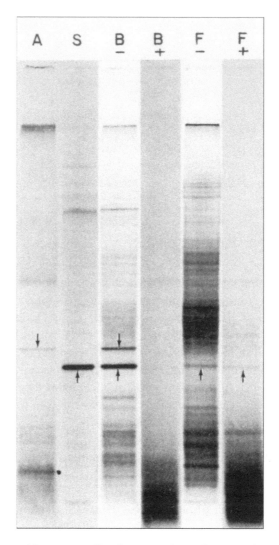

1. A figure prepared from an X-ray film showing results similar to those obtained by Günter Blobel in December 1974. The vertical columns labeled A, S, etc., are the *lanes*, and the horizontal lines in each lane (some darker than others) are the *bands* corresponding to individual proteins. The downward-pointing arrows in this figure highlight higher molecular weight potential precursors of the lower molecular weight proteins (indicated by the upward-pointing arrows). © 1975 Blobel, G., and Dobberstein, B. Originally published as figure 7 in *Journal of Cell Biology* 67 (3): 835–51.

synthesized in his experiment, and their relative positions in the vertical lanes reflected the varying molecular weights of the proteins they represented. Blobel immediately noticed that some prominent protein bands were slightly smaller than those in adjacent lanes and became very excited. The differences in size were minor, but they meant everything.

Blobel's experiment and those that followed in his laboratory and the laboratories of other scientists over the next twenty years were described as proof of the *signal hypothesis*, a proposal to explain the first step in secretion: how secretory proteins that are made in the cell's cytoplasm first find and then cross a specific membrane barrier to enter the pathway leading out of the cell. Proteins are the workhorses of cells, playing essential roles both in cell structure and as catalysts that accelerate the chemical reactions that make life possible. Proteins are large molecules, each consisting of a chain of sometimes hundreds of different amino acids linked head-to-tail into long polymers, with the amino acid composition, the amino acid sequence, and the size of the polymer varying according to the specific function of the protein. Cells such as those found in the human body make more than twenty thousand different proteins, and some of these, the secretory proteins that Blobel studied, are designed to function outside of the cell. Before proteins can leave the cell, however, they must cross the cell boundary, a barrier that separates the inside of the cell from the outside world. This barrier is a biological membrane: a thin, continuous, and almost impermeable film composed of specialized membrane proteins and fattylike substances called lipids.

The signal hypothesis states that each protein destined for secretion is synthesized with a short extension to its amino acid chain that acts as a signal to direct the protein to the location within the cell where the secretory process begins. Once there, this *signal sequence* is cut off of the secretory protein as it crosses the membrane barrier. Indeed, removal of such a signal resulted in the slightly smaller band seen by Blobel in his experiment. Blobel first proposed the idea of the signal hypothesis with his colleague David Sabatini in 1971 but only named it the signal hypothesis in key papers published in 1975, a year after his first successful experiment (Blobel and Sabatini 1971b; Blobel and Dobberstein 1975a). When Blobel won the Nobel Prize in 1999 for formulating the signal hypothesis and discovering most of the molecular machinery that secretory proteins use to find and cross the membrane barrier, newspapers described the "signals" of the signal hypothesis as "zip codes" because they are part of the protein address that enables cells to direct the proteins to their correct cellular destinations (Altman 1999).

At the time of his experiment, Blobel did not imagine that his results and the developments that followed would begin to resolve fundamental questions about the relationship of structure and function going back to the very beginning of biology. If both living and lifeless forms are composed of the same materials, how are those materials arranged to produce life? Furthermore, if attempts to investigate life render living forms lifeless and upset the life-giving arrangement of materials, how can we possibly answer the first question? Put in the context of cells, the smallest of living organisms, how can the functions of cells dependent on cell structure be explained mechanistically by the properties and chemical reactions of the cell's molecular components?

The desire to explain organisms according to mechanical principles goes back to at least René Descartes and Isaac Newton in the seventeenth century (Bechtel 2006, 2011, 2012). Although this goal was accepted by the budding field of biology in the late eighteenth and early nineteenth century, it was also perceived to be problematic because living things are purposeful; assigning functions to organisms implied an intent that was viewed as incompatible with purely mechanical explanations. Functions are embedded in the structure or *form* of the organism. Given this, it was unclear how mechanical processes could create the biological forms required for functions when the forms themselves housed the mechanical processes. The philosopher Immanuel Kant famously expressed this pessimistic view when he stated, in so many words, that there would never be a Newton of a blade of grass (Kant 1987; Godfrey-Smith 2014, 60).[2] Kant believed that natural phenomena could only be properly (i.e., scientifically) explained by mechanical principles. The dilemma was that the origins of purposeful organization in animate nature could not, in his view, be explained mechanically but that reference to this organization, this form (as a regulative principle or heuristic), was required to make sense of mechanical explanations (Richards 2000). As biology developed in the nineteenth century, this dilemma never really went away. The concept that living organisms have a purpose (referred to as teleology) became linked to vitalism, the idea that life in all its complexity is governed by principles of a distinctively holistic and purposive character and cannot be explained mechanically from analysis of its parts (Bechtel 2012). Vitalism came to have a very negative connotation among biologists as more and more about the chemistry of life began to come to light. Nevertheless, the difficulty of reconciling chemical mechanisms with the functions of whole liv-

---

2. Kant stated that "it is quite certain that in terms of merely mechanical principles of nature we cannot even adequately become familiar with, much less explain, organized beings and how they are internally possible. So certain is this that we may boldly state that it is absurd for human beings even to attempt it, or to hope that some day another Newton might arise who would explain to us, in terms of natural laws unordered by any intention, how even a mere blade of grass is produced" (1987, 282–83).

2. Edmund Beecher Wilson. Image reproduced from the website "History of the Marine Biological Laboratory" (http://hpsrepository.asu.edu/).

ing systems remained a problem. Even as late as 1925, the American biologist Edmund Beecher (E. B.) Wilson (fig. 2) remarked in the third edition of his classic text, *The Cell in Development and Heredity*:

Whether structure or function is the primary determining factor in vital phenomena is a question that has been the subject of debate for many generations of biological philosophers. As thus stated, however, the question has proven barren, for all students of the problem have, in the end had to admit that structure and function are inseparable. It is certain that vital action is not known to us apart from an organized material basis, and equally certain that vital structures exist only as products of protoplasmic activity. Thus has arisen a dilemma which belongs to the fundamental philosophy of biology and may be set aside here as practically insoluble. The fact of importance to the cytologist is that we cannot hope to comprehend the activities of the living cell by analysis merely of its chemical composition, or even of its molecular structure alone. . . . Modern investigation has . . . brought ever-increasing recognition to the fact that the cell is an organic system, and one in which we must recognize the existence of some kind of ordered structure or organization. (1925, 670)

As explained later in some detail, biologists had difficulty resolving Wilson's dilemma because investigation of the mechanistic basis of function appeared to require elimination of the cell's structure, a step that they feared might prevent meaningful insight into cell chemistry. The work of Blobel and his predecessors at Rockefeller suggested a way around this dilemma by using structure as a guide to stepwise disassembly of the cell, ultimately enabling explanation of function at the molecular level. This allowed biology to overcome the barrier that had blocked progress and arrive at a point where mechanistic explanations of life appear within reach.

It is commonly believed that the most important scientific discipline to emerge in the twentieth century is molecular biology. But is that, in fact, the case? According to a somewhat standard narrative, molecular biology arose in the 1940s and 1950s when a group of physicists and physically oriented biologists began to determine high-resolution structures of biological macromolecules, including that of DNA, and to link these, initially, to genetic phenomena (Kendrew 1967; Judson 1979; Kitcher 1984; Morange 2020). What developed from this was a scientific perspective anchored not only by "the idea that biological systems are most fruitfully investigated at the lowest possible level, and that experimental studies should be aimed at uncovering molecular and biochemical causes," something known to philosophers of biology as *methodological reductionism*, but also by the belief that the examination of molecules *alone* is sufficient to explain biological phenomena (Wimsatt 2006; Brigandt and Love 2017). By 1965, James Watson, codiscoverer of DNA structure and the embodiment of this brand of molecular biology, was confident enough of its promise to state in his textbook *Molecular Biology of the Gene*:

We see not only that the laws of chemistry are sufficient for understanding protein structure, but also that they are completely consistent with all known hereditary phenomena. *Complete certainty* now exists among essentially all biochemists that the other characteristics of living organisms (for example, selective permeability across cell membranes, muscle contraction, nerve conduction, and the hearing and memory processes) will all be *completely understood* in terms of the coordinate interactions of small and large molecules. Much is already known about the less complex features, enough to give us confidence that further research of the intensity recently given to genetics will eventually provide man with the ability to *describe with completeness* the essential features that constitute life. (1965, 67; emphasis added)

Watson's view echoed molecular biology's molecule-based investigative strategy. As stated by Alexander Powell and John Dupré, "Molecular biology did encourage a biological *Weltanschauung* [worldview] that accords priority to DNA as a causal agent within the cell, and that encourages the belief that a detailed understanding of individual molecular properties may be sufficient to account fully for cellular and organismic phenomena. . . . We think that there are a number of reasons why this position of molecular determinism was reached, but one very significant factor is the success achieved in relating the first crystallographic structures of biological macromolecules to their biochemical properties and physiological functions" (Powell and Dupré 2009, 56). Over time, this strategy became more implicit than applied, and molecular biology broadened, incorporating an array of approaches that were not necessarily consistent with the idea that studying molecules alone is sufficient. Nevertheless, at that point any biologist seeking molecular explanations of biological phenomena became, by definition, a molecular biologist, and the principle of molecular (or, ultimately, DNA) causality was rarely challenged.

Buttressed by revolutionary advances in molecular technologies, this state of affairs began to change at the end of the twentieth century when the sequencing of the human genome and analysis of the sequences revealed the enormous molecular complexity of living organisms. At that point it became clear that an approach different from one focused only on biological molecules was needed to penetrate this complexity, something more holistic and systems based (Wimsatt 2007, 309).

But what if such a strategy already existed, a strategy developed and successfully applied at the same time that Watson's reductionist molecular biology was devised and promoted? This book is about that strategy and, more specifically, cell biology, the *other* molecular biology of the twentieth century. Cell biology, like the molecular biology of Watson, is focused on explanations of cell functions at the molecular level. Crucially, however, cell biology does not consider molecules in isolation, separated from their cellular context, but includes the cellular context as an intrinsic part of its investigative strategy. This approach has its roots in the historical origins of cell biology. Cell biology is related to cytology, the field that developed in the nineteenth century after the theory that cells are the component parts of living organisms was firmly established.

Cytology shared with biology as a whole a desire to learn what separates the living from the nonliving and sought mechanistic explanations at the cellular level once it became a more experimental science in the nineteenth century. Early in the twentieth century, however, work in cytology reached a technical impasse and progress stalled, leading to the view that its approach was antiquated (Bechtel 2006; Matlin, Maienschein, and Laubichler 2018). By the time that the technical barriers were overcome after World War II, the field of genetics was thriving, launched in part by cytology's contributions to the chromosomal theory of inheritance and the rediscovery of Gregor Mendel's experiments on peas. When the structure of DNA was discovered in 1953, genetics morphed into molecular genetics and then into molecular biology. Cell biology, an interdisciplinary field that combined cytology with biochemistry and biophysics, became well established in the 1950s and 1960s. However, molecular biology was already firmly entrenched along with its gene-centric reductionist approach to the explanation of biological phenomena. Under these circumstances, cell biology was relegated to a supporting role in the modern history of biology or, in some cases, completely left out (Allen 1975; Judson 1979; Morange 2020). Molecular biology, defined as the science that sought to explain biology by tracing the pathway from gene to protein to function, was considered more fundamental. Contrary to this traditional history, I argue in this book that cell biology may be the most significant molecular biology of the twentieth century because its strategy of linking molecular events and cell structure is more effective than that of gene-centric molecular biology at providing mechanistic explanations of biological phenomena.[3] My perspective may be surprising to some readers who have grown up in the shadow of the double helix. I believe, however, that the stated primacy of molecular biology is a historical aberration caused in part by the failure of some scientists and scholars to acknowledge the significance of cell biology as an important molecular science of the twentieth century.

The work of Günter Blobel is at the center of this story because he was instrumental in extending to the molecular level the investigative strategy that makes cell biology so powerful. Blobel's achievements were the culmi-

---

3. Indeed, one can argue that many acknowledged successes of molecular biology in explaining biological phenomena rest on cell biological foundations (see chap. 11 and epilogue).

nation of a series of studies carried out mainly at the Rockefeller University (called earlier the Rockefeller Institute for Medical Research) that were initiated even before cell biology was a recognized scientific discipline. In the 1930s and 1940s, Albert Claude developed methods to reproducibly break cells open, purify their constituent parts by a technique known as cell fractionation, and then tentatively attach functions to those parts by measuring associated biochemical reactions. Throughout the fractionation process he used microscopy to link his cell fractions to the original intact cell (Claude 1948). This element of his strategy was greatly enhanced when, at about the same time, Claude and Keith Porter, one of Claude's young colleagues at Rockefeller, successfully adapted electron microscopy to their experiments. Because the electron microscope had a resolution far greater than that of the common light microscope, they soon began identifying cellular structures not previously known to exist. Among these was a system of interconnected membrane tubes in the cell cytoplasm that they named the endoplasmic reticulum. George Palade joined Claude and Porter at Rockefeller just after World War II and remained there long after they moved to other institutions. He took up the challenge of finding the function of the endoplasmic reticulum and determined that it was where the secretory process began. Blobel's experiments, by extending this previous work to the molecular level, initiated a new age of molecular studies in biology that remained grounded in a cellular context.

In the following chapters, I argue that application of the investigative strategy of Blobel and his predecessors addresses the fundamental unresolved dilemma that has bedeviled biology from its very beginning, the relationship between structure and function. Through his work and that of others, we can now link the chemistry of molecules directly to biological functions. We can now understand both the "vital structures" and "vital actions" of Wilson through the "ordered structure or organization" of the cell's molecular parts, because we examine those parts in the context of the cellular whole. Because all organisms are composed of cells, what we learn about the cell can be extended to all living organisms.

Modern biology today is awash in molecules, whether genes uncovered through high-speed sequencing or proteins inferred from "big data" analysis of those sequences. Correlations between mutated genes and diseases abound, and some claim that establishing the causes of ailments is no longer necessary because therapies can be devised without them. But what about our desire to fundamentally explain biological phenomena, to finally answer the question, what is life? Perhaps this examination of the history of cell biology and its explanatory approach will provide a helpful perspective.

The story of Blobel and his predecessors Claude, Porter, and Palade is called *Crossing the Boundaries of Life* because their work required the penetration of boundaries that were conceptual, technical, and physical. As we shall see, many cytologists were hesitant to disrupt the cell boundary because they believed that the ensuing death of the cell and loss of cellular form would prevent them from understanding the chemical mechanisms responsible for life. At the same time, they were constrained in their ability to see inside intact cells by limitations of the light microscope. Then, even when advances in biochemistry provided them with the tools to investigate chemical reactions that were characteristically cellular, they lacked the methods to systematically disassemble the cell in a way that preserved the context of the overall structure but still permitted investigation of cell functions at the molecular level. These obstacles blocked progress in cell studies until they were overcome by Claude, Porter, and Palade's development of a different approach to studying cells, a new way to create knowledge about cellular processes, a new *epistemic strategy*. Blobel then extended this strategy to the molecular level by using an experimental system containing cell fragments, providing a way to study biochemical reactions while still preserving a biological context.

To begin, I set the stage in the first chapter of part I by describing how the cell theory of the early nineteenth century became the protoplasm theory of the late nineteenth century, leading to an impasse in the study of both cell structure and function. In the second chapter, I relate how the cell boundary was identified with the cell membrane and how the cell membrane went from an operational concept among cell physiologists in the nineteenth and early twentieth century to an actual physical entity by the early 1970s. Then, in chapter 3, I introduce the breakthroughs that began to resolve the difficulties that had impeded the study of cells. Part II focuses on discoveries that led to the signal hypothesis (chap. 4), then on Blobel's work uncovering the molecular mechanisms predicted by the hypothesis (chaps. 5–8). Finally, in part III, I discuss how Blobel's work influenced concepts of biological information (chap. 9) and the use of in vitro experimental systems to study cells (chap. 10). I end with a more detailed analysis of the epistemic strategy used by cell biologists (chap. 11) and a consideration of the place of cell biology in the overall history of biology in the twentieth century.

PART ONE

The Cytologist's Dilemma

ONE

The Living Substance

The Cell as Relic

In 1896 E. B. Wilson, the most important American cytologist of the late nineteenth and early twentieth century, published *The Cell in Development and Inheritance*, a textbook based on his lectures at Columbia University (Wilson 1896) (see fig. 2). *The Cell* soon became a standard reference for biologists and continued to be well into the twentieth century through two subsequent revisions. Early in the book, Wilson presents a diagram of a generic cell that summarizes essential morphological features (fig. 3). The cell is illustrated as a rounded rectangle with a centrally located circular object, the nucleus. Within what Wilson describes as the "transparent ground-substance" that fills most of the cell are other circular or ellipsoid objects: an "attraction sphere enclosing two centrosomes," plastids, vacuoles, and numerous small objects that he calls "lifeless bodies." The remaining ground substance is filled with a netlike "reticulum" made up of lines of tiny granules, or "microsomes." In the nucleus, Wilson identifies four different internal structures, including chromatin and the nucleolus, a circular object suspended in the nuclear substance. The exterior border of the cell is not labeled; in the text Wilson remarks that cells may be surrounded by a cell wall or membrane but that this is another lifeless and optional part of the cell of no great concern (Wilson 1896, 15). Surprisingly, given the title of his book, Wilson goes on to say that use of the term "cell" is an outdated historical artifact.

> The term "cell" is a biological misnomer; for whatever the living cell is, it is not, as the word implies, a hollow chamber surrounded by solid walls. The term is merely an historical survival of a word casually employed by botanists of the seventeenth century to designate certain cells in plant-tissues which, when viewed in section, give somewhat the appearance of a honeycomb. The cells of these tissues are, in fact, separated by conspicuous solid walls which

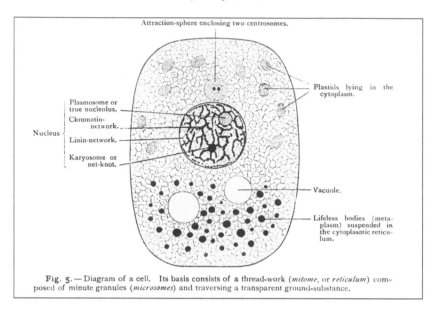

Fig. 5. — Diagram of a cell. Its basis consists of a thread-work (*mitome*, or *reticulum*) composed of minute granules (*microsomes*) and traversing a transparent ground-substance.

3. E. B. Wilson's diagram of a generic cell from his 1896 textbook, *The Cell in Development and Inheritance* (Wilson 1896). Note the absence of a labeled plasma membrane.

were mistaken by Schleiden, unfortunately followed by Schwann in this regard, for their essential part. (Wilson 1896, 13)[1]

Wilson was not alone in this opinion. Oscar Hertwig, a leading German zoologist, published *Zelle und Gewebe* (Cell and Tissues) in 1893 just when Wilson was working on the first edition of his book (Hertwig 1893). In the English translation, which appeared in 1895, Hertwig states that "[Schleiden and Schwann] both defined the cell as *a small vesicle, with a firm membrane enclosing fluid contents, that is to say, as a small chamber, or cellula, in the true sense of the word.* They considered the membrane to be the most important and essential part of the vesicle, for they thought that in consequence of its chemico-physical properties it regulated the metabolism of the cell" (Hertwig 1895, 5–6; emphasis original). In fact, not only Wilson and Hertwig, but many other cytologists at the end of the nineteenth century believed that studying the fluid contents of the cell, a substance called the *protoplasm*, was the key to deciphering the mechanisms underpinning life itself. Wilson describes the

1. In his 2017 paper, Daniel Liu was also struck by Wilson's rather surprising comment (Liu 2017).

protoplasm as "the living basis of the cell[,] . . . a viscid, translucent, granular substance, often forming a network or sponge-like structure extending through the cell body" in which are suspended a variety of "lifeless bodies" (Wilson 1896, 15).

In this chapter I describe how the cell theory was replaced by the protoplasm theory during the nineteenth century and the consequences of this change for advancements in the study of cells. Throughout the century, the primary approach to studying cells was through morphology, the microscopic examination of cellular form. By the end of the nineteenth century, the prototypical cell was considered to be nothing more than a lump of highly organized protoplasm (plus a nucleus), with life based in some unknown manner on a series of chemical reactions. At this point, the morphological approach broke down because protoplasm, despite efforts to determine its structure through the application of fixatives and stains, was essentially formless to the point that observations through the microscope no longer provided insights into its function.[2] At the same time, cytologists were concerned that disruption of protoplasm as a strategy to study its chemistry would destroy its critical organization and that any knowledge gained by this approach would not shed light on the living process. At the beginning of the twentieth century, practitioners of the new discipline of biochemistry began arguing that enzyme specificity was the organizational principle of living systems and that this principle could be studied even after cell disruption. These insights were a real advance but left cytologists without the means to link enzymatically catalyzed chemical reactions to cellular form.

The Cell Theory and the Rise of Protoplasm

The manner in which protoplasm became *the living substance* is rooted in the history of the cell theory itself. The botanists of the seventeenth century that Wilson refers to are Robert Hooke and Nehemiah Grew, both members of the Royal Society in London. Hooke, in observations with a hand lens, had noted a variety of cavities or pores in burned and petrified wood and charcoal. In 1665 he published *Micrographia* reporting findings with his more advanced, compound microscope. These included an examination of thin slices of cork in which he saw microscopic pores that he named *cellulae*,

---

2. The nucleus and chromosomes were a different matter because their structure and dynamics were easily visible in the light microscope. However, their morphological study, while contributing greatly to genetics and aspects of development, did not at this time illuminate other cell functions (see Harris 1999).

Latin for "small rooms" or "cubicles," or, in English, "cells" (Harris 1999, 5–6). The Royal Society published Grew's *Anatomy of Vegetables Begun* in late 1671, describing *bubbles* in plants that Grew initially thought reflected the gaseous product of a chemical process (Harris 1999, 9). In subsequent papers and books, most notably his *Anatomy of Plants*, which appeared in 1682, Grew extends his observations of living plants, concluding now that the bubbles are more stable features of plant structure and adopting Hooke's term *cell*, as well as *bladder*, to describe them, the latter a word employed by the Italian Marcello Malpighi in observations published at almost the same time as Grew's (Baker 1988, 109; Harris 1999, 7–9). This string of discoveries was coupled with those of Antoni van Leeuwenhoek, a Dutch amateur microscopist, who reported the first observations of an animal cell, the red blood cell, using a simple, single-lens microscope, as well as many other descriptions of microscopic *animacules*. Leeuwenhoek sent his findings as letters to the Royal Society, which published them in English summaries at about the same time as those of Hooke, Grew, and Malpighi. Many other descriptions of cells followed throughout the next century, although the idea that cells are the fundamental structural feature of all living organisms was not established until later.

By the end of the eighteenth century the diversity of the living world began to unfold. Alexander von Humboldt, the celebrated German naturalist and explorer, left for his travels to Latin America in 1799 on an expedition whose descriptions of plants, animals, and fantastic environments would inspire Charles Darwin to take a similar voyage thirty years later (Wulf 2015). The study of organismal form became an established discipline at the same time, in part through the influence of the poet and naturalist Johann Wolfgang von Goethe. In his writings Goethe sought ideal forms in nature, imagining the existence of *archetypes*, fundamental forms that captured the essence of all organisms of a particular class (Richards 2002, 408; 2018). A German physiologist proposed the term "Morphologie" in 1800 to describe the study of form as part of the systematic study of life, which he referred to as "Biologie" (Nyhart 1995, 1).[3] This interest in the form of living organisms extended to the microscopic. After Hooke and his contemporaries, observations using microscopes multiplied. An explosion of new, more detailed descriptions of cells in both plants and animals followed, and morphology became a primary means of scientific investigation.

---

3. Robert Richards points out that Goethe used the term "morphology" earlier than 1800 but did not publish it (Richards 2002).

In a lecture to the Society of German Naturalists and Doctors in Prague in September 1837, the Czech scientist Jan Evangelista Purkyně reported that a wide variety of animal tissues were composed in part by cells and that these were similar to the cells seen in plants (Harris 1999, 92). In the same year the German physiologist Theodor Schwann encountered the botanist Matthias Schleiden in Berlin, who told Schwann about his theory of cell formation in plants. Schleiden believed that "slime granules," precursors of nuclei, arise from the concentration of a homogeneous gum or vegetable gelatin and that these cytoblasts, as Schleiden calls them, then give rise to cells. Schleiden published these ideas in an article in 1838 titled "Beiträge zur Phytogenesis" (Contributions to Our Knowledge of Phytogenesis). Schwann, whose scientific mentor Johannes Müller had earlier drawn his attention to similarities between animal and plant cells (Harris 1999, 52),[4] was greatly influenced by his conversation with Schleiden. In 1839 Schwann published *Mikroskopische Untersuchungen über die Uebereinstimmung in der Struktur und dem Wachstum der Thiere und Pflanzen* (Microscopical Researches into the Accordance in the Structure and Growth of Animals and Plants) in which he adopted Schleiden's view about cell formation in plants and generalized it to animals, stating that in both cases, new cells form initially from a structureless substance (Schwann 1839). This cytoblastema, as Schwann calls it, initially produces Schleiden's cytoblasts or nuclei and then fully formed cells. Importantly, neither Schleiden nor Schwann considered this gum, gelatin, or cytoblastema to be alive but rather prebiotic raw material whose coagulation gives rise to living structures, in particular the nucleus and the cell membrane (Liu 2017). As it turned out, Schleiden and Schwann's ideas about the formation of cells were completely wrong. Nevertheless, their work, as well as that of Purkyně and others, led to the conclusion that cells are the common living components of both plants and animals, a finding that transformed the budding field of cytology.

An alternative to Schleiden and Schwann's theory of spontaneous cell formation from an amorphous gum soon arose. In 1841, the Berlin physiologist and embryologist Robert Remak published the observation that chick embryonic blood cells reproduce by dividing into two. Remak added to these findings in 1852 and by 1854 was convinced that the formation of all new cells in animal tissues is by division (Harris 1999, 116). In *Untersuchungen über die Entwicklung der Wirbelthiere* (Investigation of Vertebrate

4. Müller was one of the most important anatomists and physiologists of the nineteenth century; he spent much of his career at the Humboldt University in Berlin (Otis 2007).

Development), which appeared in 1855, Remak thoroughly critiques the ideas of Schleiden and Schwann and reviews his own work on cell formation (Remak 1855; Harris 1999, 131). Although initially skeptical, Remak's Berlin colleague Rudolph Virchow adopted Remak's views on cell division at about the same time. Virchow had established his own journal in 1847, the *Archiv für pathologische Anatomie und Physiologie und für klinische Medizin*, which he used to broadcast his views on cell division in an 1855 editorial. Here he introduced the Latin catchphrase *omnis cellula e cellula* to indicate that all new cells arise from the binary division of old cells and not spontaneously, as proposed by Schleiden and Schwann, and repeats this in his famous book *Cellularpathologie*, published in 1858 (Wilson 1896; Baker 1988, 435; Harris 1999).

While the manner in which new cells form was being worked out, other scientists focused on the liquid substance inside cells. As early as 1744, Abraham Trembley, a Swiss naturalist who was among the first to study the tiny polyp *Hydra*, a freshwater cousin of jellyfish, disrupted one of the organisms using quill pens and noticed the emergence of a sticky, stretchy substance. By 1835, Felix Dujardin, a French biologist, observed material that he named *sarcode* in a variety of organisms: "I propose to give this name to what other observers have called a living jelly, this glutinous, diaphanous substance, insoluble in water, that contracts into globular lumps, sticks to dissecting needles, and can be drawn out like mucus. It is to be found in all lower animals interposed between other structural elements" (cited in Harris 1999, 74; see also Liu 2017). Thus, in the context of the simple, unicellular organisms that he studied, sarcode is not simply the raw material that gives rise to living cells, but the living substance itself. Other observations followed. Purkyně (in 1840) and Hugo von Mohl in Germany (in 1851) independently began using the term "protoplasm" to describe the intracellular living substance. Von Mohl's findings were embedded in a somewhat confusing discussion about a structure called the primary utricle and its role in cell formation in a manner reminiscent of the theories of Schleiden and Schwann. While he did not go so far as Dujardin in considering protoplasm to be alive, he also clearly believed that it was not simply a prebiotic material (Liu 2017). At about the same time, the German botanist Ferdinand Cohn noted that the protoplasm of both plant and animal cells is very similar and likely "the chief site of vital activity" (quoted in Baker 1988, 94). Throughout the 1850s, others reported findings that supported Cohn's conclusions (Geison 1969).

In 1860, the German biologist Max Schultze went even further to claim that cell membranes, structures that featured prominently in the cell theories of Schleiden and Schwann, are not critical parts of the living cell. Instead, he states, the essential cell consists only of a "naked lump of protoplasm with a nucleus" (Schultze 1860; Harris 1999, 151).[5] Schultze arrived at this conclusion through his study of unicellular marine organisms and muscle cells in higher organisms. He observed that fingerlike cellular projections from pores in the microscopic shells (or *tests*) of Rhizopods seem to consist of nothing but protoplasm without an apparent membrane covering (Harris 1999, 150). He also saw that in higher animals mature muscle cells form from the fusion of individual cells into a large, multinuclear syncytia of continuous protoplasm, again suggesting the insignificance of the membrane. Schultze's conclusions were consistent with those of the German botanist Anton de Bary. De Bary studied slime molds that exist in different stages of their life cycle as either individual migrating cells or as plasmodia, a multicellular state in which several nuclei are present in a continuous mass of protoplasm not divided by membranes. Because the individual *swarmers* do not appear in the light microscope to have membranes, and membranes are largely absent from the plasmodia, de Bary believed that membranes are optional (Harris 1999, 150).[6] In later publications Schultze and others elaborated on these

5. A more complete statement is, "Concerning the contractility of the protoplasm, alterations of shape of the whole cells are of course impeded or completely inhibited by the presence of a stiff cell membrane. But the less perfect the surface of the protoplasm is in hardening into a membrane, the closer a cell is to the primordial membrane-less state, representing only a naked lump of protoplasm with a nucleus, the freer and less obstructed the cell can exhibit its movements. If such a cell is an organism in itself, the protean change of shape, the change of the external form of the protoplasmic lump, caused by its contractility, is most obvious." In German: "Bei dieser Contractilität des Protoplasma sind Gestaltveränderungen der ganzen Zellen durch Anwesenheit einer starren Zellmembran natürlich gehindert oder ganz unmöglich gemacht. Je weniger vollkommen aber die Oberfläche des Protoplasma zu einer Membran erhärtet ist, je näher die Zelle dem ursprünglichen membranlosen Zustande befindet, auf welchem sie nur ein nacktes Protoplasmaklümpchen mit Kern darstellt, um so freier und ungehinderter können sich die Bewegungen äussern. Ist eine solche Zelle num nun gar ein Organismus für sich, so tritt uns die proteische Gestaltveränderung, der in der Contractilität des Protoplasmaklümpchen bedingte Wechsel der äusseren Form am auffallendsten entgegen" (Schultze 1860, 299). Later in the same article Schultze emphasizes the same point: "Bear in mind my definition of the cell: 'a naked lump of protoplasm with a nucleus' and remember that I do not consider the membrane as a necessity for the concept of the cell." In German: "Ich erinnere wieder daran, dass die von mir gegebene Definition der Zelle lautet: 'ein nacktes Protoplasmaklümpchen mit Kern,' und dass ich die Membran als etwas zum Begriff der Zelle durchaus nicht Nothwendiges betrachte" (Schultze 1860, 305). Translations © Hanna L. Worliczek. Used with permission.

6. As I discuss in chapter 2, one reason that biologists were at this time unable to see membranes in some cells was the limited resolution of the light microscope, something dictated by the wavelength of visible light. Early microscopes, including those used throughout the eighteenth century, suffered from optical distortions called spherical and chromatic aberration that degraded the image (Bradbury 1967). Improved lenses and designs in the 1820s resolved both of these problems, but the resolution limit still existed. Microscopic resolution is defined as the minimal distance between two objects at which they still appear as two objects. For visible

observations, and the conclusion became clear: the living substance is nothing but protoplasm containing a nucleus; cell membranes or walls may or may not be present but in any case are unnecessary for life.

As this perspective permeated the scientific community, it also captured the popular imagination. In 1868 Thomas Huxley, a prominent British biologist and defender of Charles Darwin's ideas on evolution, gave a public address in Edinburgh that he titled "The Physical Basis of Life" (Huxley 1869). The lecture was published in a London periodical shortly thereafter and caused such a stir that it was reprinted widely. Huxley begins by indicating that he is using "the physical basis of life" as a translation of "protoplasm" for the benefit of a lay audience. Protoplasm he declares is the "one kind of matter common to all living beings," asking poetically, "What hidden bond can connect the flower which a girl wears in her hair and the blood that courses through her youthful veins?" (Huxley 1869, 5–6). Huxley speaks of protoplasm and not the cell as the "structural basis of the human body" (Huxley 1869, 10). He then goes on to comment on the chemical nature of protoplasm, concluding that it is "albumoid" or "proteinaceous" like the white of an egg (Huxley 1869, 12). With the success of his lecture, Huxley communicated the fundamental change that had swept biology: the cell theory of life had been superseded by the protoplasm theory of life.

In the same year as his lecture, Huxley reported a finding that is, from a modern perspective, bizarre. In 1857 the British ship HMS *Cyclops* returned from an expedition investigating the seafloor of the Atlantic Ocean. Among the specimens recovered was a sticky and viscous mud dredged from the bottom of the sea that was sent to Huxley for examination (Allman 1879). In an appendix included with the captain's report in 1858, Huxley describes microscopic shells provisionally called coccoliths that he believes were the remnants of single-celled organisms (Huxley 1868). Several years later, Huxley again looked at the same material, which had been preserved in alcohol. This time he reported that "the granule heaps and gelatinous matter in which [the coccoliths] are imbedded represent masses of protoplasm" (Huxley 1868, 210). Huxley concluded that this material is a form of primitive organism consisting almost entirely of living protoplasm, and names it *Bathybius haecklii* in honor of Ernst Haeckel, the noted biologist and leading advocate of Darwin in Germany. Haeckel responded with his own investigation, expressing the opinion that below 5,000 feet the ocean floor is covered

---

light, this resolution is approximately 0.1–0.2 microns, or μm (one-millionth of a meter or 10^{-6} meter). This means that objects inside cells that are smaller than 0.1–0.2 μm cannot be distinguished with a light microscope.

with a huge and formless mass of living protoplasm, something to which, as one commentator states, "no law of morphology has yet exerted itself" (Allman 1879, 724). Haeckel's enthusiasm was fueled by his own work on an organism that he called *Protoamoeba primativa* that is, he believed, the most primitive cell lacking not only a membrane but also a nucleus. Other reports appeared, including one claiming to find *Bathybius* in gutters and the sewer (Beale 1869). Before long *Bathybius* was shown to be a precipitate caused by the preservative rather than a living substance. Nevertheless, the *Bathybius* episode highlights what had become an established fact in biology: protoplasm is the fundamental living substance of all organisms, and the continuity of life is the continuity of protoplasm.

One consequence of these events was that single-celled organisms called amoebae, which appear to consist of almost nothing but protoplasm, became identified as prototypical cells (Reynolds 2008). Amoebae are motile cells that continuously modify their form as they move by projecting pseudopods (false feet) to probe different regions of their environment. The term "amoeba" or "amoeboid" refers to a generic cell with these characteristics that make up many varieties of protists and also appear as certain cell types in more complex, multicellular organisms. Most often cytologists such as Wilson and Hertwig used the species *Amoeba proteus* as a specific example, and both included drawings of it in their textbooks published near the end of the nineteenth century (Hertwig 1893; Wilson 1896; Reynolds 2008) (fig. 4). Although in Wilson's drawing a fine line demarcates the edge of the cell,[7] he does not identify or label a cell membrane. He pictures the protoplasm as somewhat granular, containing not only a nucleus but also food, water, and "contractile" vacuoles. Wilson considers the latter, however, to be transient *formed elements* of the protoplasm that do not figure as requisite components of the living substance.

Designation of the amoeba as the essential cell had important implications for the study of life. One was that more complex cells, whether free-living or parts of multicellular organisms, were thought to be built on the foundation of living amoebic protoplasm. This implied that specializations of plant or animal cells other than nuclei, including membranes, inclusions, and cilia, the hairlike projections from cells that drive motility, were not essential to life. Most significantly, investigation of these elements was considered unlikely to yield insights into the living process. Conversely, it suggested

---

7. Hertwig did label the boundary of the amoeba *Hautplasma* (plasma skin or membrane), but his views on the primary importance of protoplasm are completely consistent with those of Wilson (Hertwig 1893, 15; see fig. 2).

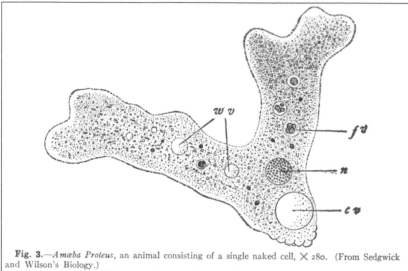

Fig. 3.—*Amœba Proteus*, an animal consisting of a single naked cell, × 280. (From Sedgwick and Wilson's Biology.)
n. The nucleus; *w. v.* Water-vacuoles; *c. v.* Contractile vacuole; *f. v.* Food-vacuole.

4. Illustration of *Amoeba proteus* from the 1925 edition of E. B. Wilson's textbook *The Cell in Development and Heredity* (Wilson 1925). An identical drawing also appeared in the first 1896 edition and in Sedgwick and Wilson's earlier biology textbook (Sedgwick and Wilson 1886).

that the answer to what did constitute life did not involve cellular form at all but was to be found within the naked bit of protoplasm constituting the amoeba, a substance superficially devoid of structure (Allman 1879, 722).[8,9]

In 1861, the physiologist Ernst Wilhelm von Brücke published an article titled "Elementary Organism" (Die Elementarorganismen) that discussed his view of cell theory (Brücke 1898). Brücke, like Schwann, had been a student

8. In an 1879 address to the British Association for the Advancement of Science titled "Protoplasm and Life," George James Allman, a botanist and zoologist, stated, "Examine it [protoplasm] closer, bring to bear on it the highest powers of your microscope—you will probably find disseminated through it countless multitudes of exceedingly minute granules; but you may also find that it is absolutely homogeneous, and, whether containing granules or not, it is certain that you will find nothing to which the term *organization* can be applied. You have before you a glairy, tenacious fluid, which, if not absolutely homogeneous, is yet totally destitute of structure. . . . And yet no one who contemplates this spontaneously moving material can deny that it is alive. Liquid as it is, it is a living liquid; organless and structureless as it is, it manifests the essential phenomena of life (Allman 1879, 722–23).

9. An early taxonomic name for "amoeba" was *Chaos*, a term that can mean a "confused unorganized state of primordial matter before the creation of distinct and orderly forms" (*Webster's Third New International Dictionary*, s.v. "chaos"); see also Reynolds 2008, 311.

of Müller in Berlin in the 1840s. By the 1850s, Brücke and other students of Müller, including Hermann von Helmholtz and Emil du Bois-Reymond, split from the morphological approach in favor of what has been termed *physicalist physiology* (Coleman 1977, 151–52; Nyhart 1995, 69). All biologists of the time, including morphologists, accepted that life had a physical and chemical basis and did not deny the value of chemistry and physics as an adjunct to morphology. Remak's textbook on microscopic anatomy, published in 1852, for example, devoted the first hundred pages to tissue chemistry (Nyhart 1995, 85). The physicalists, however, had a decidedly reductionist bias (Coleman 1977, 150–54). Instead of considering the form of the organism as a whole, they wished to elucidate the functions of living organisms, including individual cells, on the basis of submicroscopic material constituents of cells, as well as principles of physics and chemistry.

In his article Brücke wholeheartedly accepts Schultze's recent work proclaiming that the cell membrane is not an essential part of the cell. He also agrees that the protoplasm (*der Zelleninhalt*), which he describes like Huxley as resembling egg white and neither liquid nor solid, is the seat of living processes. Brücke, however, cannot accept that the complex properties associated with protoplasm can be vested in a structureless substance. Instead, he states that protoplasm, in fact, must have a very complicated structure, which he refers to as *organization* (Brücke 1898; Reynolds 2018a, 2018b). When Brücke uses the word *organization* he is not referring to cellular elements visible in the microscope but to the unknown parts of the cell interposed between those elements and molecules (Wilson 1896, 210). In the absence of good ways to get at molecules in cells (Coleman 1965), cytologists instead attempt to decipher this organization by making the invisible structure of protoplasm visible.

The Morphology of Protoplasm

In the mid-nineteenth century, plant cells were more amenable to microscopic examination than animal cells or even protists. For a specimen to be visible in the light microscope, it must be larger than the resolution of the microscope and sufficiently thin to transmit and refract the light illuminating it before it reaches the eyepiece. Plant tissues typically have a much more rigid structure than those of animals, which enables them to be cut into thin slices and observed in the living state. Furthermore, some plant cells possess chloroplasts, the green organelles responsible for photosynthesis that are easily visible in the microscope without any special preparation. In contrast, animal cells and tissues are much more difficult to observe. Individual living cells die easily due to dessication, lack of nutrients, or simply the stress caused by the process of isolation and preparation. In their fresh state, animal tissues are typically too soft to cut into thin sections. Furthermore, unstained animal cells and tissues appear translucent in the microscope because their internal structures refract light in a similar manner. Indeed, this problem undoubtedly contributed to the belief that protoplasm was a largely structureless and transparent substance. While today there are purely optical methods to introduce contrast in such specimens and enable subcellular structures to be differentiated in the light microscope, these were not introduced until the 1930s (Bradbury 1967).

To overcome these difficulties, cytologists and bacteriologists in the late nineteenth century developed methods of fixation, sectioning, and staining of tissues for microscopic observation. Fixation consists of treating living cells or tissues with strong chemicals that both kill them and, it is hoped, preserve their microscopic morphology in a form resembling the living state. Fixatives such as alcohol (wine essences) and acids were used early on. By the mid-nineteenth century more complex and rarefied mixtures were developed with the hope that they might act quickly to retard the degradation that accompanied death while also preserving more detailed structures. Among these concoctions were solutions of osmium, a rare by-product of platinum mining, and picric acid, an explosive (Liu 2016). Validation that a fixative actually preserved a cellular structure and did not create it was a significant problem (see below), requiring a kind of intellectual bootstrapping through argument and comparison of results achieved with different approaches (Rasmussen 1997; Bechtel 2006; Liu 2016; Schickore 2018). This was not always successful, and questions about the interpretation of fixed structures persisted well into the twentieth century. In Britain and especially in Germany, development of synthetic organic chemistry in the nineteenth century led to the production of various dyes (Fruton 1999). These were almost immediately used to stain cells and tissues to improve contrast, with carmine and aniline dyes employed at midcentury, followed rapidly by many others (Liu 2016). In addition to preserving structure, fixatives had the added advantage that they hardened tissues such that they could be thinly sliced. When this proved to be inadequate, techniques were developed to infiltrate samples with materials such as liquified paraffin. After hardening of the infiltrating agent, slices of specimens were cut either by hand or by machines called microtomes designed for this purpose (Bracegirdle 1978). By the end of the nineteenth century all these methods were very well developed and most obstacles to observing preserved biological specimens in the light microscope overcome.

Despite this, application of these methods to determine the morphological substructure of protoplasm did not yield definitive, consistent results. As mentioned previously, in 1896 Wilson described the protoplasm in his generic cell as a threadlike reticulum formed by chains of small granules or microsomes (Wilson 1896, 14) (fig. 3). Nevertheless, Wilson notes in the text that "regarding the precise nature of this structure opinion still differs" (Wilson 1896, 17). In certain living cells such as the eggs of sea urchins and starfish, some protoplasmic structure was visible without fixation or staining (Wilson 1899). However, in other live cells the protoplasm appeared homogeneous and unstructured. Although cytologists hoped that fixation and staining procedures might provide clear results, by the 1880s, if not earlier, it was evident that patterns seen in fixed protoplasm might be artifacts of the fixation process and not reflect the true organization of the living substance. This was clearly articulated by the cytologist Walther Flemming in a footnote in his book *Zellsubstanz, Kern und Zelltheilung* (Cell Substance, Nucleus, and Cell Division) stating that the fixative osmium causes "protoplasmic strands" and "shrunken nuclei" to appear in the alga *Spirogyra* and other plant cells, structures not visible when other fixatives are used (Flemming 1882; Liu 2016). Flemming tried to compensate for such problems by observing a diverse group of plant and animal cells prepared using a variety of fixation and staining protocols (Flemming 1882, 379; Liu 2016).

Issues of fixation and staining artifacts also arose in the controversial work of the pathologist Richard Altmann (Altmann 1890). Altmann applied a cocktail of particularly harsh fixatives and stains to various cells and observed large numbers of small, red granules (Altmann 1890; Bechtel 2006; Liu 2016). Altmann proposed that these *bioblasts*, as he called them, are fundamental living organisms capable of growth and division and that these occupy the organizational level between cells and molecules (Wilson 1896, 21). This idea was considered by many to be preposterous, eliciting the strong but polite comment by Wilson that "Altmann's premature generalization rests upon a very insecure foundation and has been received with just skepticism" (Wilson 1896, 21; see also Verworn 1899, 63). Altmann's results were further undermined in 1894 when the botanist Alfred Fischer fixed and stained a protein solution with Altmann's reagents and generated granular structures resembling bioblasts (Liu 2016).[10]

---

10. Ironically, mitochondria, actual granular constituents of the cell, were likely among the components visualized by Altmann; about a century later, DNA sequencing and other procedures demonstrated that mitochondria had indeed originated as free-living organisms that developed a symbiotic relationship with primitive eukaryotic cells.

Fischer was only one of several biologists in the 1880s and 1890s who sought to distinguish fixation artifacts from the "real" structure of protoplasm by comparison with nonliving materials. Unfortunately, these studies did not result in the acceptance of one particular pattern of protoplasmic organization. At about the same time, the zoologist Otto Bütschli inverted the logic of this approach. Instead of using nonliving materials to cast doubt on proposed structures of protoplasm, he purposely created oil-water emulsions to support his proposal that protoplasm had an "alveolar" organization (Bütschli 1894; Liu 2016). When observed under the microscope, such emulsions resembled actual protoplasm to such an extent that Bütschli was able to fool unassuming colleagues (Bütschli 1894, 85; Liu 2016). Based on this and his observations of actual cells, Bütschli concluded that protoplasm has an alveolar structure with properties of an oil-water emulsion.

Bütschli's theory was favorably received. Even Wilson, who in 1896 had been noncommittal about which protoplasmic structure was correct, wrote later, "Although in earlier papers I was inclined to regard the meshwork of the echinoderm-egg as a reticulum, further studies have left no doubt whatever, in my opinion, that in the resting cell it is in reality an alveolar structure—or, as I do not hesitate to call it, an *emulsion*—such as Bütschli has described" (Wilson 1899, 37; emphasis original). Later in the same article, Wilson tempers this conclusion by saying that "*no universal or even general formula for protoplasmic structure can be given*[,] . . . and . . . *the ultimate background of protoplasmic activity is the sensibly homogenous matrix or continuous substance* in which [alveolar, granular, fibrillar, and reticular] structures appear" (Wilson 1899, 42; emphasis original). In other words, fixation artifacts aside, protoplasm is dynamic and can appear either homogeneous or structured depending on the physiological circumstances (Liu 2016, 2017).

Chemistry and the Cytologist's Dilemma

By the end of the nineteenth century, efforts to understand Brücke's postulated organization of protoplasm using the morphological strategy had demonstrated that protoplasm was essentially amorphous in the literal sense. There was no one universal structure of protoplasm visible in the microscope with or without fixation and staining. While this represented progress, cytologists were no closer to understanding how protoplasm functioned than they had been when Huxley declared protoplasm to be the living substance in 1868. Brücke and the other midcentury physicalists had promoted chemical and physical approaches instead of morphology as an alternative way to probe living systems (Nyhart 1995). However, at that time, the chemical and physical tools needed to investigate protoplasm were not available. Gradually, this began to change.

Atomic and molecular theories of matter advanced in the first half of the nineteenth century, building on earlier breakthroughs by Antoine Lavoisier and others. John Dalton developed the concept of atomic weight and proposed that chemical compounds consist of fixed combinations of different elements whose overall molecular weight reflects the aggregate weight of component atoms (Levere 2001). At the time, many different elements, including hydrogen, oxygen, carbon, sulfur, and phosphorus were known, and elemental analysis of substances was possible. With the invention of the battery in 1800, Humphrey Davy exploited the ability of electrical discharge to decompose compounds, in the process discovering and naming additional elements, including sodium, potassium, calcium, and magnesium (Levere 2001, 87–89).

The Swedish chemist Jöns Jacob Berzelius extended Davy's work on the electrochemistry of inorganic compounds to develop an idea called electrochemical dualism that ranked atoms according to their relative electronegativity. Soon he applied his approach to the substances that make up living organisms. By the beginning of the nineteenth century, plants and animals were known to be largely composed of carbon, oxygen, nitrogen, and hydrogen, and by the 1820s, these were generally referred to as "organic" at Berzelius's suggestion (Fruton 1999, 12; Levere 2001, 96). Organic chemicals were at first not available as pure compounds and were difficult to work with (Fruton 1999, 164–65). Berzelius applied his superior laboratory skills to the isolation of organic materials and by the 1830s had purified a number of compounds and proposed molecular formulas reflecting their compositions. The French chemist August Laurent then went further to suggest that the particular arrangement of atoms in different organic compounds, not just their overall composition, is responsible for their chemical properties (Levere 2001, 101). The ideas of Berzelius and Laurent matured into the valency theory, which predicted more precisely the combinatorial possibilities of carbon and other elements, and by the 1870s the structural characteristics of organic compounds were understood.

When Schwann proposed his cell theory in 1839, he described the cytoblastema, the formative substance of cells, as gelatinous (Schwann 1847, 165). Gelatin, as well as albumin, fibrin, and casein, referred to animal and plant materials recognized since ancient times whose properties resembled egg white (Fruton 1999, 161–62). In the 1830s, Gerrit Jan Mulder named all of these *proteins*, at the suggestion of Berzelius, to indicate that they are of primary nutritional value. Mulder's analysis indicated that all classes of proteins are composed of similar amounts of carbon, hydrogen, nitrogen, and oxygen (Fruton 1999, 171–72). By the time of his 1868 lecture, Huxley concluded that all protoplasm was proteinaceous, and, as described previously, morphologists attempted to understand the structure of protoplasm by comparing fixed cells to fixed protein solutions. Nevertheless, almost nothing was known about proteins other than their elemental composition and bulk behavior when treated with organic solvents or other chemicals. As the major component of the living substance, proteins were believed to be involved in energy production, either as raw material or as catalysts for metabolic reactions (Teich 1973, 454; Fruton 1999, 166–67).[11] The idea that proteins possessed *vital energy* gave rise to the suggestion by the physiologist Eduard Pflüger in 1875 that protoplasm consists of an unstable living protein capable of polymerization into a single giant molecule (Fruton 1999, 166–67). This proposal was accompanied by speculation about the molecular structure of such a molecule, and related theories, such as the biogen hypothesis of Max Verworn, were also advanced (Teich 1973, 455–56; Fruton 1999, 168–70).

Gradually more details emerged about the actual chemical structure of proteins. Treatment of proteins with acids, which breaks them apart, led to the discovery that proteins are composed of several different types of amino acids.[12] To confront the difficulty of obtaining pure proteins for study, the organic chemist Emil Fischer applied a synthetic approach and managed to string together up to eighteen amino acids. Fischer's strategy not only shed light on the peptide bond that links amino acids in proteins but also proved that natural proteins were such polymers since his longest synthetic chain of amino acids resembled them chemically (Fruton 1999, 187–88). By the end of the nineteenth century, techniques for partially purifying proteins had advanced sufficiently to indicate that different proteins had distinct chemical properties (Fruton 1999, 191–92). These insights sowed doubt about the idea that protoplasm consisted of a giant living molecule, and biologists, reacting against such unresolved issues, preferred to consider protoplasm as more of a morphological than chemical concept.

Even with increasing knowledge of chemistry, cytologists in particular were inhibited in their ability to explore the chemical and physical proper-

11. Catalysts are substances that facilitate chemical reactions without being consumed themselves. *Oxford English Dictionary*, 2017 ed., s.v. "catalysis."

12. Amino acids are small molecules consisting of a nitrogen-based amino group on one end, a carboxylic acid group on the other end, and side chains of various chemical characteristics projecting from a carbon molecule in the middle. Most proteins are composed of some combination of twenty different amino acids. See also chaps. 4 and 5.

ties of protoplasm by the long-standing concern that disruption of cells and protoplasm not only rendered them lifeless but also destroyed the organization of the protoplasm that biologists wished to understand. Explicit statements to this effect often appeared throughout the nineteenth century, from the likes of the chemist Justus Liebig in 1842 (Coleman 1977, 149) to the cytologists Wilson and Hertwig at the end of the century.[13] In some instances such utterances reflected residual vitalism (Coleman 1977, 148) but later articulated a serious strategic concern. Citing Carl Wilhelm Nägeli, the Swiss botanist, Hertwig states:

> Protoplasm . . . cannot be placed under different conditions without ceasing to be protoplasm, for its essential properties, in which life manifests itself, depend upon a fixed organization. For as the principle attributes of a marble statue consist in the form in which the sculptor's hand has given to the marble, and as the statue ceases to be a statue if broken up into small pieces of marble, so a body of protoplasm is no longer protoplasm after the organization, which constitutes its life, has been destroyed; we only examine the considerably altered ruins of the protoplasm when we treat the dead cells with chemical reagents. (1895, 15–16)[14,15]

13. For an earlier example, Goethe wrote in *On Morphology* between 1807 and 1817, "In observing objects of nature, especially those that are alive, we often think that the best way of gaining insight into the relationship between their inner nature and the effects they produce is to divide them into their constituent parts. Such an approach may, in fact, bring us a long way toward our goal. In a word, those familiar with science can recall what chemistry and anatomy have contributed toward an understanding and overview of nature. . . . But these attempts at division also produce many adverse effects when carried to an extreme. To be sure, what is alive can be dissected into its component parts, but from these parts it will be impossible to restore it and bring it back to life" (Goethe 2016, 978).

14. In the original German: "Protoplasma dagegen lässt sich nicht in andere Aggregatzustände überführen, ohne sofort aufzuhören, Protoplasma zu sein. Denn seine wesentlichen Eigenschaften, in denen sich sein Leben äussert, beruhen eben auf einer bestimmten Organisation. Ebenso wie die hauptsächlichen Eigenschaften einer Marmorstatue in der Form bestehen, die Künstlerhand dem Marmor gegeben hat, und wie eine Statue aufgehört hat, eine solche zu sein, wenn sie in kleine Marmorsteinchen zerschlagen ist (Nägeli II 28), so ist auch ein Protoplasmakörper nach Zerstörung der Organisation, auf welcher sein Leben beruht, kein Protoplasma mehr; wir untersuchen in den abgetödteten, mit Reagentien behandelten Zellen streng genommen nur die stark veränderten Trümmer desselben" (Hertwig 1893, 15).

15. The physiologist Max Verworn expressed similar views just as dramatically: "Living substance must be killed before its chemical composition can be learned. Paradoxical as this may sound at present it is the only way by which knowledge of the chemistry of living substance can be obtained. The biting sarcasm that Mephistopheles pours out before the scholar upon this practice of physiological chemistry must be quietly endured. It is not possible to apply the methods of chemistry to living substance without killing it. Every chemical reagent that comes in contact with it disturbs and changes it, and what is left for investigation is no longer living

Given such a perceived obstacle to the study of the parts of disrupted cells using chemical and physical procedures, cytologists attempted to decipher the properties and locations of the chemical components of protoplasm in the intact cell using stains that highlighted specific chemical groups. Such approaches were of limited success (Schickore 2018, 84; see also chap. 3, below).

Colloids, Enzymes, and Specific Catalysis

While advances in organic chemistry helped to clarify the composition of protoplasm, at this point the chemical approach, like the morphological approach, did not really help explain the nature of the living process. In the absence of a clear understanding of proteins, a new twist on the emulsion proposals by Bütschli began to take hold. In 1861 Thomas Graham reported results he had obtained using what he called a dialyzer. The dialyzer consisted of parchment paper stretched across the opening of a bell jar and submerged in water. Graham compared the diffusion of different substances across the parchment membrane, finding that solutions of crystalline substances such as sugar diffuse much more rapidly than proteinaceous albumoid solutions. He concluded that protein solutions are essentially suspensions of microscopic aggregates that are too small to settle out of solution yet too large to easily pass through the membrane (Graham 1861; Fruton 1999, 196–97; Liu 2016). He called such suspensions colloids. In 1899 the British physiologist William Bate Hardy became interested in earlier work on fixation artifacts and proposed that the behavior of protoplasm on fixation resembles that of colloidal suspensions (Hardy 1899; Liu 2016, 82–88). Hardy invokes Brücke's pronouncement, made the same year that Graham's paper appeared, that protoplasm must have an organized substructure, concluding that this substructure is that of a colloid (Hardy 1899).

Among biologists, the colloidal theory of protoplasm superseded other structural models, in part because it seemed to be sanctioned by the new field of physical chemistry. Acceptance was stimulated as well by the invention in 1903 by Richard Zsigmondy and Henry Siedentopf of the so-called ultramicroscope (a kind of dark field microscope) that used side illumi-

---

substance, but a corpse—a substance that has wholly different properties. Hence ideas upon the chemistry of the living object can be obtained only by deductions from chemical discoveries in the dead object, deductions the correctness of which can be proved experimentally in the living object only in rare cases. This alone is responsible for the excessively slow advance of the knowledge of the chemistry of vital processes" (1899, 102).

nation to detect tiny particles previously invisible because they were too small to be resolved by light microscopy (Liu 2016).[16] The colloidal particles visualized in solutions and later in cells jiggled with Brownian motion, a phenomenon placed on a solid physical foundation by the work of Albert Einstein. Such observations no doubt encouraged biologists that the morphological approach to protoplasm structure might still be viable. The colloidal theory was also attractive because the total surface area of such particles provided an enormous substratum for chemical reactions, even in a volume as small as a cell.

The colloidal theory was also embraced at the end of the nineteenth century by the nascent field of biochemistry. This was particularly true among scientists interested in enzymes, the catalysts of chemical reactions within cells. The study of enzymes grew out of work on fermentation, a process known since ancient times (Fruton 1999, 117). Fermentation produces alcohol and other organic substances, mainly through the degradation of plant derived sugars. The term "fermentation" was also applied to the destruction of proteins by saliva and by substances produced in the stomach and pancreas, as well as the breakdown of urea in the urine to ammonia. Ferment became the generic name for the active agent in all these processes. Some ferments were recognized as soluble, while those associated with living organisms such as yeasts were insoluble. In the absence of techniques to purify ferments, however, these designations were largely operational and did not indicate any particular insight into the chemical nature of either the ferments or the transformations that they facilitated.

In 1876 the German physiologist Wilhelm Kühn proposed the term *enzyme* as an alternative name for ferments not associated with organisms (Fruton 1999, 148). Later, in 1897, the chemist Eduard Buchner prepared a cell-free extract from disrupted yeast cells that was able to produce alcohol from various sugars. Buchner concluded that alcoholic fermentation was a purely chemical process that could proceed completely without cells. He named the active substance zymase and indicated that it was likely a protein. Notwithstanding some claims at the time that Buchner's zymase preparations were contaminated with living protoplasm, his accomplishment suggested that living processes could be at least partially explained as a series of enzyme-catalyzed chemical reactions (Fruton 1999).

16. Dark field microscopes, including the ultramicroscope, enable the observer to *detect* particles that are below the resolution limit of the standard light microscope by light scattering, but do not increase the actual resolution. As such, there were ultimately of limited use (James 1976, 145–48).

This concept was further advanced in an important monograph by the German biochemist Franz Hofmeister in 1901, *The Chemical Organization of Cells* (*Die chemische Organisation der Zelle*) (Hofmeister 1901). As perhaps suggested by the title, Hofmeister's work attempts to add substance to Brücke's deduction that protoplasm must be more complex than it appears morphologically to enable it to carry out all the functions of the living cell (Brücke 1898). Hofmeister accepts the prevailing opinion that cellular enzymes are adsorbed onto colloidal particles. In reaction to Verworn's biogen hypothesis and earlier statements about giant living molecules promulgated by Pflüger and others, he points out that chemical reactions responsible for cell metabolism must occur in a series of interconnected steps that lead to the transformation of an initial substrate into a final product. These he believes can be accounted for by juxtaposed colloidal particles localized to specific regions of the cell. Hofmeister also recognizes that certain reaction *pathways* might be chemically incompatible with others and proposes that the cell must be divided by "impermeable partition walls," perhaps similar to some structures visible in the microscope or possibly related to proposed submicroscopic substructures postulated by various investigators (Teich 1992, 506). Several years later, however, Hofmeister retracts this proposal, indicating that advances in colloid chemistry demonstrate that such barriers are unnecessary (Fruton 1999, 158).

Hofmeister's ideas were echoed a few years later by Frederick Gowland Hopkins at Cambridge (fig. 5). Hopkins, founder of the discipline of biochemistry in Great Britain, was adamant in his rejection of the idea that protoplasm was some giant energetic molecule with properties that were indecipherable (Needham and Baldwin 1949, 33). Instead, Hopkins believed that what he referred to as *specific catalysis* by enzymes in a colloidal environment was a sufficient organizing principle to ultimately explain living processes in chemical terms, as he stated explicitly in a 1913 address.

> The highly complex substances which form the most obvious part of the material of the living cell are relatively stable. Their special characters, and in particular the colloidal condition in which they exist, determine, of course, many of the most fundamental characteristics of the cell: its definite yet mobile structure, its mechanical qualities, including the contractility of the protoplasm, and those other colloidal characters which the modern physical chemist is studying so closely. For the dynamic chemical events which happen within the cell, these colloidal complexes yield a special milieu, providing, as it were, special apparatus, and an organised laboratory. But in the cell itself, I believe, simple molecules undergo reactions of the kind we have been

5. The great British biochemist Frederick Gowland Hopkins pictured at the Marine Biological Laboratory in Woods Hole. Image reproduced from the website "History of the Marine Biological Laboratory" (http://hpsrepository.asu.edu/).

considering. These reactions, being catalysed by colloidal enzymes, do not occur in a strictly homogeneous medium, but they occur, I would argue, in the aqueous fluids of the cell under just such conditions of solution as obtain when they progress under the influence of enzymes *in vitro*. (Needham and Baldwin 1949, 150)

A significant implication of these beliefs was that the study of enzymatic reactions in extracts prepared from disrupted cells is still capable of giving insights into the organized functions of the cell as a whole. In other words, the concerns of cytologists that little certain could be learned about life's chemistry from dead cells could be circumvented by the study of extracted enzymes. This strategy was summarized by Hopkins in 1931: "One line of experiment among many others that the biological chemist has open to him, and one that has been much employed, is to make whatever possible preparations from a tissue, or from free-living cells, such that a single chemical event (one reaction or a few related reactions) characteristic of the living tissue or cell, can be isolated from the others and studied *in vitro* in respect of the material changes involved, and when possible in respect of kinetics" (Needham and Baldwin 1949, 213).[17]

Indeed, as the twentieth century progressed, Hopkins's approach, which had little consideration for the physical form of cells, a primary concern of cytologists, yielded considerable understanding of degradative and synthetic chemical transformations within cells. However, it was not successful in explaining peculiarly cellular functions such as secretory processes, cell motility, and cell division that are vested in the three-dimensional spatial organization of the cell. Progress in these areas required new tools and strategies that connected the examination of microscopic cellular form with biochemical analysis of the visible and invisible parts. Without these, the elusive linkage between biological structure and function was still out of reach, even if some understanding of individual chemical reactions carried out by cellular enzymes was achieved.

17. David Green, an American biochemist who studied in Hopkins's department in Cambridge and later became a leading investigator of oxidative phosphorylation, stated something similar in a festschrift for Hopkins: "The mastering of a particular machine requires not only a knowledge of the component parts, but also the practical ability to take the machine to pieces and reconstruct the original. . . . The study of mechanism . . . must be extremely limited in dealing with intact tissues. The variation of conditions, which is essential to studies of mechanism, must lie within the confines of those tolerated by living material. The biochemist has therefore to resort to the disorganization of the cell in order to puzzle out the mechanisms of reaction. The major discoveries of the mechanisms which cells utilize for their reactions have practically all been made by the analyses of the behavior of cell extracts and enzyme systems" (Green 1937, 185; see also Matlin 2016).

TWO

The Membrane Boundary

The Membrane Revitalized

In 1925 E. B. Wilson published the third edition of his textbook, now retitled *The Cell in Development and Heredity* (Wilson 1925). Another subtle change, but nevertheless notable, was Wilson's diagram of a generic cell (fig. 6). Unlike similar drawings in the first two editions (cf. fig. 3), this one includes a labeled plasma membrane that forms the boundary of the cell separating the living inside from the nonliving outside world. As noted previously, Wilson and other leading cytologists had earlier almost ignored cellular membranes for two reasons. First, the plasma membrane was not thought to be essential for the living state because it could not be detected by light microscopy in cells such as amoebae that were clearly alive. Second, by the end of the nineteenth century the idea that protoplasm was the key to understanding life had taken hold, leading to a reaction against the cell theory of Schleiden and Schwann that had emphasized the importance of the cell wall or membrane as an essential feature of cell identity and reproduction.

This chapter examines how membranes, the boundaries of both cells and cellular compartments, became recognized and accepted as essential parts of cells by 1925, and then rose to prominence in the post–World War II period when it became clear that membranes were essential to the functional organization of cells. As we shall see, studies of cell permeability, the process by which substances enter and leave the protoplasm of living cells, drove acceptance of the plasma membrane as a universal cell component early in the twentieth century. This work was conducted mainly by physiologists and not cytologists, although the microscope was an essential instrument. Later, biochemists and more physically oriented cytologists created structural and molecular membrane models. By the early 1970s, tools for studying membranes were in place, facilitating the molecular analysis of cellular functions in membrane-bounded cellular compartments.

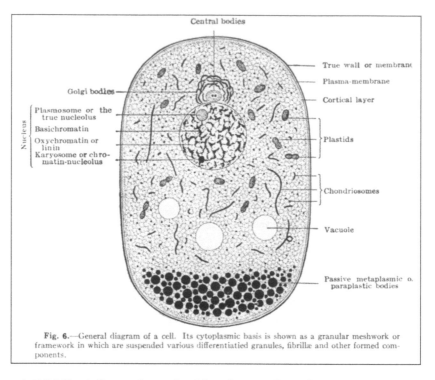

Central bodies

True wall or membrane

Plasma-membrane

Cortical layer

Golgi bodies

Plasmosome or the true nucleolus

Basichromatin

Oxychromatin or linin

Karyosome or chromatin-nucleolus

Nucleus

Plastids

Chondriosomes

Vacuole

Passive metaplasmic o. paraplastic bodies

Fig. 6.—General diagram of a cell. Its cytoplasmic basis is shown as a granular meshwork or framework in which are suspended various differentiatied granules, fibrillæ and other formed components.

6. E. B. Wilson's diagram of a generic cell from the 1925 edition of *The Cell in Development and Heredity* (Wilson 1925). Note the labeled plasma membrane as well as the "True wall or membrane" corresponding, as we now know, to the cell wall of plants.

The Permeability of Plant Cells

Wilhelm Pfeffer, who was born in Germany in 1845, is often unrecognized today for his contributions to biological membrane theory (Pfeffer 1985; Liu 2019). Even though he eventually became known for his work on plants, Pfeffer's doctorate was in chemistry, a background that proved to be an advantage (Pfeffer 1985, xiii–xiv). In 1872 he began studies of osmosis, the movement of solvents such as water across natural and artificial membranes. By this time, osmosis had been studied for nearly a hundred years, commonly using a device consisting of a pig bladder or collodion membrane stretched over the opening of a funnel that was then immersed in a solution.[1] Pfeffer's early work on osmosis was spurred by the chemist and

1. Collodion membranes are the residue produced after nitrocellulose, a modified plant product, is dissolved in ether and the ether is evaporated.

physiologist Moritz Traube's 1867 discovery that a semipermeable membrane could be produced by precipitation of potassium ferrocyanide over a solid support (Traube 1867; Pfeffer 1985; Liu 2019). Pfeffer designed an improved device by forming precipitation membranes within the walls of a clay pot (Seifriz 1936, 185; Pfeffer 1985), enabling him to make accurate quantitative measurements.

In 1877 Pfeffer published *Osmotic Investigations* (*Osmotische Untersuchungen*), summarizing his work over the previous five years (Pfeffer 1985). The first section of the book, titled "Physical Part," consisted of studies using artificial precipitation membranes; the second section, "Physiological Part," described his work with plant cells. To study osmosis in plant preparations, Pfeffer exploited the phenomenon of plasmolysis.[2] When plant cells are submerged in a hypertonic solution (often, various concentrations of the sugar sucrose),[3] the protoplasm shrinks away from the rigid cell wall as water is expelled (fig. 7). Pfeffer measured the degree to which this shrinkage correlates with different concentrations of sucrose and also determined to what extent various natural and synthetic dyes penetrated into the protoplasm.

While it had been assumed before that this shrinkage reflects a property of the protoplasm as a whole (Pfeffer 1985, 137), Pfeffer reasons that it is the surface layer of the protoplasm that is responsible for the osmotic activity. He calls this layer the *plasma membrane.*

> One may call the entire hyaline covering of the protoplasm body the surface layer [Hautschict], surface plasma [Hautplasma] or, perhaps best of all hyaloplasma [Hyaloplasma]. If so, at those places where this forms a thicker layer, there is most probably only an outer zone that is decisive for the diosmotic processes, which we observe in the protoplasm. To express this, I have decided to call this diosmotically decisive layer "plasma membrane" ["Plasmahaut" or "Plasmamembran"], and it is obviously possible that perhaps the whole hyaloplasm is a "plasma membrane," so that both terms become identical. (139)

2. Pfeffer used plasmolysis as a kind of *assay*, defined as a chemical or biological test conducted as part of an experiment.

3. The concentration of water and solutes (ions, sugars, etc.) in the interior of cells exerts a certain osmotic activity or pressure that needs to be balanced by the osmotic activity of fluids outside the cell. When the interior and exterior fluids are balanced, they are known as *isotonic*. If the external fluid has a higher concentration of solutes than the internal fluid, then it is known as *hypertonic*. Hypertonic conditions outside of a cell tend to cause water to flow out of the cell interior to try to dilute the exterior fluid, leading, in the plasmolysis assay, to shrinkage of the cell (see fig. 7).

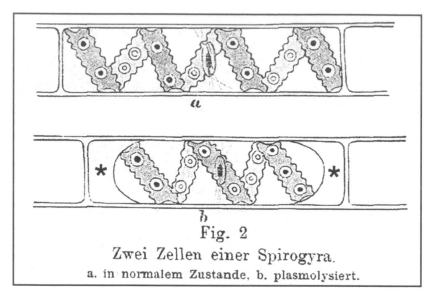

Fig. 2
Zwei Zellen einer Spirogyra.
a. in normalem Zustande. b. plasmolysiert.

7. Plasmolysis in the alga *Spirogyra*. Note the retraction of the cell membrane away from the rigid cell wall in b (asterisks). Measurement of this retraction was used as an assay for osmotic activity (Overton 1895, 162).

Pfeffer speculates that the plasma membrane is formed from proteins by coagulation of protoplasmic protein particles at the surface of the protoplasm, perhaps by a process analogous to formation of artificial precipitation membranes (141–42). He refers to the protein subunits as *tagmas*, imagining tiny diffusion pathways between them (xi, 161–62). In living cells, Pfeffer believes that these particulate "membrane formers" move from the protoplasm to allow the membrane to expand with increased osmotic pressure. In contrast, the membranes of dead cells cannot expand by this process of "intussception" and tear open. Pfeffer, like others, believes that the plasma membrane forms spontaneously when protoplasm contacts an outside aqueous solution, an attitude that was to persevere well into the twentieth century.

Although Pfeffer's plasma membrane theory was eventually accepted in general terms by more physically oriented biologists,[4] it was initially not

4. In Cowdry's 1924 *General Cytology*, both Merle Jacobs, in his chapter on cell permeability, and Robert Chambers, in his chapter on the physical structure of protoplasm, discuss Pfeffer. Jacobs states that "such a [plasma] membrane was postulated by Pfeffer (1887) to account for his osmotic results, and has been accepted as a matter of course by most subsequent inves-

taken up by many cytologists. One reason may be that his work was outside of mainstream cytology, even though his in vivo assay of osmotic activity required microscopic observation of living cells. His work with artificial membranes in particular was quantitative and mathematical and might have been difficult for cytologists without a physical science background to understand. Also, as mentioned previously, his work appeared when cytologists had already dismissed membranes as a significant contributor to the living process (Pfeffer 1985, xii–xiii).

One scientist who not only considered Pfeffer's work important, but even referred to it as "epoch-making" (Overton 1899, 97; Pfeffer 1985, xi) was Charles Ernest Overton, whose own contributions ultimately got cytologists' attention. Although British, Overton received his professional education and spent his entire career in continental Europe. In 1894 he began studies of osmotic activity in plant and animal cells, using for plants the plasmolysis assay previously exploited by Pfeffer. He summarized some of this work in a lecture in Zurich published in 1899 (Overton 1899; Branton and Park 1968, 45 52). Overton's approach was to mix chemical compounds having different molecular properties in sucrose solutions and determine if and to what extent they are able to induce plasmolysis in comparison with pure sucrose solutions. Because his measurements were conducted over specific short time periods, Overton was also able to assess how different substances affect the rate of plasmolysis. He claims in his summary that the conclusions he reports are based on ten thousand experiments conducted with five hundred different compounds!

What he finds is that, in general, the ability of a substance to cause plasmolysis (i.e., exhibit osmotic activity) depends on its relative solubility in ether, "fatty oils," or similar solvents, with those that are most soluble unable to cause plasmolysis (Overton 1899; Branton and Park 1968). For example, a solution of 7.5% sucrose causes plasmolysis within 10 seconds, while a solution of sucrose of even greater osmotic strength that also contains methyl or ethyl alcohol (both fat soluble) fails to induce plasmolysis (Overton 1899; Branton and Park 1968). Based on his experiments, Overton offers a model of cell permeability that is different from that of Pfeffer. Instead of a protein layer with pores analogous to artificial potassium permanganate membranes, he proposes that entry of substances into cells is dependent on their relative solubility in what Pfeffer had called the "diosmotically decisive

---

tigators" (1924, 150). Chambers similarly remarks that "it was Pfeffer (1890) . . . who gave significance to the cell membrane as being responsible for the semi-permeable properties of protoplasm" (1924, 252).

layer." Because Overton finds that the most permeable substances are fat soluble, he suggests that the layer is composed of lipids: "It seems to me very probable that the general osmotic properties of the cell are due to the end layers of the protoplast, which are impregnated by a substance whose dissolving properties for various compounds may well match those of a fatty oil" (Branton and Park 1968, 49).[5] From what is known about plant and animal lipids, Overton speculates that the layer is made up of cholesterol and lecithin (Overton 1899; Branton and Park 1968; Kleinzeller 1997).[6] Despite his respect for the work of Pfeffer, Overton never refers to the layer as a membrane but prefers to define it operationally (Kleinzeller 1997).

Overton's "lipoid theory" was not only noticed, but became prominent in the first decades of the twentieth century. The physiologist William Bayliss discusses Overton's ideas in his textbook, *Principles of General Physiology*, in 1918 (Bayliss 1918, 129–33). Similarly, Overton's ideas and theory are extensively described and critiqued in chapters on cell permeability by Merle Jacobs, on the physical structure of protoplasm by Robert Chambers, and on "cytological constituents" (i.e., organelles) by Edmund Cowdry in Cowdry's encyclopedic edited volume, *General Cytology* (Cowdry 1924a). Jacobs, in particular, clearly believes that Overton's model, while flawed, is the foundation on which much current research stands.

> The theory which has played a greater part in the past in discussions of cell permeability than any other is the lipoid theory of Overton. . . . The grounds on which it is based are several, of which the following are the most important: first, the non-miscibility of the protoplasm with water, which can plausibly be accounted for by the presence of a film of some sort of fatty material; second, the fact that the lipoids are substances which have a high degree of "surface activity" and which might, therefore, reasonably be expected from the Principle of Gibbs to collect at free surfaces, automatically repairing injuries, etc.; and third, the almost perfect correlation which Overton found between the lipoid solubility of hundreds of organic compounds and the ease with which they enter cells. (Jacobs 1924, 152)

5. In the original: "Wenn durch diese und zahlreiche andere Erfahrungen, auf die ich jetzt nicht eingehen kann, es mir sehr wahrscheinlich geworden ist, dass die allgemeinen osmotischen Eigenschaften der Zelle dadurch bedingt sind, dass die Grenzschichten des Protoplasts von einer Substanz imprägniert sind, deren Lösungsvermögen für verschiedene Verbindungen mit denjenigen eines fetten Oels nahe übereinstimmt, so ist es eine ganz andere Frage, gerade was für eine Substanz diese sein möge" (Overton 1899, 109).

6. Lecithin is a generic term for phospholipid. Remarkably, later work confirmed that Overton's speculation was correct.

Even Wilson, who gives membranes and cell permeability short shrift in his third edition of *The Cell*, mentions Overton's theory as "plausibly argued" without giving any specific references (Wilson 1925, 56).[7]

The "Principle of Gibbs" to which Jacobs refers relates to the thermodynamics of surface tension, a force operating at the interfacial surface between two phases. In this case, the two phases are the gel-like protoplasm and the external aqueous environment. According to the Gibbs principle, substances in the protoplasm that are capable of reducing the surface tension are energetically favored to collect at the surface (Moore 1972, 483–84). In particular, lecithin and cholesterol, Overton's proposed components of the surface layer, are poorly soluble in aqueous solutions and, if present in the protoplasm, will spontaneously collect at the interface. This application of the Gibbs principle to cells was satisfying to many cytologists not only because it was consistent with Overton's conclusions but also because it helped explain the frequent observation that a new surface film was quickly reconstituted over protoplasm escaping from disrupted cells (Chambers 1924, 258–60).

The Morphological Membrane

Even though cytologists had by 1925 accepted once again that cells were surrounded by membranes, they were slow to accept the idea that the plasma membrane might be an actual physical entity that existed apart from protoplasm, despite the development of techniques to physically manipulate the cell and its membrane. In the early twentieth century, George Lester Kite, while working at the MBL in Woods Hole, adapted a microscope-attached micromanipulator designed by Marshall A. Barber for bacteriological work so that it could be used for eukaryotic cells. Robert Chambers, who worked with Kite, gradually improved the technique to the point that he could probe, inject, and conduct microsurgery on individual living cells using glass microneedles (Kite and Chambers 1912; Kite 1913; Chambers 1918; Chambers and Chambers 1961) (fig. 8).

Chambers explored the nature of what he called the *surface film* surrounding a variety of cells. He impaled red blood cells and cells of the sea urchin ovary with sharpened needles, observing in the case of the red cells that the contents leaked out, leaving behind a "glutinous and colorless mass" (Chambers 1915, 1924). Chambers also developed a very sharp

---

7. Wilson does cite both Pfeffer and Overton in his book but not their work on cell permeability.

8. Robert Chambers (left panel) and William Seifriz (right panel) at the
Marine Biological Laboratory. Images reproduced from the website "History
of the Marine Biological Laboratory" (http://hpsrepository.asu.edu/).

micropipette capable of cleanly piercing the surface film, enabling injection
of substances directly into the protoplasm (Chambers 1922).

These sorts of experiments convinced Chambers and others that the
surface film was the permeability barrier of the protoplasm, but Chambers
seems to have never accepted the idea that it was a structure separable from
the protoplasm that was replicated along with the rest of the cell. In his
chapter in *General Cytology*, Chambers summarizes his argument: "The importance attached to the plasma membrane in the vital activities of the cell
gave support to its being considered as a definite organ. Just as Virchow had
announced the dictum *omnis cellula e cellula* and Flemming *omnis nucleus e
nucleo*, so De Vries claimed that all plasma membranes originated from a
previous membrane. Pfeffer, however, was able to show that the formation
of surface films is an innate property of any part of the protoplasm and can
be readily produced *de novo*" (Chambers 1924, 252). Even in his posthu-
mously published 1961 book, *Explorations into the Nature of the Living Cell*,
Chambers resists calling the cell boundary the plasma membrane, preferring to the end "protoplasmic surface film" (Chambers and Chambers 1961).

Whether or not the plasma membrane was a separate part of the cell appears to have been of purely academic interest to some cytologists in the first part of the twentieth century. From the perspective of cell permeability, it was sufficient to define the plasma membrane or surface film operationally in terms of what substances could or could not traverse it or how it behaved when probed with microneedles. This state of affairs was, however, challenged by the botanist William Seifriz, a slightly younger contemporary of Chambers who likely worked with Chambers at the MBL and used many of the same techniques. While Seifriz does not disagree with Chambers's conclusions about either the permeability of the surface film or the widespread notion that the surface film can regenerate, based on his studies of slime molds he argues that the surface layer is definitely a "morphological structure": "The old appellation 'naked protoplasm,' much used in reference to slime-moulds, is in important respects a misleading one. The surface layer of a plasmodium is a definite morphological structure. The membrane is very extensile, slowly contractile, and surprisingly tenacious for so delicate a layer. This superficial layer can be isolated and held by one needle while stretched to several times its length by the other" (Seifriz 1918, 309).

In a later paper, Seifriz addresses the issues even more directly.

> No topic in biophysics probably has been the subject of so much controversy as that of the plasma-membrane and its bearing on the problem of permeability. Opinions upon it differ so widely that while its very existence is questioned by some workers, others positively assert that an actual morphologically and physiologically definite surface layer can be clearly demonstrated. Thus de Vries ardently supports its existence as a morphological entity, while Kite . . . views it as a 'hypothetical' structure, and Fischer . . . calls it a 'figment of the imagination.' (Seifriz 1921, 271)

He then comments on Chambers's conception of a surface film as just another part of the protoplasm.

> Chambers's expressions for what might be construed as a plasma-membrane are 'surface film' and 'surface layer,' and the latter he makes synonymous with ectoplasm. . . . Only when the ectoplasm becomes an exceedingly thin layer, as in 'naked' marine ova, is it then even loosely comparable to a plasma-membrane. Chambers's 'surface layer' is not, even when thin, strictly a film or membrane (although he . . . occasionally uses the word membrane), for he recognizes no line of demarcation between the surface layer and the inner

plasm. The surface layer, as viewed by Chambers, is a region which 'merges insensibly' into the cell interior. (Seifriz 1921, 274)

Seifriz's conclusion that the plasma membrane is a morphologically identifiable part of the cell is based on experiments that resemble, in many respects, those of Chambers. However, Seifriz is willing to jump off what has been among cytologists an enormous conceptual cliff: the incorporation of evidence from so-called dead cells. As he states, "Where a structure cannot be readily seen in the living condition, but does, because of some physical or chemical change, become clear when dead, we are not, therefore, justified in utterly disregarding the evidence based on dead material. For example, the presence of a protoplasmic membrane in the living state cannot be indubitably established, yet the presence of the dead plasma-membrane is in some instances strikingly evident. As evidence for support of the existence of a living membrane, the presence of a dead membrane is not to be altogether ignored" (Seifriz 1921, 270). Later in the same paper he remarks, "The reactionary attitude of our science to observations on dead material is a healthy one. The criticisms directed against any evidence which purports to prove the existence of a living membrane by the presence of a dead one are legitimate. . . . But the dead membrane is to all appearances so evidently a structure having the same position and dimensions as that which can sometimes be observed on living protoplasm, that it is worthy of some consideration as evidence of the existence of a membrane about the living substance" (Seifriz 1921, 280).

How Seifriz's arguments were received among cytologists is not clear. Chambers cites Seifriz's papers in his chapter in *General Cytology* but does not explicitly discuss the issue of a morphological membrane or Seifriz's incorporation of conclusions based on "dead material." Wilson does not mention Seifriz at all (Wilson 1925). Perhaps, as stated earlier, what Seifriz considered of crucial interest was moot among cytologists. The existence of a plasma membrane distinct from the protoplasm apparently remained very much an issue even later, with arguments similar to Seifriz's appearing in work on isolated plant protoplasts reported by Janet Plowe in 1931 (Plowe 1931; Branton and Park 1968). In any case, advancement of the membrane concept did not lie in cytologists' hands but rather in those of physical chemists.

The Lipid Bilayer

The concept of colloids, tiny particles too large to pass through an artificial membrane but too small to easily settle in a solution, had by the twentieth century displaced other models of protoplasm structure (see chap. 1, above).

Part of the attraction of this model was that colloidal particles presented a huge surface area where it was believed that important biochemical reactions occurred (see Hopkins 1913). Among physical chemists, surface chemistry and its relationship to thermodynamic principles and molecular concepts became an area of great interest (Liu 2018). In Britain in the 1880s and 1890s, measurements of surface tension were made by Agnes Pockels, a self-educated chemist, and Lord Rayleigh (John William Strutt). Using a homemade instrument, Pockels found that oil placed on the surface of water reduced its surface tension, and Rayleigh, who had performed similar experiments in his bathtub, interpreted the findings in terms of the packing of molecules on the aqueous surface (Liu 2018). Some years later, Irving Langmuir, a physical chemist working at the General Electric Corporation laboratory in the United States, repeated these experiments in a more refined fashion using an improved version of the Pockels device (subsequently known as the Langmuir trough) and purified lipids. By this time, lipid molecules were beginning to be understood as amphipathic, with one end polar and the other end nonpolar or hydrophobic, and they were represented as a ball and stick, with the ball corresponding to the polar head and the stick, to the hydrophobic tail (Liu 2018). Based on his studies, Langmuir concluded that lipids on the surface of aqueous solutions were oriented such that the polar head interacted directly with water while the hydrophobic tail extended upright, away from the solution.

In 1925, Evert Gorter and François Grendel proposed a structural model for the organization of lipids in membranes (Gorter and Grendel 1925). After counting the number of red cells purified from blood samples, they estimated the surface area of individual red cells based on measurements in the microscope and calculated the total surface area corresponding to a particular number of cells. They then extracted the lipids of these cells using organic solvents and spread the lipids on water in a Langmuir trough to determine how much area the lipids occupied when packed together. Based on these measurements, they determined that the area occupied by the lipids was approximately twice the surface area of the red cells used for the extraction. From the work of Langmuir and others on the orientation of lipids at an aqueous interface, they concluded that the membrane surrounding red cells is a bilayer of lipids oriented such that "the polar groups are directed to the inside and outside" of the bimolecular layer (Gorter and Grendel 1925). While their general conclusion turned out to be right, a later reexamination of their experiments suggested that they had made fortuitous offsetting errors that enabled them to suggest a fundamentally correct model (Bar, Deamer, and Cornwell 1966).

Although James F. Danielli was responsible, with his colleague Hugh Davson, for a subsequent very influential membrane model that appeared to build on the work of Gorter and Grendel, he claimed late in his career that he was unaware of the red cell paper when he and Davson proposed their model in 1935 (Danielli and Davson 1935; Danielli 1975). Danielli, who was trained in physical chemistry at University College London by Neil A. Adam, a pioneering surface chemist, believed that the bilayer organization of lipids in membranes "flowed almost automatically from consideration of the basic properties of these molecules and would have been obvious to any competent physical chemist" (Danielli 1975). Danielli's work on biological surfaces began during postdoctoral studies at Princeton University and the MBL with E. Newton Harvey, an American biologist (Stein 1986; Stadler 2009). Harvey was interested in the physical properties of protoplasm and, in 1930, invented a microscope-centrifuge combination with Alfred Loomis that allowed him to continuously observe the deformation and stratification of cellular components during the centrifugation of living, whole cells (Harvey and Loomis 1930). In 1934, Harvey and Herbert Shapiro published a study using the cell centrifuge to measure the interfacial tension between a naturally occurring oil droplet in mackerel eggs and the remaining egg protoplasm, finding that the surface tension was much lower than expected (Harvey and Shapiro 1934). In 1935, Harvey and Danielli followed up on this work by measuring the interfacial tension between extracted mackerel oil and the "aqueous egg contents" prepared from crushed eggs. In this case they were able to make much more accurate measurements using a sensitive instrument, the du Nouy interfacial tensimeter, than was possible with the centrifuge (Danielli and Harvey 1935). When they compared the surface tension at the interface of mackerel oil and egg contents with that of oil and water, they observed that values derived from measurements using the egg contents were much lower than those of pure water, consistent with the findings in the whole egg studies. They concluded that proteins in the egg material accumulated at the surface and lowered the surface tension at the oil interface. Based on these observations, Danielli and Harvey suggested that a coating of proteins covered the polar ends of the lipid molecules of the oil at the interface (Danielli and Harvey 1935).

After returning to England, Danielli collaborated with his friend and former classmate Davson, a physiologist, to propose a model of a semipermeable cell membrane based on his work with Harvey. The model consists of a lipid bilayer coated on both polar sides with a layer of globular proteins (fig. 9). As mentioned previously, Danielli likely considered the organization of lipids into a bilayer to be self-evident; the novel aspect of the model was the addition of protein. It was the protein layer that not only helped explain the surface tension anomalies, but also the ionic permeability of membranes (Danielli and Davson 1935). With some modifications to account for more

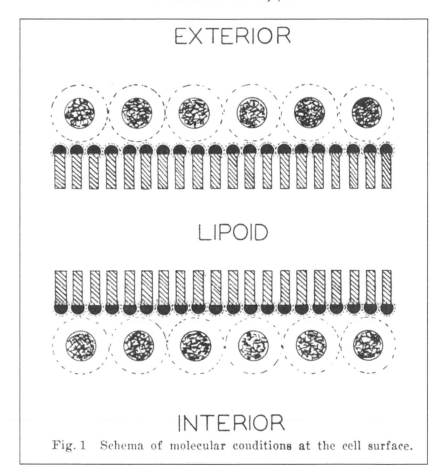

EXTERIOR

LIPOID

INTERIOR

Fig. 1   Schema of molecular conditions at the cell surface.

9. The first published illustration of the biological membrane model proposed by James Danielli and Hugh Davson (Danielli and Davson 1935, 498). The large circles represent globular proteins and the vertical sticks, lipid molecules. The dark "heads" of the lipids are polar or hydrophilic, and the crosshatched "tails" are nonpolar or hydrophobic. © John Wiley and Sons. Used with permission.

sophisticated aspects of membrane permeability, the Danielli-Davson proposal remained popular until the early 1970s (Davson and Danielli 1943, 1952; Branton and Park 1968, 6).

Biological Membrane Structure

Even with these advances, there was little direct information on the physical characteristics of membranes. With the exception of the microsurgical manipulations of Chambers and Seifriz, the plasma membrane at the cell surface could not be easily seen by conventional light microscopy, suggesting that its thickness was below the 0.1–0.2μm resolution limit.[8] If one accepted that membranes were lipid bilayers, as suggested by Gorter and Grendel and later by Danielli and Davson, then a calculation based on the size and presumed orientation of lipid molecules indicated that the membrane should be about 40Å (0.004 μm) thick (Höber 1945, 277).[9] During the first third of the twentieth century, electrical impedance measurements on concentrated suspensions of cells yielded rough estimates of anywhere from 33 to 200Å (Davson and Danielli 1943; Höber 1945, 278). This range of dimensions was not appreciably narrowed by the application of a variety of other physical techniques to the problem (Davson and Danielli 1943; Branton and Park 1968, 5–10).

The use of electron microscopy to examine cells turned out to be the breakthrough needed to advance structural studies of membranes. Electron microscopy was invented in the 1930s and first used on biological specimens in 1934 (Marton 1934; Ruska 1988; Rasmussen 1997) (for a fuller discussion, see chap. 3, below). Beams of electrons, which the electron microscope uses to illuminate specimens instead of visible light, have a much shorter wavelength than light and, consequently, a resolution that is at least several hundred times higher than that of the light microscope. By the 1950s, electron microscopy had demonstrated the existence of membranes not only on the surface of cells but also surrounding other cellular inclusions, or *organelles*. Some of these intracellular structures, including the nucleus, Golgi complex,[10] and mitochondria, were well known to cytologists because they could be seen by light microscopy, while others, such as the endoplasmic reticulum, owed their discovery to the use of the electron microscope (see chap. 3, below).

J. David Robertson stands out among other electron microscopists studying membranes in the 1950s because he synthesized observations from a variety of cells to conclude that all biological membranes have a basic

8. The membrane surrounding the nucleus was more easily visible, probably because it consists of two layers.

9. The angstrom (Å) is 10^{-10} meters. It is named after the Swedish physicist Anders Jonas Ångstöm.

10. The Golgi complex or apparatus was discovered by the Italian physician Camillo Golgi in 1898 using a histological stain that he devised. In cells it is typically located next to the nucleus. It consists of an array or stack of flattened membrane vesicles with smaller surrounding vesicles (Dröscher 1999).

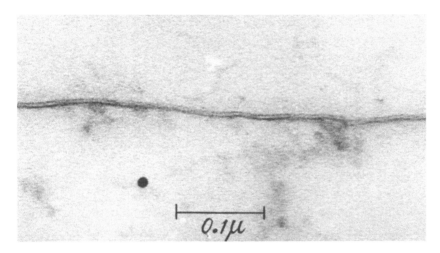

10. Robertson's unit membrane as seen in an electron micrograph of *Amoeba proteus*, the organism believed by cytologists in the nineteenth century to lack a plasma membrane (Danielli, Pankhurst, and Ridditord 1958, 260). © Elsevier. Used with permission.

structure in common. He called this the *unit membrane model* (Robertson 1959, 1981). Robertson based much of his work on observations of the myelin sheath, a thick insulating layer surrounding the axonal processes extending from nerve cells that is composed of a series of concentric membrane rings. Unlike many other investigators, Robertson used a fixative containing potassium permanganate instead of the more common osmium tetroxide because he observed that the appearance of membranes in cells fixed with osmium was inconsistent (Robertson 1959). With potassium permanganate, he visualized membranes in not only myelin but also other types of cells as two dark lines with a lighter space in between, appearing as a kind of railroad track in favorable cross sections (fig. 10). The overall thickness of the unit membrane was 75Å, with two ~20Å thick dense lines bordering the lighter 35Å thick central zone (Robertson 1959). Robertson considered his unit membrane model to be completely consistent with the general proposal of Danielli and Davson, interpreting the darkly stained lines as protein layers or other nonlipid material on the surfaces of the lipid bilayer. Given that the molecular dimensions of lipid molecules had been inferred from their chemical structure, Robertson's model strongly supported the organization of lipids into a bilayer, as suggested earlier based on principles of physical chemistry. While criticized later because it did not account for the differentiated characteristics of membranes from individual cell types or organelles (Robertson 1981), Robertson's generalized structural model remained a foundation on which further membrane studies were built.[11]

The Membrane as an Asymmetric Fluid Mosaic

The Danielli-Davson membrane model (sometimes referred to as the Danielli-Davson-Robertson model to account for Robertson's unit membrane concept) continued to be accepted into the 1970s (Singer 1971). However, there were clearly phenomena that the model had difficulty dealing with. One of these was the permeability of membranes to both lipid soluble and water-soluble substances, the latter including both ions, such as sodium and chloride, and metabolites, such as sugars, as well as water itself. These were long-standing issues. Even at the time that Overton speculated that the membrane was composed of lipids (or lipoids), an idea based on the failure of lipid soluble organic compounds to be osmotically active, Alexander Nathansohn postulated that membranes were a *mosaic* of protoplasmic particles, possibly proteins, and lipids as a way of rationalizing membrane permeability (Nathansohn 1904a, 1904b; Jacobs 1924, 155).

[The] easy permeability for liposoluble substances is not explainable by an impregnation of the plasma membrane with *Lecithin*. It has rather to be ascribed to a compound that does not have the ability to absorb water in higher quantity, therefore behaving physically similar to *Cholesterin*. But at the same time there is easy permeability for water and one has to assume that the majority of viable cells continuously exchange water with their neighbors. Furthermore, one has to assume regulatory permeability for compounds that are not liposoluble. Taken together, this forces us to conclude that this compound does not surround the plasma body in a continuous layer, but rather has a secondary role in the composition of the plasma membrane. We can imagine this composition approximately in the following way: the interspaces between the living particles of the protoplasm are filled with that lipoid substance. Liposoluble compounds could therefore always pass these interspaces without obstruction. Water and water-soluble compounds, on the

---

11. Despite the prominence of the lipid bilayer membrane model stemming from the work of Gorter and Grendel, Danielli and Davson, and Robertson, direct evidence for its existence did not come until the X-ray diffraction studies of Donald Engelman in 1971 (Engelman 1971).

other hand, would have to pass through the particles of the living substance. (Nathansohn 1904a, 642–43)[12]

Although aspects of Nathansohn's proposal turned out to be prescient, at the time his ideas were purely speculative and had little direct impact on subsequent membrane models (Jacobs 1924, 155).

When Danielli and Davson made their original proposal, they had added protein layers to the lipid bilayer not only to address the unexpectedly low surface tension found at the surfaces of cells, but also to account for aspects of membrane permeability (Danielli and Davson 1935). They speculated that the protein layer acts as a sieve to control the permeability of certain substances in conjunction with the continuous lipid bilayer (Danielli and Davson 1935; Davson and Danielli 1943). Danielli and Davson suggested that the sources of the proteins bound to the inside of the lipid bilayer were the cell contents, but it was unclear where the proteins coating the outside of the bilayer, as shown in the model, originated (see fig. 9). In 1943, when they published a monograph on membrane permeability, very little was known about the structure of proteins. Consequently, they suggested that on binding to the lipids, the proteins coated the surface in an unfolded and extended configuration (Davson and Danielli 1943). By 1958, Danielli acknowledged in a paper that "although most molecules permeate as though the cell membrane is a homogeneous lipoid layer, other substances were shown, by strictly kinetic considerations, to permeate with exceptional speed through the very limited areas of the membrane which had a high specificity towards a very limited number of molecular species" (Danielli 1958, 248). These included water-soluble molecules like glucose, glycerol,

---

12. In the original: "Suchen wir nun auf Grund all dieser Betrachtungen und Thatsachen zu einem Schlusse zu gelangen, wie die Verhältnisse thatsächlich liegen mögen, so lässt sich folgendes sagen. Die leichte Permeabilität für fettlösliche Stoffe kann nicht auf Imprägnation der Plasmahaut mit Lecithin beruhen, sondern wäre einem Körper zuzuschreiben, der nicht die Fähigkeit besitzt, Wasser in merklicher Menge aufzunehmen, der sich also in physikalischer Hinsicht dem Cholesterin ähnlich verhielte. Da aber gleichzeitig eine leichte Permeabilität für Wasser besteht, und ein grosser Theil der lebensthätigen Zellen in stetem Wasseraustausch mit den Nachbarinnen stehen dürfte, da fernerhin allenthalben eine regulatorische Durchlässigkeit für fettunlösliche Stoffe anzunehmen ist, so sehen wir uns zu dem Schlusse gedrängt, dass jener Stoff den Plasmakörper nicht in continuirlicher Schicht umgiebt, sonder nur eine secundäre Rolle im Aufbau der Plasmahaut spielt. Wir können uns diesen etwa so vorstellen, dass die Interstitien zwischen den lebenden Protoplasmatheilchen von jener fettartigen Substanz angefüllt sind. Die fettlöslichen Stoffe würden so stets ungehindert durch diese Interstitien passiren können, das Wasser dagegen, und die wasserlöslichen Stoffe müssten durch die Teile der lebenden Substanz ihren Weg nehmen" (Nathansohn 1904a, 642–43). Translation ©Hanna L. Worliczek. Used with permission.

urea, and chloride ions. In an accompanying drawing representing a revision of the original model, the lipid bilayer is shown to be penetrated by a pore formed by "polypeptide lamellae" (see fig. 2 in Danielli 1958).[13]

In 1966, two papers appeared in the *Proceedings of the National Academy of Sciences* (*PNAS*) that challenged the idea that proteins associated with the lipid bilayer were in an extended configuration layered over the polar surfaces of the lipid bilayer (Lenard and Singer 1966; Wallach and Zahler 1966).[14] Taking advantage of recently developed methods for isolating plasma membranes from cells, both studies used optical techniques based on polarized light, including optical rotary dispersion (ORD) and circular dichroism (CD), to directly examine the organization of membrane-associated proteins.

As described previously, proteins are composed of strings of twenty different amino acids linked head-to-tail by peptide bonds, yielding long polypeptide chains. In 1951, the physical chemist Linus Pauling and his colleagues proposed that polypeptide chains are often folded into two universal configurations that he called the α-helix and β-sheet that are not dependent on a precise sequence of amino acids (Pauling, Corey, and Branson 1951). Both are stabilized by a type of noncovalent bond between amino acids called a hydrogen bond. Each amino acid found in proteins has a *side chain* extending away from the polypeptide backbone that is either polar (hydrophilic) or nonpolar (hydrophobic). Polar side chains can readily associate with aqueous solutions, while nonpolar side chains prefer environments devoid of water or other charged molecules. In the α-helix the side chains of the amino acids making up the polypeptide backbone point to the outside of the helix. This means that the outer surface of the helix can be either polar or nonpolar or a combination of the two depending on the specific amino acids in the helix.

Jonathan Singer, an author of one of the 1966 papers, had been a postdoctoral fellow with Linus Pauling and, consequently, was well versed in concepts of protein folding. Using CD, ORD, and other techniques, his research group, as well as one led by Donald Wallach, an author of the second paper, found that proteins in membranes contain significant regions folded into α-helices. Other measurements suggested the possibility that

13. All proteins consist of *polypeptides*, a term that refers to the chain of amino acids linked by peptide bonds. Some proteins are made up of only one polypeptide chain, while other multisubunit proteins contain more than one polypeptide. In practice, the terms "polypeptide" and "protein" are often used interchangeably.

14. See also Ke 1965.

membrane proteins might penetrate into and possibly through the lipid bilayer (Lenard and Singer 1966; Wallach and Zahler 1966). At the time it was well known that proteins not found in membranes that are completely soluble in water tend to be folded such that hydrophilic amino acids are on the outside of the protein while any hydrophobic amino acids are deep inside the protein where they are protected from contact with water. Based on this, both groups of investigators concluded that some membrane proteins are folded into hydrophobic α-helices in distinct regions inserted into the lipid bilayer adjacent to the hydrophobic lipid tails, while parts of membrane proteins exposed to the aqueous medium on membrane surfaces are folded into hydrophilic configurations.

Five years later, Singer extended these ideas in a detailed chapter that reviewed the thermodynamics of protein folding and lipid bilayers and critically evaluated the prevailing Danielli-Davson-Robertson membrane model (Singer 1971). Singer argued that the molecular organization of membranes is not primarily determined by interactions between proteins and lipids but driven instead by the tendency of hydrophobic parts of molecules to associate with each other in an aqueous environment as a way of achieving a thermodynamically stable state. For lipids, which possess a hydrophobic fatty acid tail and a hydrophilic head, this was nothing new. In the 1930s, Danielli had appealed to such considerations to justify the inclusion of a lipid bilayer in his model (see Danielli 1975). What was new was that Singer extended such considerations to proteins, claiming that many membrane proteins are organized like lipids with distinct hydrophobic and hydrophilic regions and that the existence of hydrophobic regions in proteins causes them to be inserted deeply into the lipid bilayer. Singer proposed two types of membrane proteins that he called integral and peripheral. Integral membrane proteins are the ones with hydrophobic parts that, operationally, can only be separated from the lipid bilayer by harsh procedures that effectively disrupt the membrane, sometimes causing the integral proteins to form an insoluble aggregate. Peripheral membrane proteins, on the other hand, are, in a sense, not really membrane proteins at all but instead fairly typical water-soluble proteins that just happen to associate electrostatically with the exposed hydrophilic parts of lipids and integral membrane proteins on the membrane surface. As a protein chemist, Singer was essentially extending the principles that governed the folding of proteins in aqueous solutions to the situation where proteins needed to fold in the hydrophobic environment found in the interior of the lipid bilayer. Similar ideas also led Singer to speculate that integral membrane proteins might associate in the plane of the membrane to form multisubunit pores: "Although pores, and even protein pores, have been postulated as ion-transport mediators in membranes, simple proteins do not exhibit pores, their internal structure being quite compact for good thermodynamic reasons. . . . Subunit proteins, however, often contain narrow water-filled channels near the central axis of the aggregate, and similar membrane proteins could thereby provide specific pores across the membrane" (Singer 1971, 205). On the basis of these arguments, Singer proposed a new membrane model to replace that of Danielli, Davson, and Robertson, which he called the *lipid-globular protein model* (fig. 11).

Singer had included in his discussion new and significant sets of observations based on an electron microscopic technique called freeze etching (or, sometimes, freeze fracture). Unlike conventional electron microscopy that, at this time, relied on fixation and sectioning of cells and tissues (see chaps. 3 and 4), freeze etching is conducted by rapidly freezing biological samples to very low temperatures and, while maintaining them in the frozen state, placing them in a vacuum chamber where they are struck with a metal knife to break them apart (fracturing). At the same time, the vacuum evaporates frozen water from the surfaces of the specimens (the etching process). To visualize these samples, the fractured and etched specimens are coated with a very thin layer of carbon and platinum (or other metals) sprayed from electrodes in the vacuum chamber to create exact topological *replicas* of the biological material that can be viewed in the electron microscope. In early observations using this technique, it appeared that etching removed water from some of the surfaces exposed by fracturing but not others (Branton 1966). Further analysis suggested that the unetched surfaces were in fact the exposed *interiors* of lipid bilayers, which, because of their hydrophobicity, were not coated with water. Apparently, when frozen membranes are fractured in the vacuum chamber, they tend to split between the two halves of the lipid bilayer. Even more interesting was the observation that the inside of the lipid bilayer was studded with particles, with, in some cases, matching pits or depressions on the opposite face of the membrane. While the nature of these was unknown, one suggestion was that these corresponded to proteins inserted into the lipid bilayer (Branton 1966; Pinto da Silva and Branton 1970).

In 1972, Singer and Garth Nicolson modified Singer's original proposal, now calling it the *fluid mosaic model* (Singer and Nicolson 1972) (fig. 11). In their article, they summarized the arguments from Singer's 1971 review but then extended them based on their own experiments and recent work by L. David Frye and Michael Edidin. In 1970, Frye and Edidin published a study of the effects of cell-cell fusion on the distribution of cell surface

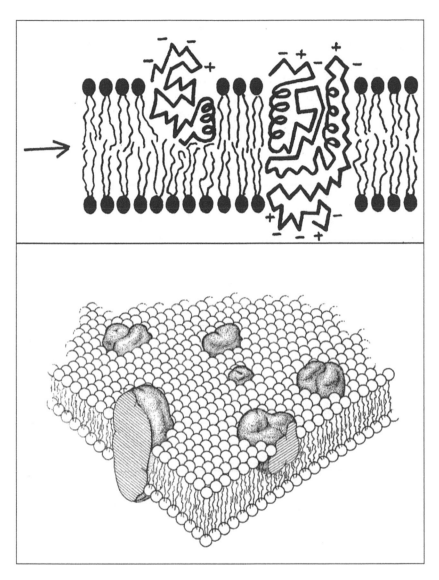

11. Singer's original lipid-globular protein mosaic model of membrane structure (upper panel) and the later fluid mosaic model (lower panel) (figures 2 and 3 from Singer and Nicolson 1972). The springlike parts of proteins in the upper panel represent α helices.
© American Association for the Advancement of Science. Used with permission.

proteins belonging to one cell or the other (Frye and Edidin 1970).[15] In their experiments, they fused cells from mice and humans and examined what happened to the distribution of mouse and human surface proteins over time as the plasma membranes contributed by each type of cell coalesced into one. Using fluorescence microscopy to separately track the surface proteins of mouse and human cells, they found that, initially, the two sets of surface proteins remained separated on the plasma membrane of fused cells. However, within forty minutes they became totally intermixed. After controlling for other possibilities, they concluded that the likeliest explanation for this was that the surface proteins had *flowed* together once the two individual plasma membranes became continuous (Frye and Edidin 1970).

Singer and Nicolson used these observations, as well as other biophysical measurements of lipid behavior in the membrane, to suggest that integral membrane proteins and lipids move independently in the lipid bilayer, as if the proteins are floating in a sea of fluidlike lipids.

> It has been shown that, under physiological conditions, the lipids of functional cell membranes are in a fluid rather than crystalline state. . . . This evidence comes from a variety of sources, such as spin-labeling experiments, x-ray diffraction studies, and differential calorimetry. If a membrane consisted of integral proteins dispersed in a fluid lipid matrix, the membrane would, in effect be a two-dimensional liquid-like solution of monomeric or aggregated integral [membrane] proteins (or lipoproteins) dissolved in the lipid bilayer. The mosaic structure would be a dynamic rather than a static one. (Singer and Nicolson 1972, 724)

They, in addition, explicitly address another issue that by this time had become an obvious implication of their studies and those of others: the proteins exposed on one surface of a biological membrane are different from the proteins exposed on the other surface. That is, membranes are *asymmetrically organized*. Some evidence for this came from their own work.

By 1972, plant proteins called lectins were known that specifically bound certain complex carbohydrates (polymers of modified sugar molecules) that are covalently linked to some membrane proteins. When lectins tagged with markers that could be seen in the electron microscope were applied to both the inside and the outside of isolated red blood cell plasma membranes, they bound only to the outside. This meant that the protein-linked complex

---

15. This work was mentioned in a footnote in Singer's 1971 article (Singer 1971) but apparently appeared too late to influence his membrane model.

carbohydrates were only on the outside membrane surface and implied that the proteins or parts of proteins exposed on the inner surface were completely different (Singer and Nicolson 1972). Singer and Nicolson also applied thermodynamic arguments to suggest that once integral membrane proteins are in place on one surface of the membrane, they are very unlikely to flip through the hydrophobic interior of the lipid bilayer to the other side. On this basis they argued that not only are biological membranes asymmetrically organized, but that this asymmetry is stable.[16]

Just as they were publishing their new membrane model, more direct evidence emerged for not only membrane asymmetry but also the existence of proteins that spanned the lipid bilayer completely (so-called transmembrane proteins). Wallach and Theodore Steck and Gary Fairbanks were studying human erythrocytes and had developed procedures to isolate the red cell plasma membrane from its cytoplasmic contents. The beauty of the system was that the membrane could be isolated as an intact and closed vesicle oriented either right side out, as in the normal erythrocyte, or inverted, such that the inside of the plasma membrane was exposed as the outer surface. When the membrane was treated with proteolytic enzymes capable of digesting proteins inserted in the membrane, they discovered that some proteins could be digested from both sides but that the degree of digestion depended on which side was treated with the enzymes. These experiments suggested not only that some proteins completely spanned the membrane but also that they were inserted in the membrane in only a single asymmetric orientation (Steck, Fairbanks, and Wallach 1971). Their results were supported by a separate study by Mark Bretscher. Bretscher, also working with erythrocyte membranes, used a radioactive chemical reagent that was not able to cross the membrane to show that a major red cell membrane protein could be labeled from both sides of the membrane. When Bretscher analyzed the parts of the protein labeled from one side or the other by peptide mapping, they were different, definitively demonstrating that the protein was asymmetrically inserted through the lipid bilayer (Bretscher 1971a, 1971b).

At this point, the fluid mosaic model of Singer and Nicolson, which was based on sound biochemical and thermodynamic arguments and supported by detailed molecular and morphological studies, completely supplanted

16. It was independently shown by others that the mixture of lipids in the inner half (or "leaflet") of the lipid bilayer is different from the mixture in the outer half (Bretscher 1972).

the Danielli-Davson-Robertson model.[17] Finally, cellular membranes, whose existence had been debated for years by cytologists and whose properties began to be deciphered by physiologists studying osmosis and transport, were on a firm footing. While the development of new, more sensitive and direct methods to analyze membrane lipids and proteins later challenged some generalizations of the fluid mosaic model, it remained the foundation of all membrane studies for the remainder of the twentieth century.

Although not emphasized at the time, the existence of stable membrane asymmetry in cells has implications beyond membrane organization per se. Asymmetry that is biologically created and maintained is a form of information in the same way that a particular array of ones and zeros representing a digitally stored document is information encoding the document. As we shall see, the recognition that *spatial* information embodied in membrane asymmetry is encoded in the signal sequences studied by Blobel is one of the most important outcomes of his work (see chaps. 5, 7, and 9).

I now return to the 1920s, when biochemistry was advancing and cytologists were contemplating new approaches to explore cellular functions. While the old obstacles to progress remained, the introduction of new people and new techniques from outside the discipline of cytology was about to make a difference.

17. The edited volume *Cell Membranes: Biochemistry, Cell Biology, and Pathology*, published in 1975, included chapters written by both Danielli and Singer. In his chapter, Danielli states, "Whether and to what extent the details of this [Danielli's] 1950's model of the cell membrane have been confirmed by subsequent research will be discussed extensively by other contributors to this volume. . . . What seems incontestable, however, is that the cell membrane does consist of a lipid-protein 'mosaic' of the sort originally suggested—the lipids serving to isolate the cell from its environment, with the liquid lipid bilayer acting as a two-dimensional solvent for the macromolecules" (Danielli 1975, 9). In saying this, Danielli not only attempts to claim that his earlier membrane model had features similar to the fluid mosaic model described by Singer in this same volume; he also acknowledges in a backhanded fashion that the fluid mosaic model has replaced his. For his part, Singer states at the end of his contribution, "The model of membrane structure discussed in this chapter, after a few years' gestation in the scientific literature, has emerged and been widely accepted in the last year or so, a time in which many different kinds of experimental results were obtained that could be explained by the model" (Singer 1975, 44).

THREE

Breakthroughs

The Cell and Its Parts

As the twentieth century entered its second quarter, studies of the cell had advanced, but fundamental issues remained. Cytologists were stalled at the cell boundary, unwilling to break it open and unable to see much inside it with their light microscopes. Edmund Cowdry's 1924 edited volume, *General Cytology*, appeared with contributions from many of the leading American biologists (Cowdry 1924a). E. B. Wilson, who wrote a historical introduction to Cowdry's book, almost simultaneously published the third edition of his classic textbook, now greatly expanded to reflect important advances that linked the dynamics of chromosomes during cell division with the new field of genetics (Wilson 1925). Both books were benchmarks in cell studies but for different reasons. Wilson's text, even with significant additions, was essentially a review and revision of past accomplishments in cytology since the cell theory had emerged from the work of Schleiden and Schwann. As described previously, Wilson now accepts the ubiquity of cell membranes and even suggests that they may surround vacuoles in the protoplasm (Wilson 1925, 55). Statements about the functions of most protoplasmic formed elements, however, remain vague and inconclusive. By this time, biologists had largely abandoned efforts to discover a definite morphological structure of protoplasm, and Wilson concedes that protoplasm is essentially a submicroscopic colloid, a concept more compatible with emerging ideas of protoplasmic chemical organization favored by physical chemists and biochemists (Hopkins 1913, 78; Wilson 1925; Liu 2016, 2017; see also chap. 1, above).[1]

---

1. By the time of Wilson's book, the term *cytoplasm* was supplanting *protoplasm* as the name of the cell's interior contents, excluding the nucleus. As Wilson notes, protoplasm originally referred to the "cytosome" and not the nucleus. Later it began to be used to describe the cell

Cowdry's book, on the other hand, looked forward, attempting to predict how a multidisciplinary approach might finally get at the details of cell function. Individual chapters dealt with cell chemistry, permeability, "reactivity," and physical properties (Cowdry 1924a). Others highlighted the new technique of cell culture and the more traditional topics of fertilization and differentiation. Two chapters focused on chromosomes and Mendelian theory. Notably, only Cowdry's own rather unilluminating chapter highlighted the morphological approach, coupling descriptions of mitochondria, the Golgi apparatus, and something called the "chromidial substance" with functional speculations and statements of ignorance (Cowdry 1924b; see also Maienschein 2018).

Indeed, by the 1920s, the use of light microscopy alone to discover new things about the cell had largely run its course.[2] Certainly, there were incremental advances in optics, but the resolution limit remained an impediment. Chemical dyes helped to highlight various intracellular structures. With the exception of chromosomal studies, however, little new was being discovered. Certain stains with known chemical characteristics seemed promising as a means to probe cell chemistry and the organization of chemical reactions and molecules within the intact cell. Yet significant advances with these techniques remained few and far between (Claude 1946a, 15; Danielli 1953, 1–15).

Almost all of the investigative strategies highlighted in Cowdry's book shared one characteristic: their subjects were the intact and often living cell. Studies of permeability and reactivity described by Merle Jacobs and Ralph Lillie depended on monitoring how whole cells responded to mechanical, electrical, thermal, radiant, chemical, and osmotic stimuli (Jacobs 1924; Lillie 1924; Maienschein 2018), while the main strategy for probing the physical properties of the cell described by Robert Chambers was to touch, impale, bisect, and inject living cells with microsurgical implements (Chambers 1924). Whether because of the lack of methods or philosophical hesitancy, many biologists seemed unable to overcome the belief that had overshadowed cytology for decades: results obtained with disrupted dead cells could not provide insights into living processes. Biochemists,

---

substance as a whole, including the nucleus, and the term "cytoplasm" took on the original meaning of protoplasm (Wilson 1925, 22).

2. One can argue that progress continued to be made by means of light microscopic observations of recognizable particulate constituents of cells, such as the insightful studies of Bowen on the role of the Golgi apparatus in secretion. However, even this work was somewhat undermined by later claims that the Golgi was a morphological artifact (Bowen 1926; Palade and Claude 1949b).

on the other hand, had no such restrictions. After accepting the idea that enzyme specificity was not dependent on the intact cell, they pursued a strategy where full destruction of cell structure was acceptable as long as enzyme activity was preserved and began to decipher metabolic pathways. Their now formless cell, however, bore little resemblance to the cell cytologists had studied throughout the nineteenth century.

A Literal Breakthrough

By the 1930s, strategies for working with broken cells that still retained some structural organization began to emerge. One can argue that Johann Friedrich Miescher's isolation of nuclei in 1871 and Otto Warburg's preparation of metabolizing particles from liver cells in 1913 used such an approach (Claude 1948; Bechtel 2006, 128; Moberg 2012, 29; Veigl, Harman, and Lamm 2020). However, Robert R. Bensley and his collaborators at the University of Chicago, in particular Norman L. Hoerr, were the first to develop a systematic method for the isolation of cell parts. At first glance, Bensley seems an unlikely person for such an advance (fig. 12). Bensley, whose long career at Chicago began in 1901, was primarily an anatomist and histologist, although one who attempted to decipher function by combining histological staining reactions with morphology (Hoerr 1957). In 1932, Bensley began the work that led to the separation of cell parts by attempting to deal with a well-known problem in the application of chemical fixatives and stains to cells and tissues.

> The investigation of the distribution of chemical substances in cytoplasm has been beset with extraordinary difficulties. When reagents are applied to the living cell or the surviving cell the results are not easily interpreted, because the reagent may penetrate slowly and slowly achieve a considerable concentration in the cell. Moreover, the reagent is acting not on the individual constituents of the cell whose distribution is sought, but upon the cytoplasm as a whole, and the result is a composite one depending on the interaction of all these several factors. Moreover, the excitation of cytoplasm by even minute quantities of reagents employed in microchemical investigations may so disturb the equilibrium between several phases of cytoplasmic structure that the distribution of substances in the cell as a whole is completely changed. (Bensley and Gersh 1933a, 208)

To circumvent these problems, Bensley, with his graduate student Isidore Gersh, revived a procedure for fixing cells by freeze-drying rather than

12. Robert Bensley. Image reproduced from the website "History of the Marine Biological Laboratory" (http://hpsrepository.asu.edu/).

by the application of chemicals (Gersh 1932; Bensley and Gersh 1933a, 1933b). Pieces of fresh tissue were rapidly frozen in liquid air and then dried in a vacuum for twelve hours before being embedded in paraffin. After this, the embedded material was cut into sections for microscopy, as with conventional histological procedures. Bensley believed that such preparations preserved not only cellular form but also the distribution of chemical substances found within the living cell. Bensley and Gersh used this procedure to investigate the chemical characteristics of the cell by extracting the sectioned material with a variety of reagents and treating it with various enzymes while at the same time monitoring any changes in the microscope. The results yielded a series of papers in 1933 and 1934 titled "Studies on Cell Structure by the Freeze-Drying Method" (Bensley 1933; Bensley and Gersh 1933a, 1933b, 1933c; Bensley and Hoerr 1934a, 1934b).

One part of the cell on which Bensley and his student Hoerr focused was mitochondria, a subject that had been of great interest to Bensley earlier in his career. Mitochondria, one of the few organelles in cells readily visible in the light microscope, were discovered in 1890 by Altmann and later named mitochondria by Carl Benda to reflect their appearance as either threads or granules (Bechtel 2006, 80–81). In sections of *Amblystoma* (salamander) liver,[3] Bensley and Hoerr obtained inconsistent results when attempting to solubilize mitochondria from paraffin-embedded sections with salt solutions, acids, or digestive enzymes (Bensley and Hoerr 1934a, 253). They then tried to extract mitochondria from rough, hand-cut sections of freeze-dried material that had not been embedded in paraffin. While this yielded some additional information, the amount of material provided by *Amblystoma* was insufficient for their experiments. They then tried preparations of finely ground freeze-dried liver and macerated fresh liver from both guinea pigs and rabbits. In the fresh preparation, they observed in the microscope that a number of mitochondria had been released from liver cells into suspension. Realizing that isolation of mitochondria might give them sufficient material for chemical characterization, they devised a protocol to separate mitochondria from larger cell fragments and nuclei by centrifugation at low speed and then to concentrate and purify them by a series of high-speed centrifugation steps interspersed with washes to remove soluble protein contaminates (Bensley and Hoerr 1934a, 1934b).[4] While these and

---

3. Bensley and Hoerr use the name *Amblystoma*, but others consider the spelling *Ambystoma* correct (Scott 1916; Bensley and Hoerr 1934a).

4. "Centrifugation" refers to a process in which samples of material are spun at high speeds in a *centrifuge* to separate them by size and/or density. The largest and densest components

the few subsequent studies of this type that followed yielded only chemical compositional and not functional insights, their conceptual break with the past was substantial because they were now working with material isolated from broken cells.

In 1929, before Bensley's lab began its series of studies, Albert Claude arrived at the Rockefeller Institute for Medical Research in New York City from Belgium to join the cancer research laboratory headed by James Murphy (fig. 13). Murphy was a protégé of Peyton Rous, a Rockefeller scientist who in 1910 had discovered a *filterable agent* that transmitted a form of cancer called sarcoma between chickens. This agent was much later shown to be a virus, but at the time of its discovery its exact nature, other than a size small enough to pass through filters, was unknown. By 1925, Rous lost interest in the agent, but Murphy continued to study it on his own (Moberg 2012, 15–16).

Claude's involvement in cancer research had begun in Europe soon after he received his MD in 1928, motivated in part by his mother's death from breast cancer (Moberg 2012, 19–20). Notably, Claude's background in medicine and pathology did not likely inculcate him with the biases of the American cytology community, perhaps leaving him with a more open mind.[5] In Murphy's lab he began to try to isolate the cancer-causing agent by any available means. Murphy did not favor the theory that the agent was a potentially living virus but instead thought it might be a kind of enzyme or mutagen. Consequently, Claude's initial purification efforts resembled those for enzymes, including chemical precipitation of material from tumor extracts and collection of the precipitate by centrifugation. In 1935, after becoming aware of British studies suggesting that the Rous sarcoma agent could be concentrated from filtrates by centrifugation without precipitation or other chemical procedures, Claude began to employ a centrifuge to separate fractions from disrupted normal and tumor cells (Moberg 2012, 25).

Centrifuges had been used for research for more than a century, although the instruments and their applications were not very sophisticated (Beams 1943). By the twentieth century, a variety of simple centrifuges were available (Elzen 1988; Creager 2002). Biologists used these to not only separate

---

move first to the bottom of the tubes holding the samples, with other components distributed within the tubes along the central axis of the centrifuge according to their size and density. See discussion below.

5. As Carol Moberg notes, Claude never went to high school in his native Belgium. During World War I, he participated in underground intelligence activities for the British. After the war, he was able to enter medical school without a high school diploma or an examination as a reward for his service (Moberg 2012, 19–20).

13. Albert Claude at the Rockefeller Institute for Medical Research. Photo courtesy of the National Library of Medicine. Used with permission of Philippa Claude.

and purify particulate extracts and preparations, but also spin intact cells to observe how the intracellular contents stratified in response to the centrifugal field (Harvey and Loomis 1930; Bensley and Hoerr 1934a, 1934b; Beams 1943). In 1924, Theodor Svedberg, a colloid chemist, constructed the first modern centrifuge in Sweden, which he called an ultracentrifuge, to determine the size distribution of colloidal particles and proteins (Elzen 1988; Creager 2002). Svedberg's was an analytical instrument with an optical system that enabled the separation to be monitored during the centrifuge run and was not designed to purify proteins in any quantity. Shortly thereafter, Emile Henriot and Emile Huguenard in Belgium designed the centrifuge that was used in 1935 for sedimentation of the Rous sarcoma agent in Britain (Elzen 1988; Moberg 2012). In the United States in the 1920s and 1930s, Jesse Wakefield Beams and his student Edward Graydon Pickels designed air and electrically driven centrifuges with the important modification of a vacuum chamber that increased the stability and reduced the heating of the spinning rotor (the container holding samples) due to friction with the air (Elzen 1988; Creager 2002). In 1935, Pickels was recruited to the International Health Division of the Rockefeller Foundation, with laboratories located at the Rockefeller Institute, where he built preparative scale centrifuges for virus purification (Elzen 1988; Creager 2002; Moberg 2012).[6] By the 1940s, the term "ultracentrifuge" had been extended to include any instrument that operated at very high centrifugal forces.

The centrifuge that Claude used for his early work was a common, unsophisticated commercial model manufactured by the International Equipment Corporation. Claude modified it with an optional pulley and belt to substantially increase its speed and operated it in a refrigerated "coldroom" to keep it from heating too much while running (Claude 1937; Moberg 2012, 25–26). In early experiments, Claude worked with a filtrate derived from water-extracted tumor tissue. Instead of simply centrifuging the extract at high speed for a long time to concentrate the infectious material, Claude used a series of short centrifuge runs at varying speeds to more carefully characterize the sedimentation behavior of the agent, and even to estimate its size. As Claude proceeded, he kept track of how much of the infectious activity could be isolated in individual fractions sampled at different stages and compared this to the total infectious activity of the initial filtrate. This

6. Pickels left Rockefeller in 1946 to help found the Specialized Instrument Corporation (Spinco) to manufacture what became the most successful and widespread analytical (the Model E) and preparative (the Model L) ultracentrifuges (Elzen 1988; Moberg 2012). Spinco eventually became part of Beckman Instruments.

quantitative accounting allowed him to estimate how much activity was recovered in the fractions in relation to the total activity in the starting material. Such a rigorous approach was to become an important distinguishing characteristic of his work.

In addition to measuring infectious activity, Claude examined the fractions in the microscope. Like the British investigators working on the Rous agent, he observed tiny particles in the extract visible by dark-field microscopy but not by conventional light microscopy, suggesting that the particles were below the resolution of the light microscope. His conclusion from these studies was that differential centrifugation was vastly superior to any chemical procedures for isolating the infectious activity or agent (Claude 1937; Moberg 2012, 25–26).

As the work proceeded, Claude began omitting the filtration step from his preparation of extracts to increase the yield of infectious activity and conducted chemical analyses of the concentrated material, noting the presence of particles containing protein and ribonucleic acid (RNA) (Claude 1937, 1938a, 1938b, 1939).[7] He continued to believe that some of these particles might be related to the transmissable agent. In early 1940, Claude published a paper comparing the particles derived from tumor extracts to similar extracts prepared from normal tissues (Claude 1940). To his apparent surprise, Claude notes that there are no differences between the particulate composition of the normal and tumor extracts: "The material isolated from these various sources is strikingly similar and the purified fractions obtained from mouse embryo and mouse tumors have been found to possess many of the physical and chemical characteristics already described for the homologous fractions previously obtained from chick embryo and chicken tumor I" (Claude 1940, 77). He then goes on to suggest that the particles are kinds of formed elements of the cell, possibly mitochondria (Claude 1940). Claude's work had now converged with that of Bensley, who he cites in this paper for the first time.

At this point, Claude began to focus more and more on the fractionation of a variety of cell types, from liver to lymphocytes, both normal and tumor derived. He concluded that similar particles consisting of a lipid-ribonucleoprotein complex are present in all cells (Claude 1941). As he refined his techniques, his results brought him into conflict with those of the Bensley laboratory. Claude reported that fractionation of guinea pig liver, which Bensley and Hoerr had also studied, yields three fractions: large

---

7. Nucleic acids are polymers of nucleotides based on either the sugar ribose (ribonucleic acids [RNA]) or deoxyribose (deoxyribonucleic acid [DNA]).

"secretory granules," small particles, and even smaller particles that are only sedimented by two hours' centrifugation at high speed (Claude 1941). Bensley and Hoerr had concluded that the large granule fraction consisted of mitochondria (Bensley and Hoerr 1934b). Claude finessed this difference by suggesting that his secretory granules originated as mitochondria (Claude 1941). Two years later, Claude was invited to a symposium in honor of Bensley organized by Hoerr (Hoerr 1943a). In his published symposium paper, Hoerr gently critiques Claude's findings, saying that he believes that Claude's secretory granules are their mitochondria (Hoerr 1943b). Claude, however, does not relent in his presentation or in a subsequent paper, concentrating not only on his secretory granules but also on his small particles, which he has now named *microsomes*, adopting E. B. Wilson's term and generic definition, "a small granule of undefined nature" (Wilson 1925, 32–33; Claude 1943a, 1943b; Moberg 2012, 31).[8]

In 1946, Claude published a pair of detailed papers on a refined strategy for fractionating cells. The first paper describes procedures for breaking open liver cells derived from either rat or guinea pig livers (Claude 1946a). The chilled tissue is first coarsely broken up by forcing it through a one millimeter mesh screen and then ground in a mortar for 5 minutes. When viewed in the microscope, this "liver suspension" contains small tissue fragments and individual cells, as well as nuclei released from broken cells, granules of varying sizes, and fat globules. To produce what Claude calls a "liver extract" from this material, he centrifuges it three times for 3 minutes at the very low speed of 1,500 x g,[9] each time discarding the sedimented material (called the *pellet* because it collects as a mass at the bottom of centrifuge tubes). The final extract is largely devoid of whole cells, free nuclei, and tissue fragments. Claude calculates that this extract contains about half of the starting liver tissue on a dry weight basis.

In the second paper, Claude describes fractionating the extract by centrifuging it at higher speeds, illustrating his strategy in a diagram (Claude 1946b) (fig. 14). He first spins the extract in the centrifuge for 25 minutes at

8. It is important to emphasize that Wilson's microsomes were visible in the light microscope while Claude's were submicroscopic, detectable optically only in the dark-field microscope (Claude 1941). Hence it is very unlikely that Wilson and Claude were referring to identical entities even though Claude borrowed Wilson's term.

9. This indicates that the material is placed in a centrifugal field that is 1,500 times stronger than gravity. Use of this unit of measurement is independent of the type of centrifuge employed. Centrifugal force indicated as multiples of the gravitational force is related to the speed of the centrifuge in revolutions per minute (rpm) and the radius of the centrifuge rotor measured from the center.

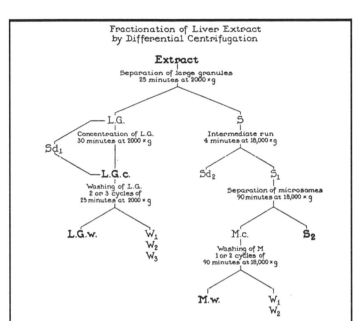

Fractionation of Liver Extract by Differential Centrifugation

Extract
Separation of large granules
25 minutes at 2000 × g

— L.G.
Concentration of L.G.
30 minutes at 2000 × g
Sd₁

S
Intermediate run
4 minutes at 18,000 × g

— L.G.c.
Washing of L.G.
2 or 3 cycles of
25 minutes at 2000 × g

Sd₂

S₁
Separation of microsomes
90 minutes at 18,000 × g

L.G.w.   W₁
W₂
W₃

M.c.
Washing of M
1 or 2 cycles of
90 minutes at 18,000 × g

S₂

M.w.   W₁
W₂

TEXT-FIG. 1. The diagram illustrates the procedure of fractionation described in the text. The liver fractions especially investigated were: (1) the large granules, concentrated (L.G.c.); (2) the large granules, washed (L.G.w.); (3) the microsomes, washed (M.w.); and (4) the supernate (S₂).

TABLE I

Fractionation of Mammalian Liver by Differential Centrifugation
Solid Content and Proportion of Various Fractions in Rat Liver Extract

	Fraction	Volume of original extract	Volume of fraction	Dry weight of fraction	Total dry weight of fraction	Amount of fraction in extract		Proportion of liver pulp in extract
		cc.	cc.	mg. per cc.	mg.	mg. per cc.	per cent	per cent
E	Extract..................			27.1		27.1	100.0	47.5
S	Large granules removed....			23.5		23.5	86.7	
S₁	Intermediate sediment (Sd₂) removed...............			21.7		21.7	80.1	
S₂	Microsomes removed.......			17.3		17.3	63.8	
L.G.c.	Large granules concentrate..	315	25	41.7	1043	3.3	12.2	
L.G.w.	Large granules, washed.....	189	15	34.7	521	2.8	10.3	
M.c.	Microsomes, concentrate....	168	15	48.2	723	4.3	15.8	
M.w.	Microsomes, washed........	140	10	38.8	388	2.8	10.2	
Sd₁	Cell debris removed from L.G.c.................	315	5	12.2	61	0.2	0.7	
Sd₂	Intermediate sediment (L.G. and M., mixed).........	252	18	23.6	425	1.7	6.3	
W₁	First washings from L.G.c...	189	35	2.3	81	0.4	1.5	
W₂	Second washings from L.G.c.................	189	35	0.4	14	0.1	0.4	

14. Albert Claude's cell fractionation scheme. Upper panel: flow chart. Lower panel: quantitative analysis of total mass distribution in individual fractions (the "balance sheet").

2,000 x g and divides the resulting pellet and "supernate" (the liquid above the pellet, referred to today as the supernatant). He then resuspends the pellet in a saline solution and spins it under similar conditions several additional times to wash it. Claude calls this fraction the large granular fraction. Examined microscopically, the fraction consists of visible granules ranging in size from 0.5 to 1 or 2 μm in diameter.[10] To deflect the critique of his claim in previous studies that the large granule fraction consists of secretory granules, Claude accepts that it consists of both mitochondria and secretory granules (Claude 1946b; Moberg 2012, 28–31). Claude then centrifuges the supernatant at high speed (18,000 x g) for an extended time to yield a microsomal pellet and washes the pellet several times. He describes this pellet as a "soft, jelly-like material, completely transparent and dark amber in color" (Claude 1946b, 73). Claude refers to the microsomes as colloidal when resuspended and estimates their average size based on the speed of centrifugation required to sediment them as 0.1 μm, too small to be seen in the conventional light microscope but visible in dark field as "refractile bodies constantly agitated by Brownian motion" (Claude 1946b, 73).

Claude carries out an extensive chemical analysis of the fractions. As he does so, he carefully monitors the distribution of the dry mass in the original extract, attempting to account for every bit of the initial material (see fig. 14). He reports that the large granules are composed of mainly proteins and lipids but also ribonucleotides, the constituents of RNA. Microsomes consist of more lipid than protein, along with some ribonucleotides. Alluding to other work, some of which is unpublished, Claude suggests that cytochrome-linked enzymes have been found in the mitochondrial part of the large granule fraction, but none of the enzymes found in the overall liver extract appear to be associated with microsomes (Claude 1946b).

Overall, the fractionation methods and results that Claude reports in these papers are not just an extension of the earlier work of Bensley and Hoerr or simply an improvement on Claude's own previous attempts. Rather, he describes a fully integrated strategy for breaking open cells and characterizing the components of the cell interior both morphologically and chemically. While Bensley and Hoerr focused on one cellular component at a time, Claude included almost everything at once in his scheme. Although he discarded the nuclei, not including them in his analysis, he accounted quantitatively for everything else, setting the stage for the assignment of specific functions to the individual isolated parts of the cell. This approach,

10. In his paper, Claude uses the designation "μ" for micron, or 10^{-6} meter. The correct symbol is now μm. Claude also uses mμ for 10^{-9} meter, now called nanometer (nm).

later referred to as the *balance sheet*, set Claude apart from the others and became crucial when the enzymatic constituents of fractions were identified and studied (Hogeboom 1951).[11]

While Claude was not successful in associating specific enzymes with particular cell fractions in the studies published in 1946, other work was already under way that would change this. George Hogeboom was a biochemist who had joined Murphy's lab in 1941 and began using cell fractionation to help isolate enzymes from mouse melanoma cells (Bechtel 2006, 178). Along with many other scientists at Rockefeller and elsewhere, Hogeboom was obligated to conduct research on war-related projects, and in 1943 he was sent to Florida to do this work. When he returned in late 1945, he began collaborating with Claude and another Rockefeller biochemist, Rollin Hotchkiss, to localize the enzymes cytochrome oxidase and succinoxidase to the cell fractions from rat liver generated by Claude's fractionation scheme (Hogeboom, Claude, and Hotchkiss 1946; Bechtel 2006, 178–79; Moberg 2012, 80–82). Using Claude's careful, quantitative approach, they concluded that the enzymes were mainly localized to the large granule fraction composed mostly of mitochondria (Hogeboom, Claude, and Hotchkiss 1946). Before they could publish their results, they learned that similar findings had been obtained by Walter Schneider, a graduate student of Van R. Potter's at the University of Wisconsin (Schneider 1946a, 1946b). Potter, also a biochemist, had encountered Claude at the Bensley symposium in 1942, where he told Claude that he was attempting to purify oxidative enzymes. Claude encouraged him to try cell fractionation and provided him with detailed protocols that proved crucial to Schneider's success.

After completing his thesis with Potter, Schneider moved to Rockefeller in 1946 on a postdoctoral fellowship to work with Hogeboom, while Hotchkiss moved on to other unrelated research (Moberg 2012). They were joined by a new investigator, George Palade (fig. 15). Palade was a Romanian physician and anatomist who immigrated to New York City in 1945. Although trained originally as a clinician, while still in Romania Palade

---

11. The first published reference to the "balance sheet" may be in a paper by George Hogeboom, one of the biochemists who worked with Claude in early fractionation studies. In a 1951 paper he states, "It is essential to analyze both the original whole tissue and all the fractions obtained from it. By this means it is possible to draw up a balance sheet, which serves as a test of the validity of the analytical method, enabling one to detect the presence of inhibitors or activators or the possible role of more than one cell component in a given biochemical reaction" (Hogeboom 1951, 640–41). As late as the 1970s, the importance of the balance sheet was emphasized in any discussion of cell fractionation by scientists in the Laboratory of Cell Biology at Rockefeller that Claude helped build.

15. George Palade. Photo courtesy of the National Library of Medicine.

became more interested in research as a way of expanding medical knowledge and began studying the kidneys of dolphins obtained from the Black Sea. However, with the chaos that accompanied the end of World War II, Palade realized that carrying on his research in Romania was, at least for the moment, impossible. With the recommendation of his mentor in Romania, Grigore Popa, he arranged a visiting investigator position with Robert

Chambers at New York University and managed to leave Romania with his family (Farquhar 2012) (see fig. 8). In Chambers's lab, Palade learned the techniques of micromanipulation and tissue culture and worked on renal cysts obtained from a culture of the chicken mesonephros (an embryonic precursor of the kidney). He also accompanied Chambers to the MBL in Woods Hole for three months where he was involved in a project using sea urchin eggs.[12] At some point Palade heard a seminar by Albert Claude and, impressed by Claude's electron micrographs (see next section), managed to move to Rockefeller in October 1946 (Farquhar 2012).

Claude initially asked Palade to work on the isolation of the Golgi apparatus by cell fractionation. However, he soon began working with Hogeboom and Schneider to improve methods for the isolation of mitochondria. In both Schneider's work and that of Hogeboom, Claude, and Hotchkiss, the mitochondria that they isolated did not morphologically resemble those seen in whole cells. They were round instead of elongated and failed to stain with the vital dye Janus Green, indicating that they were not enzymatically active. In some cases, they were even depleted of internal contents. The problem was thought to be related to the solutions used to fractionate cells and isolate mitochondria. Claude had previously tried an alkaline solution and later switched to isotonic saline, but this failed to preserve mitochondrial morphology or staining activity. Hogeboom, Schneider, and Palade attempted to improve the isolation procedure by employing a homogenizer designed by Potter and Conrad Elvehjem at Wisconsin to disrupt the liver tissue and also by modifying the centrifugation steps, using a new refrigerated centrifuge with a vacuum chamber designed by Josef Blum in the Rockefeller instrument shop (Potter and Elvehjem 1936; Moberg 2012). The most important change turned out to be the use of 0.88 molar sucrose solutions instead of saline, an idea attributed to Palade by Schneider many years later (Moberg 2012, 84–85). Under these conditions, the mitochondria not only retained their expected shape but also stained readily with Janus Green (Hogeboom, Schneider, and Pallade 1947, 1948; Bechtel 2006; Moberg 2012). The use of sucrose in cell fractionation not only revolutionized the way that mitochondria were isolated, paving the way for future studies of oxidative phosphorylation, but also turned out to be a technical breakthrough in the isolation of other cellular components that was exploited in subsequent studies.

12. Letter from George Palade to Grigore Popa, January 18, 1947, Palade Archives. Courtesy of the National Library of Medicine.

Penetrating the Structureless and Optically Empty Protoplasm

Embedded in Claude's published description of his fractionation results are references to three recent studies from his lab using a new technology: electron microscopy. While he does not describe his findings in any detail, he does indicate that the electron microscope is generating some new insights (Claude 1946b, 70). As mentioned previously, Max Knoll and Ernst Ruska invented the electron microscope in Germany in the 1930s based on developments in cathode ray technology and the realization that visible light and streams of electrons, which have properties of both particles and waves, follow similar optical principles (Ruska 1988; Rasmussen 1997). Almost simultaneously, projects to develop electron microscopes were initiated in a variety of other countries, both in Europe and in North America (Rasmussen 1997, 26). By 1939, the Siemens company in Germany produced the first commercial instrument under Ruska's guidance, but with the outbreak of war, the microscope was unavailable in most countries.

While the electron microscope is much more complex than the light microscope, its operational principles are analogous. Light is replaced by electrons accelerated at high voltage, and glass lenses are replaced by electromagnets. Images are formed when specimens scatter electrons and are visualized on a fluorescent screen and then recorded on photographic plates or film. As with the light microscope, resolution of the electron microscope is limited by the wavelength of, in this case, the electron beam. Based purely on this principle, the electron microscope has a theoretical resolution limit that is thousands of times greater than that of the light microscope. In practice, the actual resolution was at the time much less but still more than one hundred times better than the light microscope (Meek 1976).

When used to study the cell, this resolution is sufficient to see cellular parts much smaller than mitochondria, the smallest component observed by both Bensley and Hoerr and Claude in the light microscope. Nevertheless, when the electron microscope was taken up by biologists, its use was complicated by inherent properties of the instrument. The electron beam must operate in a vacuum to avoid electron scattering by air molecules, and objects to be imaged must therefore be stable in the vacuum chamber, precluding observations of living cells. Furthermore, even though the electron beam is powerful, it still cannot easily penetrate biological specimens of the thickness used in light microscopy. As a consequence, most of the early biological samples that were imaged with the electron microscope were thinly spread and dried suspensions of bacteria and viruses (Rasmussen 1997; Moberg 2012, 53).

The first commercially available American electron microscope was the RCA EMB (also called the Model B) designed at the Radio Corporation of America in Camden, New Jersey, and available in 1941. The motivation for production of this instrument may have been to promote RCA's television technology, although interest from other American firms in the production of electron microscopes, including Kodak and General Electric, suggests some expectation of a lucrative market (Rasmussen 1997, 31). RCA's effort began in 1938 when Vladimir Zworykin, head of electronic research, hired Ladislaus Marton. Marton was a Belgian physical chemist who in 1932 had built an electron microscope in Brussels and used it to produce crude images of plant tissue (Marton 1934). Marton's first microscope at RCA, the Model A, was functional but difficult to use. Zworykin next hired James Hillier, a Canadian who also had experience in microscope design. Hillier soon developed the improved Model B, and Marton, who did not get along with Zworykin, left RCA for an academic position at Stanford. To encourage use of the microscope by biologists, Zworykin invited scientists from the University of Pennsylvania and the Rockefeller Institute campus in Princeton, New Jersey to try it. In 1940, he obtained funding from the National Research Council to form a committee of scientists to coordinate research and recruited Thomas Anderson, a physical chemist and experienced investigator, as a postdoctoral fellow to coordinate several projects. At the same time, RCA began selling the Model B to both academic and industrial laboratories (Rasmussen 1997; Moberg 2012).

Claude may have first heard about the electron microscope from a letter and manuscript he received from Marton in the mid-1930s (Moberg 2012, 53–54). In 1940, Claude learned that other scientists were working with Anderson at RCA to visualize viruses, including the Rous sarcoma agent. By 1942, some of this work was published to great acclaim in the *Journal of Experimental Medicine*, edited by Peyton Rous at Rockefeller (Green, Anderson, and Smadel 1942; Moberg 2012, 54). When Claude went to the symposium for Bensley in Chicago at about the same time, he met Francis O. Schmitt from the Massachusetts Institute of Technology. Schmitt had purchased an RCA EMB in 1941 with a Rockefeller Foundation grant to study protein fibers, among other things. On learning of this, Claude sent Schmitt some isolated chromatin to examine. While this collaboration never bore fruit, Claude apparently realized that electron microscopy was a technology worth pursuing (Rasmussen 1997; Moberg 2012).

One of the microscopes produced by RCA was sold to the Interchemical Corporation located in Manhattan near the Rockefeller Institute. Interchemical obtained the instrument to measure particle sizes in ink and paint. In

May 1943, Claude was contacted by Henry Green, head of physics at Interchemical. Green had read an article of Claude's in *Science* on the constitution of protoplasm and asked Claude if he could provide them with some of his cellular particles to examine in the electron microscope (Claude 1943a; Moberg 2012, 55). Green was apparently encouraged to contact Claude by Albert E. Gessler, head of research at Interchemical, after a close friend of Gessler's died of cancer. Gessler hoped that working with Claude would advance cancer research. Fortuitously, Interchemical also employed Ernest F. Fullam, a technician who had learned to operate the RCA Model A from Marton at the University of Pennsylvania, and Claude was able to recruit him to the project (Moberg 2012, 55).

Initially, Claude sent a variety of samples to Fullam, with mixed results. By early 1945, he and Fullam were able to produce publishable micrographs of isolated mitochondria, but sample preparation was a major unresolved issue (Claude and Fullam 1945; Moberg 2012, 57). The isolated mitochondria were stabilized by fixation with osmium tetroxide and other fixatives typically used in conventional microscopy and then dried on tiny stainless steel screens, or *grids*, coated with a plastic film called Formvar (Claude and Fullam 1945). Such grids fit into the sample holder used to insert specimens into the microscope's vacuum chamber. When viewed in the electron microscope, the mitochondria, with a diameter less than one micron, were still too thick for the electron beam to penetrate and showed up mainly as a series of round black objects with little internal detail evident. Attempts to circumvent such problems in a subsequent study by making very thin sections of guinea pig liver with a microtome designed by Fullam and Gessler were only modestly successful (Claude and Fullam 1946; Moberg 2012).

In 1944, Keith Porter began working with Claude on the electron microscopy project. Porter was a young Canadian biologist who joined Murphy's lab as an assistant in 1939. He was an expert in nuclear transplantation into amphibian eggs and set out to use his knowledge and technical skills to investigate interactions between the nucleus and cytoplasm in cancer. Unfortunately, Porter's progress in the first few years at Rockefeller was slow because of his persistent tuberculosis infection. At Murphy's instigation, Porter learned to grow chick embryo cells in culture with the goal of injecting Claude's particles into the cells, and a collaboration with Claude ensued (Moberg 2012). After joining Claude on his trips to Interchemical, Porter became familiar with the challenge of obtaining biological samples thin enough for the electron microscope and came up with a possible solution using cultured cells.

Porter realized that cells in tissue culture attach and spread out over the surface on which they are growing. Because of this, the entire cell, particularly near the periphery of the cytoplasm, is very thin. Porter reasoned that this region might be thin enough for the electron beam to penetrate. To get the spread cells into the electron microscope, Porter grew them on glass cover-slips coated with Formvar and fixed them with vapors of osmium tetroxide. He then cut out a small circle of Formvar with cells on it underwater to keep the plastic film extended and captured it by placing a microscope grid under the circle and then removing it from the water. After drying the sample, he examined it on Interchemical's electron microscope with Claude and Fullam (Porter, Claude, and Fullam 1945).

Porter and his colleagues located a single cell in a space between the bars of the grid. They were astounded by what they saw and worked into the night taking images (Moberg 2012, 60). At the magnification used (1,600X), the entire cell may not have been visible in a single microscope field, possibly requiring them to take a series of overlapping micrographs to capture everything (fig. 16).[13] In the paper reporting their findings, published in March 1945, they reassembled this cell as a photomontage and also presented other enlargements and higher power images of parts of other cells (Porter, Claude, and Fullam 1945). This was the first time that an entire eukaryotic cell was seen at such high magnification. Porter, Claude, and Fullam focused in the paper on describing technical aspects and observations, being careful to not overinterpret their findings because they were concerned that some details might be artifactual (Moberg 2012, 61).

In the nuclear region the cell was too thick to make out much detail, although they observed "some elements of high density" that they speculated might be part of the Golgi apparatus, a part of the cell visible by light microscopy with a special stain. Closer to the periphery they easily recognized mitochondria by their filamentous shape and also made out for the first time a "delicate lace-work extending throughout the cytoplasm," described elsewhere in the text as a "lace-like reticulum" (Porter, Claude, and Fullam 1945, 238, 246). They concluded the paper with a modest statement: "The electron micrographs disclosed details of cell structure not revealed by other

13. A number of years after Claude returned to Belgium, Palade requested from Claude copies of the original figure, possibly for a review that he was preparing. When Claude sent the images he noted that on July 6, 1944, he, Porter, and Fullam took a total of thirteen pictures of the cell shown in their paper and then used these images to construct the montage that was published. Originally, they made a montage composed of cutouts from four separate images of the cell but then, for publication, made a separate montage composed of cutouts from five separate images so that the details of the nucleus were more visible. In the original montage, which Claude also provided to Palade, the center of the cell that includes the nucleus is nearly opaque (Palade Archives, Courtesy of the National Library of Medicine).

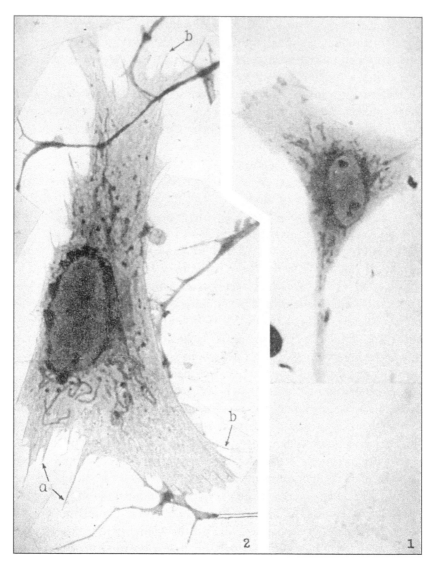

16. The first electron micrograph of an entire eukaryotic cell taken by Keith Porter, Albert Claude, and Ernest Fullam, labeled "2" in the figure. Note that the image is a montage of several smaller images that have been trimmed and mounted together. The "lace-like reticulum" is visible in the periphery of the cell's cytoplasm. The image to the right of the electron micrograph (labeled "1") is a light micrograph of a cell prepared identically to the one shown in the montage. Magnification: 1,600X. © 1945 Porter, K., Claude, A., and Fullam, E. Originally published in the *Journal of Experimental Medicine* 81 (3): 233–46.

methods of examination" (244). Over the next few years, as the technique of biological electron microscopy improved, these parts of the cell that had never been seen became the subjects of intense investigation. Along with cell fractionation, which had circumvented cytologists' fear of working with broken cells, electron microscopy now began to eliminate another obstacle to progress, the limited resolution of the light microscope.

The issue of sample thickness remained a problem for several years. In standard histological procedures in place since the nineteenth century, tissue samples were embedded in paraffin wax and then sectioned on a microtome with a steel knife (Bracegirdle 1978). Using this traditional approach, investigators were unable to cut sections sufficiently thin for electron microscopy. Early microtomes designed for electron microscopy operated on the principle that thinner sections could be obtained by cutting at high speeds; while somewhat successful, these instruments were difficult and even fearsome to operate, and the high speeds tended to distort and tear the specimens (Moberg 2012). In 1946, the Rockefeller instrument maker Josef Blum worked with Claude to design a low-speed microtome with important innovations. Among these was the ability to carefully adjust section thickness and a bypass mechanism that caused the arm holding the specimen to withdraw from the knife on the return stroke to avoid damage to the face of the sample block (Claude 1948; Moberg 2012). As more and more laboratories attempted electron microscopy, other developments followed. These included the use of plastic embedding media that were much harder than paraffin and thus amenable to thinner sectioning and the adoption of knives made from freshly broken plate glass (Moberg 2012). In 1953, Porter and Blum reported the design of an improved *ultramicrotome* based on the earlier Claude-Blum design. The new instrument was capable of dependably cutting sections that were 25 to 50 nm thick and easily penetrated by the electron beam. The design was commercialized as the Model MT-1 by the Ivan Sorvall company of Connecticut and remained in use in some laboratories as late as the 1990s (Porter and Blum 1953; Moberg 2012).

Among the different fixatives initially used to preserve biological samples for electron microscopy, osmium tetroxide turned out to be the most satisfactory (Porter, Claude, and Fullam 1945). One reason for this was that osmium served not only as a fixative but also a stain that highlighted cellular membranes and other structures due to deposition of the electron-dense osmium molecule. A problem with osmium, however, was that its effectiveness as a fixative was variable. With thin specimens, osmium could be used as a vapor, but with improvements in sectioning, blocks of tissue were fixed by immersion in osmium solutions. These were prepared in water with the pH adjusted to neutrality. In 1952, Palade reported that when such osmium solutions contacted tissue they caused acidification that interfered with fixation (Palade 1952b). To remedy this situation, Palade buffered osmium solutions to pH 7.3–7.5 (slightly alkaline) with a mixture of veronal (a barbituate) and acetate so that the pH remained constant as fixation proceeded (Palade 1952b; Meek 1976). This modification, which may appear minor from a more recent perspective, had a major effect on the quality and consistency of tissue fixation for electron microscopy and was widely adopted throughout the nascent community of electron microscopists. Even with the development of the fixative glutaraldehyde in 1964, use of Palade's osmium formulation continued (Sabatini, Bensch, and Barrnett 1964).

Because the electron microscope revealed details in the cell that had never been seen before, deciding what was "real" as opposed to an artifact of sample preparation proved a challenge. In Porter's original paper with Claude and Fullam, they addressed this issue by comparing light and electron micrographs of cells identically fixed with osmium and dried (fig. 16). They argued that the overall disposition of the cells was the same and that parts of the cell such as the nucleus and mitochondria that were visible in both the light and electron microscopes appeared similar (Porter, Claude, and Fullam 1945). Later, as they further developed their sample preparation procedures, they attempted to monitor cells by light and dark-field microscopy during the fixation and drying processes to see if visible cellular structures were altered (Rasmussen 1997, 110). In the early to mid-1950s, and even later, disagreements continued about the existence of certain cellular structures seen in the electron microscope. One of the most famous of these was between Palade and Fritiof Sjöstrand over the internal structure of mitochondria (Palade 1952a, 1953; Sjöstrand 1956; Rasmussen 1997). While both believed that every effort should be made to ensure that electron micrographs represented as much as possible structures found in the living cell, Sjöstrand felt that micrographs could ideally be interpreted literally, even to the point of yielding insights into molecular function (Rasmussen 1995, 1997; Matlin 2016). Palade, on the other hand, thought that even the best electron micrographs were inherently limited because the harsh techniques used to prepare samples undoubtedly changed the cell. Instead, he believed that consilience between different and diverse techniques was required to ultimately validate conclusions reached from examination of micrographs alone (see chap. 11). Palade's conviction became a key element in the later success of the Rockefeller group in deciphering cellular functions.

Consolidation

In 1949, shortly after presenting a Harvey Lecture summarizing his achievements,[14] Claude left Rockefeller to return to Belgium. Murphy retired in 1950. At about the same time, Hogeboom also left Rockefeller together with Schneider for positions at the National Institutes of Health in Bethesda, Maryland (Moberg 2012). This left Porter and Palade devoid of leadership and without a trained biochemist. The situation was nonetheless soon stabilized by the formal creation of the Laboratory of Cytology by Herbert Gasser, director of the institute, and the promotion of first Porter and then Palade to associate members (Moberg 2012). Although functioning in part to provide electron microscopy services to other investigators, the laboratory would soon grow and establish an international reputation.

The enormous impact of Claude's work chiefly with Porter, Palade, and Hogeboom cannot be overstated. When he began breaking open cells and separating their contents by differential centrifugation, Claude overcame the prejudice among many cytologists against using dead cells to find the functions of living cells.[15] In an interview in 1985, Porter said that "when [Claude] started tearing cells apart, taking pieces out and examining them, everybody who called himself a decent cytologist or cell biologist was at him[,] . . . [asking,] what was the good of doing that, breaking up that gorgeous structure? And he got a lot of bad, bad press from that" (cited in Moberg 2012, 38). Claude's advantage was perhaps that he was not a cytologist and never bought into the belief that may have obstructed progress in cytology.

Claude was also not a biochemist. However, soon after he devised his scheme for separating the cytoplasm into three fractions, he began to link specific enzymatic activities to those fractions. He was fortunate that

14. The Harvey Lectures are a prestigious invited lecture series held annually in New York City and sponsored by the Harvey Society, which publishes a monograph of the lectures. The Harvey Society was founded in New York in 1905 by a group of prominent physicians and laboratory investigators (Bearn and James 1978).

15. Not everyone was convinced. In his 1953 book, *Cytochemistry*, James Danielli wrote, "Amongst the most common techniques employed today, particularly by biochemists seeking to make a contribution to cytology, are techniques involving the disintegration of cells into fragments, and fractionation of the fragments so formed. . . . It is hoped that methods of this type will isolate granules, mitochondria, nuclei, and chromosomes, in a condition which is closely similar to, if not identical with, the state of those bodies in intact cells. It would undoubtedly be of the greatest value if it were true that cell organs could be isolated in this way. But so far there has been an almost complete lack of proof that the bodies isolated are in the same condition as in intact cells" (Danielli 1953, 7–8).

Murphy had hired Hogeboom and that an encounter with Potter led Schneider to Rockefeller. By the time of his seminal 1946 papers, Claude completely understood that the hope among cytologists that cytochemical reactions would yield understanding of the chemical organization of the cell was unlikely to be fulfilled.

> The intracellular topography of biochemical functions constitutes one of the major problems of cytology, and one that has benefitted the least from the microscopical technique. With the successful application of staining to the study of cell morphology, the hope was entertained that specific color tests could be used under the microscope to determine the distribution of enzyme systems within the cell. Unfortunately, most of the color tests involve chemical reactions incompatible with the life of a cell and almost invariably it is found that essential cell structures have been severely damaged or completely destroyed by the procedure. (Claude 1946a, 15)

Claude's approach to cell fractionation was not only to break open the cell, separate its parts, and analyze the chemical constitution and enzymatic activities localized in those parts, but, more important, to do this quantitatively. In this way he could say that when an enzyme is concentrated in a particular part of the cell, it says something about the function of that part, a proposition realized when oxidative enzymes were localized to mitochondria (Hogeboom, Schneider, and Pallade 1947, 1948). The implications of this were recognized in a report by the director, Gasser, published in 1951: "What is going on inside the cell is a microcosm of complexity. The cytoplasm is filled with particles, and something about the nature of the particles is now emerging. The particles may be thought of as intracellular organs. . . . One wants to know, step by step, the chain of chemical transformations peculiar to each particle, and then the relationships of each chemical system within the cell to the others in the cell" (cited in Moberg 2012, 106). With the findings of Claude and his colleagues, the early twentieth-century speculations of Hofmeister and others on the chemical organization of the protoplasm were seen in a new light. The cell consisted of a series of what came to be called organelles, each of which sequestered a defined chemistry.

At the same time that Claude broke open cells, he did not lose sight of cellular morphology. As he stated in his 1948 Harvey Lecture, "In the presentation of the results it would be difficult to separate the biochemical work from the morphological observations since the microscope has constantly served as a guide or check for the chemical and biochemical studies" (Claude 1948, 123). The microscope was not only used by Claude to determine the effectiveness of his cell breakage techniques, but also to relate the particles released from the cells back to the intact cell as a way of validating his results. This was most evident in the mitochondria work with Hogeboom, Schneider, and Palade when a premium was placed on isolating mitochondria that accurately resembled those in the intact cell *and* reacted identically with Janus Green (Hogeboom, Schneider, and Pallade 1948). In the course of his fractionation studies, Claude identified micro-somes, particles that could be sedimented in the centrifuge but not seen in the light microscope. At this point, his efforts with Porter and Fullam to look at cells and fractions in the electron microscope began to bear fruit, shattering a second obstacle to the advance of cell studies, the resolution lim-its of the light microscope.

In sum, what Claude and his collaborators created was a new epistemic strategy, a new way to learn about how cells work that depends on the in-tegration of disparate techniques. The strategy consists of disruption of cells and separation and characterization of their parts by fractionation and biochemical analysis while at the same time monitoring this process step by step with the microscope so that the parts can always be related to the whole. With the development of systematic cell fractionation and electron microscopy, Claude, Porter, and Palade crossed both the physical bound-ary of the cell, the cell membrane, and the technical boundary of micro-scope resolution, overcoming two of the major obstacles to the study of cell structure and function. Most significantly, by integrating morphology and fractionation, they also found a way to biochemically analyze cell func-tions while preserving the cellular context, an approach that contrasted with that of biochemists, who destroyed cell structure to access enzyme activities (Green 1937; Matlin 2016, 2018). These advances prepared the ground for molecular explanations of biological phenomena at the cellular and subcel-lular level, developments that would remain unrealized for the time being.

PART TWO

From Cells to Molecules

FOUR

The Endoplasmic Reticulum

Basophilia

The next step toward a molecular cell biology required detailed focus on the structure and function of one specific part of the cell. Of the various possibilities, the "lace-like reticulum" was perhaps the most intriguing structure seen by Porter, Fullam, and Claude in their first examination of a cell in the electron microscope, both because it had not been seen before and because it was located in areas of the cell formerly considered to be optically empty (see fig. 16). This suggested that it was smaller than the resolution of the light microscope (Porter, Claude, and Fullam 1945; Porter 1955). Claude, of course, was hoping to see microsomes in the electron micrographs that resembled those that he had isolated from fractionated cells and therefore tended to focus more in his comments on observed particles than on reticular structures (Porter, Claude, and Fullam 1945; Claude, Porter, and Pickels 1947; Claude 1948). In any case, it was not clear early on whether the reticulum was an artifact of fixation and drying procedures. At this point, even speculations about its function were premature, although results using other approaches began to provide clues.

In the nineteenth century, when various dyes began to be used to highlight intracellular structures, cytologists observed diffuse regions of the cytoplasm in secretory cells that were stained by basic dyes, suggesting that cellular material localized in those regions was *basophilic*; that is, the material itself was acidic (Garnier 1900; Porter 1953; Schickore 2018). In detailed studies of glandular tissues, Charles Garnier referred to basophilic structures seen in the cytoplasm as the *ergastoplasm*, which he illustrated as somewhat thick, fibrous bodies (Garnier 1900). Later investigators lumped ergastoplasm and other cytoplasmic basophilic components together as an indistinct chromidial substance, so named because it stained similarly to chromosomes (Cowdry 1924b). In general, chromidia were not visible in living cells and were regarded by some as fixation and staining artifacts (Cowdry 1924b; Claude 1948; see also Haguenau 1958).

In the 1930s and 1940s, Jean Brachet in Belgium and Torbjörn Caspersson in Sweden began investigating the distribution of nucleic acids in cells. Although their approaches were dramatically different, the strategies they employed were essentially advanced forms of traditional cytochemistry in the sense that they were originally based on imaging of whole cells rather than disruption of cells followed by chemical analysis. As Caspersson noted in a 1947 review of his work:

> The study of the chemical composition of the individual cell and its internal structures involves two principal difficulties, one is the smallness of the objects and the other is that the structure of the cell must not be impaired—preferably one should investigate the cell in its living state. Under such conditions it is, of course, too much to hope for quite general methods; we have to limit ourselves to studying certain substances with specially suitable properties. (1947, 127)

After the discovery of nuclein by Friedrich Miescher in the 1860s and 1870s, the existence in cells of two types of nucleic acids, RNA and DNA, was not recognized until the twentieth century (Fruton 1999; Veigl, Harman, and Lamm 2020). Stains devised at about that time enabled the distribution of nucleic acids in cells to be determined, particularly when staining was combined with specific enzyme treatments. Brachet began localizing nucleic acids with the Feulgen reaction for DNA and the Unna formulation of methyl green and pyronin, the latter of which stains DNA green and RNA red (Brachet 1960; Burian 1996). He found that while DNA was confined to the nucleus, RNA was concentrated in both the nucleolus, an intranuclear structure, and the cytoplasm. Brachet confirmed these results by digesting cell sections with crystallized ribonuclease (RNase), an enzyme that destroys RNA, demonstrating that staining attributed to RNA disappeared (Brachet 1960). Caspersson came to similar conclusions through the use of what was essentially a microscope illuminated by ultraviolet light. Because nucleic acids absorb at ultraviolet wavelengths, he was able to quantitatively determine their distribution in cells, finding considerable absorption in the cytoplasm (Caspersson 1947). Although he could not distinguish DNA and RNA on the basis of this alone, he deduced that the cytoplasmic component was RNA because Feulgen staining of DNA was confined to the nucleus. This conclusion was strengthened by Brachet's almost simultaneous results (Caspersson 1947; Brachet 1960). Both investigators also found that cells with the largest amounts of RNA in the cytoplasm were those that were very active in protein synthesis, providing the first such linkage many years before the roles of RNA in the mechanism of protein synthesis were understood (Rheinberger 1997; Bechtel 2006). Brachet even developed these ideas into a precursor of what later became known as the central dogma that postulated the flow of information in cells from DNA to RNA and then to proteins (Crick 1958, 1970; Burian 1996).

Albert Claude's chemical analysis of large particles and microsomes isolated by cell fractionation had concluded that both contained RNA (Claude 1939, 1943b, 1944). Although he gradually became aware of the results of Caspersson and Brachet, he independently concluded that the RNA in microsomes corresponded to the basophilic staining in the cytoplasm. To demonstrate this he centrifuged whole, intact cells and then fixed and stained them. Under these circumstances, the contents of the cells contained within the cell membrane separated along the axis of centrifugation into areas containing nuclei, mitochondria, secretory granules, and an amorphous zone with no particular structure (Claude 1943a, 1943b). On further centrifugation, the amorphous zone separated into two areas, one of which was brightly stained with basophilic dyes. Because the stained material did not contain any morphologically recognizable components when viewed in the light microscope but was nevertheless sedimentable in the centrifuge, Claude concluded that this basophilic element and the well-known basophilia found in many other cell types corresponded to the microsomes isolated during cell fractionation. He also stained cell fractions, confirming that microsomes stained differently from the large particles in the cell and were basophilic. By 1948, Claude was confident in concluding that not only were microsomes the basophilic component of the cytoplasm, but also that the basophilia was caused by RNA. While familiar at this point with the correlations made by Caspersson and Brachet between RNA and protein synthesis, he did not accept their findings without reservation (Claude 1948).

After Claude returned to Belgium, Porter followed up on his first work with the electron microscope, culminating in a study titled "Observations on a Submicroscopic Basophilic Component of the Cytoplasm" (Porter 1953; see also Porter and Thompson 1948; Porter and Kallman 1952). By the time that this work was done, Porter had refined his procedures for working with cultured cells and presented them in detail in his study (Porter 1953). To substantiate his earlier finding of a lacelike reticulum, he observed a variety of tissue culture cells by both light and electron microscopy. In most cases, cells were fixed with osmium vapors as before, and some were stained with either hematoxylin or toluidine blue to visualize cytoplasmic basophilia in the light microscope. Porter also observed living cells in culture with phase contrast and dark-field microscopy (Porter 1953). Although it required some semantic gyrations, Porter argued that the *submicroscopic* reticulum seen in the electron microscope is, nevertheless, also visible in the light microscope, supporting his claim that the reticulum is indeed an actual cell structure and not an artifact: "The hematoxylin-staining material of the cytoplasm is finely divided and has the general form of the reticulum of the electron microscope image. This indicates that the reticular system corresponds to the basophilic substance of the cytoplasm" (Porter 1953, 740). Porter then attempted to link his observations to those of Claude by examining osmium fixed, isolated microsomes in the electron microscope, concluding that the "highly variable, partly amorphous component . . . displays enough similarity to vesiculated strands to convince one that it is made up, in part at least, of elements of the reticulum" (Porter 1953, 740). Although he had previously introduced the term (Porter and Thompson 1948; Porter and Kallman 1952; Palade 1956a), in this paper Porter formally christened his structure the *endoplasmic reticulum* (ER) to indicate that, so far, it was best described as a "complex reticulum of strands" located predominantly in interior parts of the cytoplasm (Porter 1953, 747).

At this point, Palade began to take over the work on the ER from Porter. Although the 1953 paper was not the last that Porter published on the topic, he became more interested in the development of the field of biological electron microscopy and its application to a diverse set of biological systems, publishing in the early 1950s articles on protozoa, muscle, and connective tissue proteins. Upset with the difficulty of getting their largely morphological studies accepted in leading journals like the *Journal of Experimental Medicine*, as well as the poor quality of reproduced electron micrographs when the papers were finally published, Porter also took time to spearhead the founding of a new journal published by the Rockefeller University Press. Initially called the *Journal of Biophysical and Biochemical Cytology* when it appeared in 1955, it became the *Journal of Cell Biology* in 1962 (Porter and Bennett 1981; Bechtel 2006, 260–61; Moberg 2012). Palade, on the other hand, began to focus on the ER and its function in cells.

A Particulate Component of the Cytoplasm

In the first of what became a series of papers on the ER, Palade, while still collaborating with Porter, compared its appearance in individual cultured cells and in cells from tissues (referred to as *in situ*, i.e., in its original place). The first group, cultured cells, was prepared by Porter's original protocol of fixation with osmium vapors and drying and viewed in their entirety in the electron microscope. Tissues, on the other hand, were excised as small blocks and fixed by immersion in Palade's buffered osmium, dehydrated and embedded in plastic, and cut into thin sections using the new Porter-Blum ultramicrotome (Porter and Blum 1953; Palade and Porter 1954). In whole cells, the ER appears similar to what was reported before. In sections, however, it is seen as a clearer though more complicated structure because the sections are thinner than the diameter of the strands seen in whole cells. As a result, the overall morphology of the ER is less apparent and is visualized as a series of cross-sectioned circular or oblique vesicular profiles, some of which they describe as *cisternae* to reflect what they believe is the intersection of a series of tubes into hollow, flattened areas. To consolidate their observations, they also devised methods to embed and section cultured cells, some of which had been grown from the same tissues examined in sections of tissue blocks. This allowed them to directly compare whole cell observations with sectioned cells and tissues. They concluded that the ER seen in tissue sections is identical to that seen both previously and in this study in whole cells (Palade and Porter 1954). Shortly after the publication of this article, Palade extended these conclusions in a second paper by examining sections from forty different mammalian and avian cell types (Palade 1955b). He established that the ER is present in nearly every cell and consists of an interconnected three-dimensional network of membrane-bounded profiles approximately 50 to 300 nm in diameter, usually with a lightly stained, homogeneous content (Palade 1955b, 576).

In the paper with Porter, Palade notes briefly that in some cells there appear to be "small granules" attached to some membranes of the ER, rendering these membranes "rough surfaced" while other membranes are "smooth surfaced" (Palade and Porter 1954, 646). In a description of a sectioned kidney cell, Palade elaborates: "Note the roughness of the outside surface of the membrane limiting the vesicular and tubular elements of the reticulum and the agglomeration of small dense granules around these elements and in contact with their membrane. In a few places, where the section apparently exposes the limiting membrane over a relatively large surface, . . . the small granules appear to be disposed in linear series" (Palade and Porter 1954, pl. 58 caption).[1] In the subsequent, more comprehensive study of the ER, Palade again mentions the particles associated with the membranes, calling

---

1. In fact, these details are hard to see in the reproduced electron micrograph, an example, no doubt, of Porter and Palade's complaint that journals, in this case the *Journal of Experimental Medicine*, do a poor job of printing the micrographs.

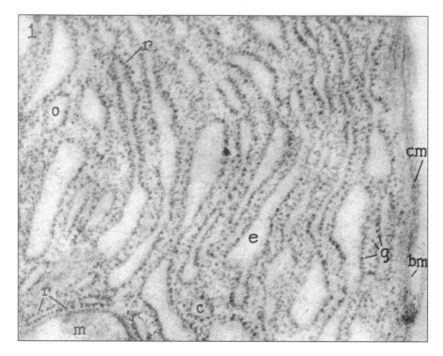

17. Palade's endoplasmic reticulum of the "rough surfaced variety." The cell membrane is indicated by *cm*, a mitochondrion by *m*, granules by *g* (i.e., ribosomes), and rows of granules by *r*. Elongated (*e*), oval (*o*), and circular (*c*) profiles of the endoplasmic reticulum are also labeled. © 1955 Palade, G. Originally published as figure 1 of plate 11 in the *Journal of Biophysical and Biochemical Cytology* 1 (1): 59–68.

the overall structures "profiles of the rough surfaced variety," or rough ER (Palade 1955b, 579).[2] However, by this time he had published a separate paper focused on this "small particulate component of the cytoplasm" in the first issue of the new *Journal of Biophysical and Biochemical Cytology* (Palade 1955a; see also Palade 1956c).

Palade had originally presented his findings to a meeting of the Electron Microscope Society of America in 1953, and the new paper expands on his earlier observations with a series of highly enlarged electron micrographs that more clearly show the particles than in previous publications, reflecting the improved reproductions in the new journal (fig. 17). His subjects are

2. In an interview more than forty years later, Palade continued to refer to ER "of the rough surfaced variety."

mainly secretory cells, such as those of the exocrine pancreas[3] and parotid gland, because of their very elaborate ER, but some other cell types are included as well. In pancreas cells in particular, where the cytoplasm is filled with parallel layers of ER, the membrane surface facing the cytoplasm is encrusted with hundreds of particles, alternating with particle-free zones corresponding to the interior lumens of the tubular structure (Palade 1955a) (fig. 17). Palade notes that in oblique or "grazing" sections that display flat membrane surfaces (referred to as *en face*), the particles assume patterns that resemble "rosettes, spirals, and circles" (Palade 1955a, pl. 13 caption). Palade and others had earlier noted that the ER extends as a double membrane around the nucleus (Watson 1955, 257). Now he supports that contention by noting that particles appear to be associated with only the membrane facing the cytoplasm and not with the membrane surface exposed to the interior of the nucleus (Palade 1955a).

Although Palade clearly believes that the particles are attached to the membrane, he is careful to place the word attached in quotation marks throughout the paper to emphasize that actual physical attachment has not been demonstrated. At the end of the paper, he engages in a careful and elaborate argument concerning the relationship between cytoplasmic basophilia and the particles. He notes that others have isolated small RNA-rich particles by centrifugation that they referred to as ultramicrosomes or macromolecules, suggesting that the RNA content of typical microsomes might be due to the particles attached to microsomal membranes rather than the membranes themselves (Palade 1955a, 64–65). He then extends his argument using his own observations of cells from the intestinal crypt. These cells, Palade found, have a "poorly developed" ER but high concentrations of apparently free particles in their cytoplasm. Since the regions with the particles in these cells are known to be basophilic, his observations support the idea that it is the particles and not the ER membrane that are responsible for the basophilia (Palade 1955a, 65). Palade now believes that the particles should be considered a component "distinct" from the ER membrane (Palade 1956b, 91).

Over the next few years, Palade's results are extended by others to not only the mammalian cells that formed the core of his observations but also

---

3. The exocrine pancreas synthesizes digestive enzymes and secretes them into the small intestine. The endocrine pancreas, whose secretory cells are intermixed with the endocrine pancreas in mammals in so-called Islets of Langerhans, synthesizes and secretes hormones such as insulin and glucagon into the bloodstream.

plant cells and microorganisms (Roberts 1958). When combined with the earlier conclusions by Brachet and Caspersson that cytoplasmic RNA has something to do with protein synthesis, it is natural to link the particles to protein synthesis as well, even if the linkage is circumstantial (Roberts 1958, vii). In early 1958, the topic of the first symposium of the new Biophysical Society is "Microsomal Particles and Protein Synthesis." Although Palade gives an invited paper, the bulk of the presentations are biochemical and biophysical characterizations of the particles. In the introduction to the published proceedings of the conference, the editor, Richard B. Roberts, reports that in response to the "semantic difficulty" that participants encountered about what to call particles from different sources a consensus was reached to use the name *ribosome* (Roberts 1958, viii).[4]

The discovery of ribosomes also provided a means to definitively link microsomes from fractionated cells with the ER seen in whole cells. In studies of fractionated liver and pancreas, both of which have substantial amounts of rough ER, Palade and his new collaborator, Philip Siekevitz (see below), reported that isolated microsomal vesicles also appeared to be covered on the outside with particles (Palade and Siekevitz 1956a, 1956b). They correctly surmised that these vesicles were derived from the rough ER during homogenization, stating:

> It is assumed that the breaking down of the reticulum into microsomes is not due to mechanical tearing but rather to a generalized pinching-off process taking place upon cell injury. This assumption is supported by the following findings: (*a*) the limiting membrane of the microsomal vesicles is usually continuous,—fragments with broken, torn, or open ends are not encountered; (*b*) a fragmentation of the ER into apparently independent vesicles is frequently encountered in cells undergoing cytolysis and in cells damaged by trimming at the periphery of tissue blocks. If the mechanical factors involved in tissue homogenization are directly responsible for the fragmentation of the network, then it follows that the broken fragments "heal" rapidly and thus form closed vesicles. (Palade and Siekevitz 1956b, 686)

With this observation, the stage is set for a functional analysis of the ER using isolated *rough microsomes*, the vesicular derivatives of the rough ER seen in intact cells.

---

4. In some circles, the particles had been referred to as "Palade-somes."

Microsomes and Protein Synthesis

Almost nothing was known about the biochemical mechanism of protein synthesis at the end of the 1940s. Since the nineteenth century, chemists had been aware that proteins are polymers of amino acids linked together head-to-tail with peptide bonds, but how the polymers were constructed was unknown. One idea was that protein synthesis is just the reversal of proteolytic reactions that severed peptide bonds. Another thought was that short peptides with just a few amino acids might be preassembled in some way and then linked together to create full-size proteins (Loftfield 1957). Because synthetic reactions require the input of energy, biochemists such as Fritz Lipmann proposed the involvement of phosphorylated high energy intermediates (Siekevitz and Zamecnik 1981; Rheinberger 1993, 1997).

After World War II, the availability of radioactive amino acids made new sorts of experiments possible, because even trace amounts of radioactivity could be easily detected. At Massachusetts General Hospital (MGH) in Boston, Robert Loftfield, a physical organic chemist, devised improved techniques for the chemical synthesis of the amino acids glycine and alanine containing the radioactive isotope ^{14}C (Rheinberger 1997). Paul Zamecnik, a physician at MGH interested in the role of protein synthesis in cancer, added radioactive alanine to slices of normal rat liver and rat liver tumors called hepatomas to compare the amounts of incorporation of the amino acid into the tissues. Initial results suggested that the rate of incorporation in the hepatoma slices exceeded that of normal liver. Because incorporation required oxygen, Zamecnik and his colleagues took advantage of results from Lipmann's laboratory, a neighbor at MGH, showing that the chemical dinitrophenol (DNP) uncoupled the production of phosphorylated compounds from oxygen consumption. When added to the liver slices, DNP inhibited incorporation into the hepatoma-derived material but not the normal liver (Rheinberger 1997, 46–47). Nancy Bucher, another investigator studying liver regeneration at MGH in the same unit as Zamecnik, noted that incorporation was higher in slices of regenerating rat liver than in normal liver, weakening the connection between increased protein synthesis and cancer (Rheinberger 1997, 49). More significantly, it was not clear where any of the experiments with the liver slice system would lead. Because metabolism of ^{14}C-alanine into compounds other than proteins by liver slices was possible, it was very difficult to link incorporation of the radioactive amino acid directly to protein synthesis (Siekevitz and Zamecnik 1981; Rheinberger 1997). Other strategies were needed.

In 1949, the biochemist Elizabeth Keller joined Zamecnik's lab as a research fellow and tried an alternative approach to the liver slice system to study incorporation. Keller injected radioactive amino acids directly into rats and, at various times after injection, removed their livers and homogenized and fractionated them using the techniques developed by Schneider, Hogeboom, and Palade at Rockefeller (Hogeboom, Schneider, and Pallade 1947, 1948). She found that shortly after the injection radioactivity was concentrated in the microsome fraction, from which it later disappeared (Keller 1951; Rheinberger 1997). At about the same time, Philip Siekevitz obtained a postdoctoral position with Zamecnik at MGH. Siekevitz had completed his PhD degree in biochemistry with David Greenberg at the University of California, Berkeley. Greenberg's lab was one of several other than Zamecnik's studying radioactive amino acid incorporation into tissue slices (Rheinberger 1997, 48; Moberg 2012, 137–49). Based in part on Keller's results, Siekevitz set about developing a completely cell-free system to study incorporation of radioactive amino acids.

Siekevitz began by adding radioactive alanine to rat liver homogenates together with ATP and a metabolic intermediate capable of being oxidized to provide further energy to the system and incubating the mixture for up to forty minutes. At various times, he took samples and fractionated the homogenate into microsomes, mitochondria, and nuclei, resulting in a supernatant free of particulate sedimentable material. As with Keller's in vivo experiment, Siekevitz observed that the greatest incorporation was into the microsomal fraction (Siekevitz 1952). Significantly, after precipitating proteins and other components from the incubation, Siekevitz was careful to demonstrate by chemical analysis that the radioactivity incorporated was in fact peptide bonded to protein and not associated with proteins through other linkages or in compounds trapped in the precipitate. In subsequent experiments, he separated the fractions and recombined them in different combinations. He then incubated these mixtures with radioactive alanine and the other metabolites and measured incorporation of radioactivity into protein. Siekevitz found that the greatest incorporation occurred when he combined microsomes with mitochondria and that it depended on phosphorylated high energy compounds provided by the mitochondria because inhibitors such as DNP blocked protein synthesis (Siekevitz 1952; Moberg 2012).

Although Siekevitz soon left MGH for a second postdoctoral position with Van Potter at the University of Wisconsin, where he worked on mitochondrial function, Zamecnik's group continued refining the in vitro system and made a number of major contributions to understanding the mechanism of protein synthesis (Rheinberger 1997). In 1950, while Siekevitz's experiments were under way, Zamecnik described the liver homogenate as "a biochemical bog in which much effort is being expended to reach firm ground" (Zamecnik 1950, 663; Siekevitz and Zamecnik 1981; Rheinberger 1997). With the success of the fractionated system, at least a toe had reached firm ground, with solid footing ahead. The impact of the system was, however, not only on the study of protein synthesis but also on establishing the functions of the ER, a morphologically recognized part of the cell. This was because the experiments of both Keller and Siekevitz demonstrated that microsomes, fragments of the ER, were clearly a site of protein synthesis. As Hans-Jörg Rheinberger notes, "The reconstitution assay linked the construction of a *metabolic* space to the *topological* space of fractionation" (Rheinberger 1997, 71; emphasis original). Others would soon reposition the latter in the context of the whole cell.

An Integrated Approach

By 1949, Hogeboom and Schneider, the only two biochemists working in the Laboratory of Cytology at the Rockefeller Institute, had moved to the National Institutes of Health (Bechtel 2006; Moberg 2012). As the characterization of the ER proceeded, Palade realized that a purely morphological approach was insufficient to investigate the function of the ER or the particles associated with it (Palade 1955a, 65; Moberg 2012, 133). In 1950, Porter had asked Potter at Wisconsin, Schneider's former mentor, if he might recommend someone to replace Hogeboom and Schneider, but he did not identify any candidates (Moberg 2012). When Palade contacted Potter again in 1954, Potter recommended Siekevitz, who was hired immediately (Moberg 2012). Given the presumed linkage between rough microsomes and protein synthesis, the choice of Siekevitz, who had pioneered the study of protein synthesis in a cell-free fractionated system, was fortuitous if not strategic.

Palade and Siekevitz quickly embarked on what they called "an integrated morphological and biochemical study" of microsomes derived from rat liver (Palade and Siekevitz 1956a).[5] The starting point was a careful morphological description of the whole tissue, the tissue homogenate, and the cell fractions isolated from the homogenate. They took great care to prepare

---

5. In an article written when Palade won the Albert Lasker Basic Research Award in 1966, he seemed to suggest that his earlier morphological and biochemical work on mitochondria in collaboration with Hogeboom and Schneider in the 1940s and 1950s was his first integrated study (Palade 1966).

identically both the intact tissue and cell fractions for electron microscopy so that their morphologies could be compared directly. They observed that "almost all the structures described in the cytoplasm *in situ* can be recognized in sections of pellets prepared from the liver homogenate" (Palade and Siekevitz 1956a, 178). These similarities persisted as they isolated microsomal vesicles. In parallel with morphological observations, they carried out chemical analyses of the isolated microsomes, measuring protein, RNA, and phosphorus derived from phospholipids. They also measured the activity of the enzyme diphosphopyridine nucleotide (DPNH)-cytochrome c reductase that Hogeboom had found was associated with liver microsomes (Hogeboom 1949).[6]

Palade and Siekevitz next attempted to dissect isolated microsomes chemically, apparently modeling their approach in part on the one published in early 1955 by John Littlefield in Zamecnik's lab (Littlefield et al. 1955). The microsomes were "aged" by incubating them for periods at different temperatures or treated with Versene, a compound that strongly binds divalent cations such as magnesium and calcium,[7] the detergent deoxycholate (DOC), and the enzyme RNase (Palade and Siekevitz 1956a). In each case, they tracked the release of protein, RNA, phospholipid phosphorus, and DPNH-cytochrome c reductase activity from the microsomes by pelleting them in the centrifuge and examining both the supernatant and the pellet both chemically and in the electron microscope. From their analysis, Palade and Siekevitz began to establish the functional identity of microsomes. As described previously, they concluded that microsomes were fragments of the ER that were sealed, osmotically active vesicles with retained, electron-dense content "imprisoned" inside (Palade and Siekevitz 1956a, 192). The disposition of particles on the outsides of the vesicles also suggested the preservation in isolated rough microsomes of the same functional organization found in the intact rough ER. Treatments of the rough microsomes with Versene and RNase released or degraded particles and RNA from the microsomal membrane, supporting the conclusions that most microsomal RNA was found in the particles and that the membranes

---

6. Enzymes such as DPNH (now NADH, or nicotinamide adenine dinucleotide)-cytochrome c reductase catalyze specific chemical reactions. The amount of enzyme present in a particular fraction (called activity or *specific activity*) is calculated as the rate of product creation by the catalyzed reaction divided by the amount of protein in the fraction. Fractions containing high concentrations of a particular enzyme have a high specific activity because a lot of product is created by a relatively small amount of protein.

7. Versene is now usually referred to by its chemical name, ethylene diamine tetraacetate (EDTA).

and particles were separate entities. DOC treatment, on the other hand, dissolved the vesicle membrane, solubilizing most protein, phospholipid, and DPNH-cytochrome c reductase activity and releasing apparently intact particles that were recovered by centrifugation.

The integrated morphological and biochemical approach used by Palade and Siekevitz was a refinement of the epistemic strategy pioneered by Albert Claude when he first began monitoring cell fractionation with the microscope while maintaining a balance sheet to track the distribution of protein, RNA, and other chemical components in his fractions in relation to the amounts found in his starting homogenates. However, while Claude's microsomes referred to "a small granule of undefined nature" (Claude 1943b, 1943a), for Palade and Siekevitz this was no longer the case. Now, their microsomes corresponded to a definite identifiable part of the cell, likely associated with specific functions.

Immediately after completing their study of liver, Palade and Siekevitz switched to the guinea pig exocrine pancreas as the subject of their experiments. They chose the exocrine pancreas because it contains very large amounts of rough ER arrayed in stacked layers in the cytoplasm (see fig. 17), as well as so-called zymogen granules believed to be involved in secretion of a variety of digestive enzymes synthesized by the cells (Palade and Siekevitz 1956b). After conducting an initial set of experiments that yielded results similar to those obtained from liver, they embarked on a more detailed and systematic multiyear study of protein synthesis in the exocrine pancreas.

To begin, Siekevitz and Palade surveyed the tissue morphologically and refined their procedures to isolate cell fractions enriched in nuclei, zymogen granules, mitochondria, and microsomes (Siekevitz and Palade 1958b). Their chemical and enzymatic analysis determined that both RNase and TAPase (trypsin-activated protease)[8] are concentrated in the zymogen fraction, confirming what they referred to as the "classical hypothesis" about the storage site for the enzymes in the cell (Siekevitz and Palade 1958b). In a second paper, they focused their attention on microsomes (Siekevitz and Palade 1958c). In their initial examination of the pancreas, they had noted that the ER in situ contains amorphous content of varying density that differs from cell to cell (Palade and Siekevitz 1956b, 674). Now, in an attempt

8. The pancreas secretes a variety of protease precursors that are activated when cleaved by the protease trypsin. Trypsin itself is also secreted by the pancreas as inactive trypsinogen, and then activated by existing active trypsin. A form of the enzyme RNase, which Palade and Siekevitz used as an added, exogenous reagent to characterize the microsomes, is, coincidentally, also synthesized and secreted by the pancreas, something that would later cause problems (see later section of this chapter and chap. 5).

to synchronize the production of secretory proteins by the pancreas, they starved guinea pigs for forty-eight hours and then fed them for one hour before removing the pancreas. As they anticipated, the hormonal signal triggered by feeding stimulated the pancreas to immediately synthesize large amounts of digestive enzymes. When they examined the pancreas in the electron microscope, they observed "large accumulations of intracisternal granules in the ER," in comparison to the relatively empty ER seen in the pancreas from starved animals (Siekevitz and Palade 1958c, 311). Isolated microsomes contained inside both visible granules and increased amounts of both RNase and TAPase activities. Siekevitz and Palade then treated microsomes with low concentrations of the detergent DOC, and separated the resulting suspension into heavy and light fractions by centrifugation. In the electron microscope, the heavy subfraction appeared to consist mainly of the granules also seen inside untreated microsomes, corresponding to the amorphous material found in the intact ER, while the light subfraction was mostly ribonucleoprotein particles separated from the microsomal membranes. The granules contained concentrated amounts of RNase and TAPase activities. Surprisingly, however, the particle subfraction was also enriched in these activities (Siekevitz and Palade 1958c).

As they discussed their results, Siekevitz and Palade speculated that they had observed the initiation of a "new secretory cycle" and that the granules that they saw in the ER after feeding are "precursors of the zymogen granules" that "represent the intracellular transport, packing, and storing of already finished products" (Siekevitz and Palade 1958c, 316). In other words, the secretory products are synthesized in association with the rough ER and then transported through the cell to the zymogen granules where they are stored prior to their release from the cell.[9] Furthermore, they proposed that "significant steps in the synthetic process precede formation of intracisternal granules and can be reasonably connected with more stable structures in the ER such as the limiting membrane and the attached ribonucleoprotein particles" (Siekevitz and Palade 1958c, 316). They concluded the paper by suggesting that "the particles and subjacent membrane are the loci of enzyme formation" (Siekevitz and Palade 1958c, 317).

In a follow-up study, Siekevitz and Palade injected radioactive amino acids into guinea pigs and followed the appearance of the radioactivity in cell fractions over time (Siekevitz and Palade 1958a). They observed that

9. In separate experiments not described here, Palade and his colleagues definitively showed that this postulated secretory pathway, with some additional elaborations, is correct, as he summarized in his Nobel Lecture in 1975 (Palade 1975; Matlin and Caplan 2017).

radioactivity first appeared in microsomes and later, as the amount in microsomes declines, in the zymogen and mitochondrial fractions. They then treated the microsomes with DOC and fractionated them into smaller components, finding that radioactivity appeared in ribonucleoprotein (RNP) particles within one minute after injection and quickly declined as it appears in the microsomal content subfraction. Their results were consistent with earlier, more limited findings by Littlefield and others (Littlefield et al. 1955). From this they concluded, "The findings presented here are compatible with our general hypothesis according to which the digestive enzymes are synthesized by the attached RNP particles, transferred across the limiting membrane into the cavities of the ER, segregated temporarily . . . into intracisternal granules, and finally packed and stored in the form of mature granules" (Siekevitz and Palade 1958a, 565). With this, Siekevitz and Palade proposed a route by which proteins leave their site of synthesis in the cytoplasm, the ribonucleoprotein particles (named ribosomes that same year), and enter membrane-enclosed compartments that will eventually leave the cell. They were, however, frustrated by looking at "the sum of a diversity of simultaneous synthetic processes," and resolved to examine the synthesis and transport of a single protein (Siekevitz and Palade 1958a, 565).

The protein that Siekevitz and Palade decided to study is chymotrypsinogen, the inactive form of chymotrypsin, one of the major trypsin-activated proteases secreted by the pancreas. As before, their approach was to inject a radioactive amino acid into recently fed guinea pigs, remove the pancreas at different short time points, and fractionate it. The difference in this case, however, is that they would then base their conclusions on chymotrypsinogen purified from the cell fractions. Preliminary electron microscopy conducted at the beginning of their study verified the identity of the fractions. The subsequent analysis of chymotrypsinogen synthesis was a veritable tour de force because the amount of radioactive chymotrypsinogen synthesized within the time limits of their experiments is minuscule, and the protein must be purified from individual cell fractions using conventional biochemical procedures that are time-consuming and inefficient (Siekevitz and Palade 1960).[10] They were fortunate to have access to sophisticated chromatography techniques to purify chymotrypsinogen that had been developed

---

10. Siekevitz and Palade state that "It follows that only a small fraction of the incorporated leucine is used for α-chymotrypsinogen synthesis; that the extent of labeling is not necessarily the same for all enzymes produced by the gland . . . ; and finally that substantial losses of α-chymotrypsinogen are probably incurred during chromatography" (Siekevitz and Palade 1960, 626).

at the Rockefeller Institute (Hirs, Moore, and Stein 1953). Nevertheless, to get sufficient material, they had to use up to ten or twelve animals for each time point. This means that they could look only at one time point per experiment, of a total of five time points between one and forty-five minutes examined. Overall, radioactive labeling, cell fractionation, enzyme purification, and analysis of enzyme activity and incorporation of radioactivity required about a week for each of the time points, without accounting for preliminary experiments and repetitions.

Their findings from these experiments verified their previous conclusions but with greater time resolution. One minute after injection of radioactivity, the largest amount of newly synthesized chymotrypsinogen is found in the microsomes, and most of that in the particles (ribosomes) that could be released from the microsomes by DOC treatment. After two and a half minutes, some of the chymotrypsinogen in the particles has shifted into the microsomal contents, and after three minutes radioactivity begins appearing in the mitochondrial[11] and zymogen granule fractions, peaking in the latter at forty-five minutes, the last time point examined (Siekevitz and Palade 1960, table II). From these results, they concluded that "α-chymotrypsinogen is synthesized in or on the attached RNP particles [ribosomes], to be subsequently transferred to the cavities of the ER and finally concentrated and stored in the zymogen granules" (627). They also stated that "the attached RNP particles are an important, possibly unique, site for α-chymotrypsinogen synthesis" (627). There is one puzzle. Some of the labeled chymotrypsinogen was also found on "free" RNP particles isolated by centrifugation from the original "post-microsomal" supernatant, that is, particles not released from microsomes by DOC. Because both the free RNP particles and those attached to the microsomes are associated with chymotrypsinogen, Siekevitz and Palade considered it unlikely that they are "physiologically distinct cell components" (628). Instead, they speculated that the free particles need to attach to the microsomal membrane before they complete the synthesis of the enzyme. Time will prove their statements prescient.

Siekevitz and Palade were certainly not the only ones studying the synthesis of proteins in the pancreas or in other tissues. Indeed, by 1956, several laboratories, most prominently Zamecnik's, had already provided evidence that ribonucleoprotein particles and not the microsomes per se were the site of protein synthesis. Furthermore, the Zamecnik lab's use of the detergent

---

11. Based on later results, it is clear that chymotrypsinogen was not actually associated with mitochondria but with other membranous compartments in the same fraction, possibly elements of the Golgi apparatus.

DOC was critical to the work of Siekvitz and Palade (Keller, Zamecnik, and Loftfield 1954; Littlefield et al. 1955; Loftfield 1957; Rheinberger 1997, 94).[12] Siekevitz and Palade were, however, the only ones conducting such elaborate *integrated* studies that combined biochemical analysis with correlated morphological observations. On a conceptual basis, the key difference between their work and that of others was that they were not interested in protein synthesis as a biochemical problem. Certainly, increased knowledge about the mechanism of protein synthesis provided by research groups like Zamecnik's were useful to them for both analysis and interpretation of experiments. However, what they wanted to know was how the *cellular* task of synthesizing a specific group of proteins and getting them to intracellular sites where they could be secreted was accomplished. For this, their combination of morphology and biochemistry was essential because the biochemical activities that they examined remained rooted in defined parts of the cell.

The chymotrypsinogen experiments focused attention on the interaction between the RNP particles, or ribosomes, as they began to be called after 1958 (Roberts 1958), and the cytoplasmically disposed membrane surface of the ER as the key first cellular step in the process of protein secretion. It seemed clear that ribosomes were where protein synthesis begins and continues and that the association of ribosomes with the membrane led to the deposition of the synthesized secretory proteins into the interior of the ER. But how?

A Vectorial Process

By the mid-1960s, a number of advances had begun to clarify the biochemical mechanism of protein synthesis. The Cambridge biochemist Fred Sanger's sequencing of the protein insulin in 1951 had definitively established that proteins consist of unique sequences of amino acids connected by peptide bonds (Sanger and Tuppy 1951a, 1951b). The coupling of this

---

12. One oft-cited study linking particles to protein synthesis in the pancreas was by Vincent Allfrey, Marie M. Daly, and Alfred E. Mirsky published in 1953 (Allfrey, Daly, and Mirsky 1953). Mirsky was a prominent scientist at the Rockefeller Institute. Although Allfrey and his colleagues used a form of cell fractionation to isolate what they call microsomes, they do not reference a single paper by Claude, Porter, Hogeboom, or Palade. The likely reason for this was competition between the groups at Rockefeller. In 1954, when Mirsky learned that Palade and Siekevitz were planning to work on protein synthesis in the liver and pancreas, he tried to get the Rockefeller director to block their work, to no avail (George Palade to Philip Siekevitz, July 28, 1954, Palade Archives, Courtesy of the National Library of Medicine).

information with what was known about DNA structure and protein synthesis then led Francis Crick to propose the *sequence hypothesis* stating that the sequence of nucleotides in DNA corresponds to the sequence of amino acids in proteins (Crick 1958). In the late 1950s, Zamecnik's lab discovered a form of RNA that transfers amino acids to the ribosome, called transfer RNA (tRNA) (Siekevitz and Zamecnik 1981). At about the same time, evidence began accumulating, particularly from bacterial systems, that another type of RNA, called messenger RNA (mRNA), is synthesized or *transcribed* from DNA and carries the encoded sequence information to ribosomes. There, the sequence information is *translated* into proteins using amino acids coupled to specific tRNAs (Siekevitz and Zamecnik 1981). During the process of protein synthesis, ribosomes remain attached to the mRNA they are translating. When protein synthesis is robust, a single mRNA can attach to more than one ribosome such that the staggered synthesis of multiple copies of the same protein can proceed simultaneously. Such a combination of one stringlike mRNA with multiple ribosomes is known as a polyribosome or *polysome*.

In the same period, research in the Rockefeller laboratory, which was now referred to more often as cell biology than cytology, both broadened and deepened. The Rockefeller Institute became the Rockefeller University in 1955 and began to accept graduate students, some of whom found their way to Porter and Palade's operation. Other postdoctoral fellows and visiting scientists also joined the laboratory and pursued an array of projects distinct from the work on protein synthesis and secretion (Moberg 2012, 242). Siekevitz and Palade and some of these new collaborators began characterizing the composition and biogenesis of the ER as a way of getting at possible functions that were either related or unrelated to the issues investigated so far (Ernster, Siekevitz, and Palade 1962; Dallner, Siekevitz, and Palade 1966a, 1966b). Siekevitz also collaborated with Yutake Tashiro, a Japanese scientist, to show that rat liver ribosomes were composed of a single large and small subunit, a structure similar to that of bacterial ribosomes, whose characterization was more complete than in eukaryotes (Tashiro and Siekevitz 1965a, 1965b).

Some of this work was designed to provide a foundation for the next step in deciphering the process of secretory protein synthesis localized to the rough ER. In describing their hypothesis on the synthesis and transfer of chymotrypsinogen to the interior of the ER, Siekevitz and Palade state that "the final proof remains to be obtained by studying *in vitro* amino acid incorporation into a digestive enzyme by isolated cell fractions" (Siekevitz and Palade 1960, 627). Their first steps to accomplish this were taken in

1960 at the same time that the chymotrypsinogen study was under way. Together with the graduate student Jack Kirsch, they demonstrated that ribosomes detached from guinea pig liver rough microsomes with DOC can incorporate radioactive amino acids into protein (Kirsch, Siekevitz, and Palade 1960). While not a novel finding, it was both a satisfying culmination of their previous work and a critical step forward (Palade 1966).[13] Despite this, in 1962, Palade wrote that their attempts to achieve protein synthesis with pancreas ribosomes had been unsuccessful (Palade, Siekevitz, and Caro 1962, 37).

The painstaking efforts of Siekevitz, Palade, and their collaborators to achieve a breakthrough in understanding how secretory proteins were synthesized and sequestered by the ER began to yield significant results by late 1965. One key study, published in 1966, resembled the analysis of chymotrypsinogen synthesis six years earlier in that it was based on in vivo labeling and fractionation of the guinea pig pancreas, along with electron microscopic examination of fractions. However, it differed in being focused on the synthesis of amylase, a pancreatic secretory enzyme that digests complex sugars, and on the refined subfractionation of rough microsomes with the detergent DOC. Amylase was an attractive choice because it could be purified from fractions in one step by precipitation with the natural glucose polymer glycogen. Subfractionation was carried out with different concentrations and preparations of DOC worked out in part in Siekevitz's previous collaboration with Lars Ernster, a visiting Swedish biochemist (Ernster, Siekevitz, and Palade 1962; Palade 1966), and employed linear density gradients of sucrose to isolate the subfractions inspired by another collaboration (Dallner, Orrenius, and Bergstrand 1963; Dallner, Siekevitz, and Palade 1966a, 1966b).[14]

After isolating rough microsomes by conventional fractionation, Siekevitz treated them with either low or high concentrations of DOC and separated them on a density gradient ranging from 10% to 30% sucrose.

13. In a review written when Palade was awarded the Lasker Award in 1966, Palade stated, "From our point of view, the first phase of our project was completed with the Kirsch paper. . . . We had started with the discovery of a new particulate component by electron microscopy; we had isolated the attached and free particles from microsomes and the cell sap, respectively; we had characterized them biochemically as ribonucleoprotein particles; we had demonstrated their ability to synthesize proteins when properly supplemented. This was the type of inquiry we had in mind; the characterization of newly discovered structures not only in terms of their morphology, but also in terms of their chemistry and function" (Palade 1966, 146–47).

14. Siekevitz had a close personal relationship with Ernster, having spent a sabbatical period with him in the late 1950s working on mitochondria (Moberg 2012, 140).

With low concentrations of DOC, amylase appeared in both a light fraction sedimenting near the top of the gradient and a heavy fraction sedimenting further down and a pellet on the bottom of the tube. By electron micros-copy the light fraction consisted of ribosomes connected together in chains (presumably polysomes), while the heavy fraction and pellet consisted of ribosome-covered microsomal vesicles containing dense content in their lumens (Siekevitz and Palade 1966). When the heavy fraction was "osmoti-cally shocked," which presumably broke open the closed microsomal vesi-cles, and then centrifuged again at high speed, the resulting pellet at the bottom of the tube contained vesicles with attached ribosomes that had lost their contents and had reduced amounts of amylase activity (see Siek-evitz and Palade 1966, fig. 4). Similarly, when the microsomal vesicles with attached ribosomes from the pellet were treated with high concentrations of DOC and resedimented, they still appeared by electron microscopy as vesicles with bound ribosomes but with loss of some of the material within the vesicles. The critical observation was that, in both cases, microsomal vesicles with bound ribosomes that had lost their content had the highest concentration of radioactively labeled amylase.[15] From this Siekevitz and his colleagues concluded:

> When microsomes are gradually disrupted by detergent treatment, ribosomes begin to be detached from the membranes, and those detached most eas-ily are those having low specific radioactive amylase . . . Even when disrup-tion of the microsomes is stepped up by high-DOC treatment, there remains on microsomal membranes a small percentage . . . of ribosomes which are resistant to detachment by detergent, and contain amylase of much higher specific radioactivity . . . [I]t is clear that the presence of newly synthesized, completed protein . . . on the ribosomes coincides with a firmer attachment of the ribosomes to the membrane. One of the factors responsible for this situation might be that part of the amylase molecule is still firmly bound to the ribosome, while the rest of it is already anchored to the membrane of the ER. (Siekevitz and Palade 1966, 527–28)

They conceded that an alternative explanation might be that the mRNA for amylase that is associated with the actively synthesizing ribosomes might itself be responsible for membrane binding but did not discuss this idea further (Siekevitz and Palade 1966, 528).

---

15. The measure for amylase that is highly radioactive is *specific radioactivity*, which is the ratio of radioactivity in amylase to amylase enzymatic activity.

One reason that Siekevitz and Palade might have favored the conclusion that the amylase molecule is responsible for the strong attachment of some ribosomes to the membrane is that other studies in their laboratory leaned in the same direction. At about the same time that the in vivo amylase study was under way, Colvin Redman, a postdoctoral fellow with Siekevitz and Palade, had succeeded in synthesizing radioactive amylase in an in vitro system derived from pigeon pancreatic rough microsomes (Redman, Siekevitz, and Palade 1966). He tried a system derived from the pigeon pancreas because it had earlier been successfully used by Fritz Lipmann's laboratory in protein synthesis studies modeled after those of Zamecnik (Weiss, Acs, and Lipmann 1958).[16] What was key is that the pigeon pancreas has one hundred times less RNase, an RNA degrading enzyme synthesized by the pancreas, than the guinea pig pancreas. This made the survival of endogeneous mRNAs coding for pancreas proteins such as amylase more likely (Redman, Siekevitz, and Palade 1966). Redman incubated the microsomes and radioactive leucine, together with other factors required for protein synthesis, for periods from one to ten minutes. He then isolated amylase using glycogen from either the total incubation mixture or ribosomes or dissolved microsomes separated by treatment with high concentrations of DOC. The dissolved microsomes included both the solubilized microsomal membrane and the interior microsomal contents. Redman observed that after one minute of incubation the highest amount of amylase radioactivity was in the ribosome fraction. At later times ribosome incorporation remained high, but increasing amounts of radioactive amylase appeared in the DOC soluble fraction, suggesting that newly synthesized amylase was being transported into microsomal vesicles from ribosomes, a conclusion consistent with both the previous experiments by Siekevitz and Palade on chymotrypsinogen synthesis and contemporaneous experiments on amylase synthesis in vivo (Siekevitz and Palade 1960; Redman, Siekevitz, and Palade 1966; Siekevitz and Palade 1966).

By 1966, the Rockefeller group, soon called the Laboratory of Cell Biology,[17] had become fully established. Soon after completion of a new research building, South Laboratory, in 1959, Palade, Porter, Siekevitz and their by now large number of associates moved there from their original quarters in the third basement of the Theobald Smith Building (fig. 18). Despite this,

16. Siekevitz and the biochemist and Nobel laureate Fritz Lipmann overlapped at MGH during the time that Siekevitz was working on in vitro protein synthesis with Zamecnik. By 1966, Lipmann had moved to Rockefeller, and his laboratory was in the same building just upstairs from the cell biology group.

17. According to Carol Moberg, Vincent Allfrey and Alfred Mirsky formed a joint laboratory at Rockefeller in 1963 that they also called the Laboratory of Cell Biology (Moberg 2012, 406).

18. The Rockefeller Laboratory of Cell Biology ca. 1960. Upper row, from the left: Tom Ashford, unidentified, Peter Satir, Philip Siekevitz, George Palade, unidentified, David Spencer Smith, Steven Wissig, Jack Kirsch, Don Young, Myron Ledbetter. Middle row: unidentified, unidentified, unidentified, unidentified, Lucien Caro, unidentified, Mary Bonneville. Lower row: Rachel Maggio, "Sherbie" (Porter's administrative assistant), Marilyn Farquhar, Keith Porter, unidentified, unidentified, unidentified. Identifications by Peter Satir. Photo courtesy of the National Library of Medicine.

Porter left for Harvard in 1961, leaving Palade in charge of the unit with Siek-evitz as his senior partner. In 1966, Palade received the Albert Lasker Award for Basic Research, a formal acknowledgment of his achievements (Palade 1966). At about the same time, the fruitful collaboration between Palade and Siekevitz, while still continuing, began to change. Siekevitz no longer focused on microsomal protein synthesis and ribosomes but instead on bio-genesis and differentiation of both the ER membrane and, in a completely new project, chloroplast membranes. Palade, with new collaborators, began to concentrate on steps in pancreatic secretion beyond the ER. It is unclear if this apparent divergence by Siekevitz was the result of changing interests or if he considered the synthesis and transfer of proteins to the cisternae of the ER to be, for him, sufficiently settled. There may have been other factors. Both of the studies on amylase synthesis were funded by Siekevitz's grant, but Palade remained the senior author (Redman, Siekevitz, and Palade 1966; Siekevitz and Palade 1966). Siekevitz was promoted to full professor in 1966 and may have wanted to be more independent of Palade.[18] Furthermore, Siek-

18. When Rockefeller was the Rockefeller Institute, investigators were promoted as institute "members." When it became a university, investigators were promoted along the typical profes-sorial ranks found in other academic settings.

evitz, who had always been interested in the sometimes fraught relationship between science and the government, became even more interested as the Vietnam War unfolded, and his energies began to be focused outside of the laboratory (see, e.g., Siekevitz and Nagin 1966). This kind of activism was likely not supported by Palade. In any case, the microsomal protein synthesis project shifted to a new, younger faculty member, David Sabatini.

Sabatini was an Argentine MD who learned electron microscopy while conducting research with Eduardo D. P. De Robertis, a well-known cytologist and the author of a 1946 cytology textbook that was translated into English and became in later editions one of the first to introduce students to cell biology (De Robertis, Nowinski, and Saez 1948). To improve his knowledge of biochemistry, Sabatini obtained a Rockefeller Foundation Fellowship to join Palade's laboratory beginning in January 1961. When the time came, however, Porter had not yet departed from Rockefeller, and there was insufficient space for Sabatini. Because Sabatini could not change the initiation date of his fellowship, Palade arranged for him to spend six months at Yale in Russell Barrnett's laboratory (Sabatini 2005; Moberg 2012, 165–67). To fill his time, Sabatini tested various aldehydes other than formaldehyde, a commonly used tissue preservative, as fixatives for electron microscopy. He discovered that one of them, glutaraldehyde, was very effective at preserving cell structure. When his discovery became known and published, he got the attention of the electron microscopy community as a young scientist to watch (Sabatini, Bensch, and Barrnett 1964; Moberg 2012).[19] When he arrived at Rockefeller, he began working on the nature of the ribosome-membrane interaction and soon joined the graduate program as a PhD student.

Sabatini's first work at Rockefeller, published in 1966, followed up on the results of Redman and Tashiro, Siekevitz's collaborator (Redman and Sabatini 1966; Sabatini, Tashiro, and Palade 1966; Sabatini 2005). As described previously, Redman successfully developed an in vitro system that was capable of transferring a polypeptide chain in the process of synthesis, known as a *nascent* polypeptide, into the lumen of isolated microsomal vesicles. Tashiro used analytical centrifugation to study the dissociation of mammalian ribosomes into large and small subunits and demonstrated with Siekevitz that some newly synthesized polypeptides were associated with the large ribosomal subunit, something already known in bacterial systems (Tashiro and Siekevitz 1965a, 1965b). With Tashiro, Sabatini combined an analysis of the dissociation of membrane-bound ribosomes using

19. Glutaraldehyde, usually in combination with osmium tetroxide, one of the original fixatives for electron microscopy, is still very commonly used.

centrifugation with observations of rough microsomes by electron micros-copy. For this, Sabatini used the negative staining technique in which thin sections are flooded with an electron-dense dye that penetrates accessible spaces in the sample, revealing the impermeable parts as light areas against a darker background. With this procedure, Sabatini showed that ribosomes are bound to the microsomal membrane by the large subunit, most likely oriented such that an axis normal to the membrane passes through the mid-dle of both the large and small ribosomal subunits (Sabatini, Tashiro, and Palade 1966) (fig. 19). In addition, Sabatini labeled proteins in vivo prior to isolation of rough microsomes and dissociation of ribosomes, demon-strating that newly synthesized polypeptides remained associated primarily with membrane-bound large subunits after release of small subunits, an extension of the earlier results of Tashiro and Siekevitz.

In the other paper, Sabatini and Redman used a modification of Red-man's in vitro protein synthesis system using guinea pig liver microsomes to study the fate of newly synthesized polypeptides. As with Redman's earlier study, they added radioactive amino acids to isolated rough microsomes and incubated them under conditions compatible with continued protein synthesis. In this case, however, they prematurely terminated protein syn-thesis using the antibiotic puromycin. Puromycin inhibits protein synthesis because it structurally resembles a tRNA molecule coupled to an amino acid (Yarmolinsky and De La Haba 1959; Allen and Zamecnik 1962). When added during protein synthesis, it enters the ribosome and is covalently bonded to the distal end of the nascent polypeptide, causing the unfinished polypeptide to be released from the ribosome. To their surprise, Redman and Sabatini observed that the prematurely terminated polypeptide synthe-sized by membrane-bound ribosomes was transferred through the micro-somal membrane like a complete protein and accumulated in the lumen (Redman and Sabatini 1966).[20]

When the combined results of these two papers were considered, Sa-batini and Redman were able to propose a model for what Redman had earlier called the *vectorial* transfer of proteins through the microsomal mem-brane (Redman, Siekevitz, and Palade 1966).

> The model provides for a central channel or space within the large [ribo-somal] subunit, separated from the cytoplasmic matrix by the 30S [small] subunit, but continuous with the cisternal space through a permanent or

---

20. Günter Blobel later cited this puromycin experiment as a major influence on his thinking.

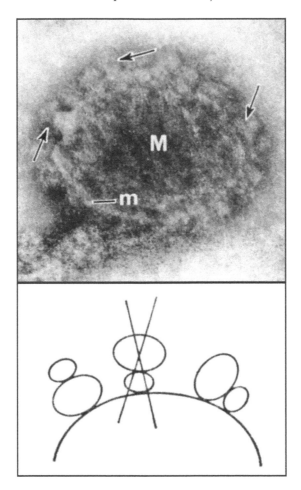

19. David Sabatini's analysis of the orientation of bound ribosomes on the microsomal membrane. The upper panel is an electron micrograph using negative stain with the arrows pointing to individual ribosomes. The lower panel is a diagram indicating the two most likely and one unlikely orientation of the bound ribosomes. M indicates the overall circular microsomal vesicle, and m indicates the microsomal membrane (Sabatini, Tashiro, and Palade 1966, plate I, image a, and figure 11). © Elsevier. Used with permission.

intermittent discontinuity in the microsomal (ER) membrane to which the 47S [large] subunit is attached. In the model, the vectorial (unidirectional) character of polypeptide release is explained by structural restrictions. The peptides being synthesized are assumed to grow within the central channel of the large ribosomal subunit (47S) in an environment which is (or can be made) continuous with the cisternal space through a discontinuity in the ER membrane. As visualized at present, the transfer mechanism relies primarily on release from the large ribosomal subunit and on structural restrictions at the ribosome-membrane junction, and hence it is nondiscriminatory and possibly passive. (Redman and Sabatini 1966, 614)

While the mechanism by which the ribosome attached to the membrane was not explicitly investigated, they stated that "it seems plausible to consider that strong attachment of 'active ribosomes' is related to passage of the protein through the membrane," with one possibility being that "the product of protein synthesis on the ribosomes is what makes them stick to the membrane" (Sabatini, Tashiro, and Palade 1966, 523–24). At this point, with the departure of both Redman and Tashiro and the absence of further direct involvement by Siekevitz, progress stalled until others were added to the team.

The accomplishments of Sabatini and his colleagues fell along a continuum of successive findings obtained by Claude, Porter, Palade, and Siekevitz. Sabatini and his collaborators initiated their experiments with isolated rough microsomes, but their strategy, nevertheless, was a direct extension of the integrated approach pursued earlier by Palade and Siekevitz. Through coordinated electron microscopy, cell fractionation, and biochemical analysis, Palade and Siekevitz had determined that microsomes, an isolated part of the cell first described by Claude, were identical with the ER, a structure originally seen in the whole cell by Porter and Claude. Now, with the cellular context of microsomes established, Sabatini was able to work directly with microsomes at higher resolution. Even now, Sabatini continued to integrate electron microscopy and biochemistry as he probed the functions of the ER through its surrogate, microsomes. In this way, he set the stage for the next step of inquiry into the relationship between the structure of the ER and its function in secretion: investigation of molecular mechanisms. Now, the boundary of the cell that Claude and Porter had originally crossed using cell fractionation and electron microscopy had led to a new boundary in the cell, the membrane of the ER and microsomes, and its traversal by Palade, Siekevitz, and Sabatini. Gradually, the epistemic strategy devised by Claude and refined by Palade and Siekevitz was inching closer and closer to an explanation of the first step in the secretory process at the molecular level.

FIVE

The Signal Hypothesis

Günter Blobel

In 1966, Günter Blobel, a new postdoctoral fellow, arrived in the Rockefeller cell biology laboratory with the intention of working with Siekevitz on the attachment of ribosomes to the ER membrane. His path to Rockefeller was quite circuitous. Blobel was born in Silesia, a part of Germany that is now in Poland, before World War II. Near the end of the war in 1945, his large family of two sisters and four brothers left Silesia with their mother in an effort to escape the Russian advance, arriving finally in the vicinity of Dresden, a center of romantic and baroque culture. A third sister, the eldest and a nurse, was killed in a bombing attack. His father, who was serving in the German army, did not rejoin the family from a prisoner of war camp in Finland until the 1950s. Soon after arriving in the Dresden area, Blobel visited the city, which had not to that point been significantly damaged during the war and was amazed by its architecture and beauty. Within days he witnessed Dresden's firebombing and destruction from outside the city, an event that made an intense and lasting impression.

His family remained in the small town of Freiberg near Dresden, where Blobel grew up in the midst of music and Goethe's poetry, visiting Weimar, Goethe's home, on a school culture trip. However, as the children of bourgeoisie, according to the East German authorities (his father was a veterinarian), he and his siblings were not eligible for higher education. At that time, in the 1950s, the border with West Germany was porous and, one after another, Blobel and all of his siblings and parents migrated to the west. Blobel left home at the age of seventeen and a half, arriving in Frankfurt on Goethe's birthday, and, with the support of his brothers, prepared to enter the university. That Frankfurt had been the birthplace of Goethe was important to Blobel, since he imagined that he was retracing in reverse the path of the poet-scientist who had served as a source of solace and beauty for him in otherwise drab postwar East Germany. The kind of tragic dichotomy of the heart and the head that Goethe embodied became almost a personality trait of Blobel's that motivated and inspired him.

In Frankfurt, Blobel somewhat arbitrarily decided to attend medical school, which one enters in Germany immediately after high school. Because of his relative youth (West German medical students were at least eighteen or nineteen years old when they began) and the unwillingness of authorities to accept his East German educational credentials, his entry was obstructed until he completed a series of examinations. One area where he struggled was, surprisingly, German language and literature. In an exam he was asked to interpret the eighteenth-century drama *Emilia Galotti*, which tells the story of the romance between a daughter of the bourgeoisie and a son of the nobility. Blobel described it as an example of class struggle, an interpretation labeled nonsensical East German propaganda by his examiners. Nevertheless, Blobel persevered and was admitted to medical school.

After six years of medical school and a two-year internship, Blobel found himself more interested in the causes of disease than treating patients. One of Blobel's brothers, who by now was a professor at the University of Wisconsin in Madison, suggested that he get proper research training and helped arrange a spot for him in graduate school at Wisconsin. He arrived sight unseen and entered Van Potter's laboratory in the McCardle Cancer Center, a choice that, while unknown to him at the time, provided him with a crucial future connection (fig. 20). In that period, Potter's involvement in experimental science waxed and waned, in part because of his broad and passionate interests outside the lab, although he always maintained a small research group (Trosko and Pitot 2003). In particular, when Blobel joined his lab in 1962, Potter had become involved in a long-term struggle with the city of Madison to build Monona Terrace, a lakeside center originally designed by Frank Lloyd Wright, a friend of Potter's, and redesigned by Wright just before his death in 1959. Consequently, Potter provided little direction to Blobel, simply instructing him to follow up on findings in bacteria by trying to purify mRNA from rat liver, even though the lab had little to no experience in this area. Blobel soon became an expert in cell fractionation, publishing one of his first papers with Potter on the isolation of nuclei, a line of research that he would follow in parallel with his studies of secretion for his entire career (Blobel and Potter 1966).

Ultimately, Blobel's attempts to isolate mRNA led him to polysomes, the active array of ribosomes on mRNA that assembles during protein synthesis, and then to the problem of the so-called ribosome cycle. Ever since Palade had described ribosomes, it was recognized that some ribosomes are free in

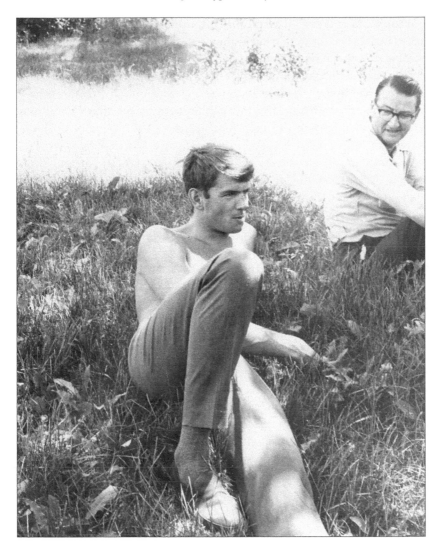

20. Günter Blobel picnicking with his graduate adviser Van Potter (right) in the environs of Madison, Wisconsin, early 1960s. Photo courtesy of the Rockefeller University. Used with permission of the Rockefeller University and Carl Blobel.

the cytoplasm while others are bound to the surface of the ER membrane (Palade 1955a). One hypothesis (which was ultimately confirmed) was that there is no important difference between the two sets of ribosomes and that those that are bound cycled with the free population. As Blobel delved into the subject, he began reading the work of Siekevitz and Palade and their collaborators at Rockefeller on bound ribosomes and decided to go there when he finished his degree rather than return to Germany. Fortuitously, as was previously described, Potter's lab had long been a source of biochemists for the Rockefeller group. Among these was of course Siekevitz, who Palade had recruited in the 1950s. Potter called Siekevitz, and in 1966 Blobel was on his way to New York. His time with Potter revealed Blobel's ambition and drive; after starting a project from scratch in 1962, he published nine papers, including three in 1967 in the prestigious *Journal of Molecular Biology*. One of these dealt with the "interaction of ribosomes and membranes," a topic that became his life focus and crowning achievement (Blobel and Potter 1967a, 1967b, 1967c).

After Blobel began at Rockefeller, he soon recognized that Siekevitz was no longer actively working on the ribosome-membrane problem and that Palade was the driving force of the laboratory. Palade, however, paid little attention to him, and Blobel struggled to get started on a project. What he was good at was purifying ribosomes from rat liver, and when he gave his preparations to Palade to examine in the electron microscope, Palade complied and produced dozens of micrographs. At this time, David Sabatini had just been appointed assistant professor after completing his PhD and was continuing his work on ribosome-membrane interactions. When he realized what Blobel was doing, he considered it an infringement on his territory. Nevertheless, after some conflicts, they agreed to work together.

It is possible that Sabatini recognized that Blobel was a useful partner. In his two major papers on ribosome-membrane interactions, Sabatini had benefited from collaborations with Redman and Tashiro, who were experts, respectively, on in vitro protein synthesis and ribosome isolation (Redman and Sabatini 1966; Sabatini, Tashiro, and Palade 1966). Blobel, with a more biochemical orientation and extensive experience in cell fractionation and ribosome isolation, may have filled a technical gap in Sabatini's group. In any case, the two had complementary but very different personalities. Sabatini was older than Blobel, and his scientific education was much broader due to his experiences in Argentina and in the PhD program at Rockefeller. Some regarded him as possibly more knowledgeable than even Palade, and he seemed to have never forgotten anything that he learned. On the other hand, his approach to experimental research was somewhat passive. He depended mainly on technicians to conduct experiments, preferring to remain in his office, where he liked to engage in intense discussions about research ideas in front of the blackboard. Blobel, in contrast, was a gregarious and even manic presence in the laboratory, constantly conducting experiments and doing preparations with his own hands. Years later, when other investigators were asked where Blobel's laboratory space was located during this period, they replied that he worked everywhere.

The cell biology laboratory at Rockefeller was much more cohesive than a typical academic department. Rockefeller University considered the organization of research at the institution as based on the European model, namely, very hierarchical. Instead of departments, there were formal laboratories headed by a senior professor who dictated what other members would work on, regardless of their rank. In the case of the Laboratory of Cell Biology, Palade was in charge, but he was largely a benevolent force. Certainly, the individuals appointed to faculty positions represented Palade's interests, but once they had their own operations, he did not micromanage their work. He seemed to view his role as one of standard setting and education and often read and provided extensive stylistic and substantive comments on manuscripts prior to their submission to journals for publication. The laboratory had a weekly seminar series in its small library, with Palade inviting speakers from far and wide. Attendance at the seminars was expected but not forced, and frequently, when other members of the laboratory wondered why the work of the speaker invited by Palade was of interest, Palade would provide a background and summary of the speaker's field that convinced them that the work was important and relevant.

The laboratory was located primarily on the fifth floor of the South Laboratory building. There were some small individual laboratories, such as Sabatini's located at the end of the hall. However, most major equipment was distributed in different rooms along the corridor. As one would expect from a laboratory founded on cell fractionation and electron microscopy, centrifuges and electron microscopes dominated, with several microscopes located in a suite of rooms at the opposite end of the corridor from Sabatini's lab. Centrifuges were of various types and generations, and included old Spinco models, often called washing machines because of their size and configuration, and more modern ultracentrifuges manufactured by the International Equipment Company and Beckman Instruments. The centrifuges were noisy and were so heavily used that they were constantly running, producing a high-pitched whine from their spinning rotors and a rumble from compressors and vacuum systems, sounds that provided a kind of background music for all other activities.

The Collaboration

By fall 1968, Blobel and Sabatini were actively collaborating, and results started to appear in print in 1970. In a pair of papers published in the *Journal of Cell Biology*, they examined the degree to which nascent polypeptides are protected from proteolytic digestion during synthesis on free ribosomes or when the ribosomes are bound to the ER membrane.[1] They reported that during synthesis on free ribosomes, the entire nascent polypeptide chain, with the exception of a thirty-nine amino acid peptide at the carboxy-terminal end, is degraded by added proteases when digested at 0°C (Blobel and Sabatini 1970). This peptide is protected inside a channel traversing the interior of the large ribosomal subunit, as had been previously postulated (Redman and Sabatini 1966; Malkin and Rich 1967). In contrast, when membrane-bound ribosomes are treated with proteases, larger protein fragments are protected (Sabatini and Blobel 1970). These results suggested that membrane-associated ribosomes in the process of protein synthesis bind to the membrane so tightly that the nascent chain emerging from the large subunit is inaccessible to proteases. Furthermore, they indicated that the microsomal vesicles are so tightly sealed that their contents are not subject to proteolytic digestion. Indeed, when the vesicles were made permeable by the addition of small amounts of detergent, then the larger polypeptides were degraded. Much later, the use of proteases to determine if a protein had entered the lumen of a microsomal vesicle would prove an essential assay.

Blobel and Sabatini's use of detergents was nothing new. Biochemists had employed detergents for years to clarify tissue extracts, often without any clear understanding of how they worked. Chemists, on the other hand, knew a great deal about mechanisms of detergent action because most detergents had originally been synthesized for industrial processes. In the early 1970s, biochemists began exploring the extensive chemical literature to help them use detergents more intelligently in their experiments, particularly when the experiments involved biological membranes (Tanford 1974; Helenius and Simons 1975). Among the most commonly used detergents were DOC, exploited by Zamecnik's lab and by Siekevitz and Palade in the 1950s and 1960s to subfractionate microsomes (see chap. 4), Triton X100 (TX100), and sodium dodecyl sulfate (SDS). Both DOC and TX100, which were employed by Blobel and Sabatini in their experiments, are considered mild detergents because they do not completely destroy membranes or un-

---

1. Proteins or polypeptides are synthesized by ribosomes reading an mRNA from the 5′ end to the 3′ end of the polymer (referring to specific locations on ribose residues at either end of the RNA molecule). The first amino acid incorporated in the new protein forms the amino- or N-terminus and the last the carboxy- or C-terminus, so called because the initial amino acid has a free amino group while the final amino acid has a free carboxy group. The term *nascent* refers to polypeptides that are being actively synthesized.

fold (denature) proteins when used in low concentrations. In contrast, SDS is a strong detergent that disrupts cellular structures and converts unfolded proteins into negatively charged "rigid rods"(Reynolds and Tanford 1970b). The latter property turns out to be very useful in conjunction with techniques for separating protein mixtures (see below).

In the period of their collaboration, the Sabatini and Blobel laboratory consisted of three postdoctoral fellows, Mark Adelman, Takashi Morimoto, and Gert Kreibich, and one graduate student, Dominica (Nica) Borgese. Another investigator associated with the lab, Yoshiaki Nonomura, would shortly return to Japan. All these individuals formally worked under Sabatini's direction but in reality worked closely with both Blobel and Sabatini. Blobel had been promoted to assistant professor at Rockefeller in 1969. Sabatini was in the process of being promoted to associate professor but, for a variety of reasons, was already considering leaving Rockefeller for a new position. The emphasis of active projects in the laboratory was on ribosomes and, to a lesser extent, mRNA (Blobel 1971a, 1971b; Blobel and Sabatini 1971a; Blobel 1972; Morimoto, Blobel, and Sabatini 1972a, 1972b). The relationship of these studies to the problem of ribosome-membrane interactions was not always evident. Because it was by now clear that cytoplasmic proteins are almost exclusively synthesized on free ribosomes and secretory proteins on bound ribosomes (see Hicks, Drysdale, and Munro 1969), it is probable that Blobel and Sabatini were looking for key differences between free and membrane-bound ribosomes evident either morphologically or biochemically.

The work of Adelman and Borgese addressed these issues more directly. Adelman optimized procedures for releasing active ribosomes from microsomal membranes, demonstrating definitively that ribosomes interacted with the microsomal membrane through both a salt-labile electrostatic linkage and through the nascent chain, as suspected earlier (Adelman, Blobel, and Sabatini 1973; Adelman, Sabatini, and Blobel 1973). Borgese's work focused on a detailed analysis of ribosome subunit circulation between the bound and free populations, leading to the conclusion that, in vitro at least, small ribosomal subunits were released from bound ribosomes more readily than large subunits. This was consistent with Sabatini's earlier finding that ribosome-membrane interactions are mediated by the large subunit (Sabatini, Tashiro, and Palade 1966) (fig. 19) but also with the idea that some large ribosomal subunits might possess a component that directed them to the membrane. However, when Borgese compared the protein composition of purified ribosomal subunits derived from free and bound ribosomes using the relatively new technique of SDS–polyacrylamide gel electrophoresis, there were essentially no differences between the free and bound populations (Borgese, Blobel, and Sabatini 1973).[2]

Information

Blobel and Sabatini's concentration on ribosomes, in particular, structural aspects of the ribosome-membrane interaction (Sabatini et al. 1971), reflected an underlying assumption in the early 1970s about how ribosomes synthesizing secretory proteins find their way to the ER membrane. This assumption, whether stated explicitly or not, was that ribosomes specialized for the synthesis of secretory proteins *direct* themselves to the ER membrane. This was not an unreasonable idea since some ribosomes bound to the ER membrane in the absence of protein synthesis (Sabatini et al. 1971; Adelman, Sabatini, and Blobel 1973; Borgese et al. 1974).[3] Furthermore, preliminary experiments reported by Sabatini at a meeting suggested that the protein polyphenylalanine synthesized in vitro from the synthetic mRNA polyuridylic acid (polyU) "was transferred to the microsomal contents." Because the polyU mRNA was both synthetic and coded for an artificial polypeptide and not a secretory protein, this result seemingly implied that a protein's mRNA or the protein itself did not determine if the ribosomes translating it bound to the membrane (Sabatini et al. 1971, 128).[4] Never-

2. The technique of SDS–polyacrylamide gel electrophoresis, developed in the 1960s, was a breakthrough in the analysis of complex protein mixtures. The detergent SDS binds to proteins and converts them into rigid rods with a net negative charge proportional to the protein molecular weight (Reynolds and Tanford 1970a, 1970b). When such solubilized mixtures of proteins in SDS are "loaded" onto gels of polyacrylamide and an electric current is applied, all proteins migrate through the pores of the gel toward the positive pole because of their negative charge and are separated according to molecular weight (Maizel 2000). The technique is simple, and multiple experimental samples can be run simultaneously, resulting in the resolution of complex protein mixtures into a series of bands on the gel in a few hours. Each band corresponds to an individual polypeptide (see fig. 1).

3. In Nica Borgese's paper published in 1974 but based on work conducted earlier at Rockefeller, she states that "if large subunits could attach to microsomal membranes independently from protein synthesis [which she observed], it would be unlikely that the nascent chain itself is the only factor involved in the recognition mechanism directing polysomes, engaged in the synthesis of defining products, to become membrane bound. . . . An alternative mechanism could operate by which specific initiation complexes would recognize membrane-bound large subunits. The latter process could be mediated by modified initiation factors, capable of recognizing both messenger RNAs and membrane-bound large subunits" (Borgese et al. 1974, 578). Borgese's results raised the possibility that there might be a proteinaceous ribosome receptor on the microsomal membrane.

4. In this experiment, published in a symposium report, proteases were not used to show that polyphenylalanine is actually inside microsomal vesicles rather than just stuck to the outside of the membrane (Sabatini et al. 1971).

theless, all their intense efforts to study the morphology of free and bound ribosomes in the electron microscope, to carefully disassemble and reassemble subunits derived from free and bound populations, and to examine the molecular composition of the two groups had uncovered no significant differences.[5] Gradually, this absence of evidence pushed them to think more about the role of the nascent polypeptide chain itself in ribosome-membrane attachment.

When Blobel assumed a faculty position in 1969 he began to apply for his own research funding. One desirable avenue was a so-called Research Career Development Award from the National Institutes of Health (NIH). The RCDA was attractive not only because it was a source of salary support but also because it represented a peer-reviewed endorsement of an investigator's promise. At some point in 1969 or 1970, Blobel submitted an RCDA application describing his proposed work on bound ribosomes. Contained within the application was a novel proposal for how ribosomes synthesizing secretory proteins are directed to the ER. Blobel hypothesized that there is a common sequence at the 5' end of mRNAs for secretory proteins that, when translated, facilitates the attachment of the ribosome-mRNA complex and the growing polypeptide chain to the ER.

Shortly after submitting his proposal, Blobel, who wanted to establish his scientific identity and independence from Sabatini, attended a meeting on biomembranes held in the Smoky Mountains of Tennessee on April 5, 1971. Despite its remote location, the meeting was anything but obscure, with many leading membrane investigators in attendance. Blobel's presentation was brief, and the write-up, which was published in its original typescript form the same year, was only three pages long.[6] Although his presentation was blandly titled "Ribosome-Membrane Interaction in Eucaryotic Cells," it contained the same revolutionary proposal outlined previously in his grant application and was illustrated with a hand-drawn figure (Blobel

5. Borgese found one unique polypeptide in large ribosomal subunits derived from free ribosomes but was not sure what to make of the observation (Borgese, Blobel, and Sabatini 1973). In a paper from a 1974 meeting, published in 1975, Sabatini reported other protein compositional differences between free and bound ribosomes, suggesting "that several protein bands in both subunits may correspond to factors involved either in the recognition of specific mRNAs and/or in regulating the binding of selective polysome classes to endoplasmic reticulum membranes" (Sabatini et al. 1975). It is not clear if Sabatini's group, by that time located at New York University, followed up on this observation.

6. Blobel recalled that he was very nervous about his presentation and rehearsed it extensively while walking on forest trails in the Smoky Mountains. When it was time for his talk, Palade introduced him and, in his remarks, gave away Blobel's most important ideas, somewhat deflating his presentation.

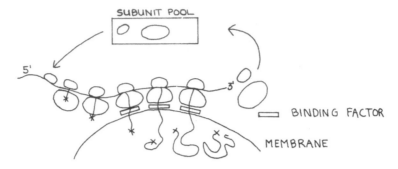

21. The published model of what later became the signal hypothesis from Blobel and Sabatini's 1971 symposium paper (Blobel and Sabatini 1971). Used with permission of Günter Blobel.

and Sabatini 1971b) (fig. 21). In the figure, the "X" drawn at the amino terminus of nascent polypeptides emerging from actively synthesizing ribosomes is described as the primary element—the information—responsible for directing active ribosomes to the membrane. The existence of such an X-factor on the growing polypeptide chain that directs the ribosome-mRNA complex to the membrane is a completely new idea. As stated in the published paper:

> This . . . implies that the information as to whether a particular mRNA is to be translated by free or membrane-bound ribosomes lies in the mRNA itself. This and other features of protein biosynthesis . . . can be incorporated into a tentative scheme. . . . All mRNAs to be translated on bound ribosomes are assumed to have a common feature such as several codons near their 5′ end, not present in mRNAs which are to be translated on free ribosomes. The resulting common sequence of amino acids near the N-terminal of the nascent chains or a modification of it (indicated by X) would then be recognized by a factor mediating the binding to the membrane. This binding factor could be a soluble protein, which recognizes both a site on the large ribosomal subunit and a site on the membrane. (Blobel and Sabatini 1971b, 194)

According to contemporary accounts, this proposal, which had first appeared in Blobel's grant application, originated from vigorous exchanges between Blobel and Sabatini in front of Sabatini's office blackboard. At the time, the proposal was not supported by any direct data and was later described by Blobel as a "fantasy."[7] The write-up itself, which did not cause much of a stir in the scientific community, was kindly but derisively referred to as "the blurb" by other members of the laboratory.[8]

Blobel's description of the sequence at the 5'end of the mRNA (the end first translated into protein) as *information* was a novel use of the term. Prior to this, invoking information encoded in the sequences of nucleic acids had referred to inherited genetic information in DNA and the genetic code (Kendrew 1967; Crick 1970; Jablonka 2002). The information that Blobel refers to has nothing to do with genetics or even the function of individual proteins bearing this information once it is translated into a protein sequence. Instead, this is *spatial* information that indicates the destination of the protein in the cell, in this case, the ER. That is, although this information is encoded just like other sequences that give rise to proteins, it can only be decoded by existing topographic and asymmetric structures in the cell, structures that are inherited independently of the genes in the nucleus (see chap. 9).

Clues from Cambridge

While Blobel and Sabatini were developing the model depicted in the 1971 paper, Timothy Harrison was conducting his doctoral research with George Brownlee in the Laboratory of Molecular Biology (LMB) in Cambridge, United Kingdom (fig. 22). Brownlee was an expert in RNA sequencing and assigned Harrison, in collaboration with the immunologist and biochemist Cesar Milstein, a project to purify the mRNA of immunoglobulin light chains.[9] At the time, it was exceedingly challenging to isolate mRNAs. Even though the plasmacytoma and myeloma cells (derived from cancers of the immune

---

7. In an interview many years later Sabatini mentioned that the observation that the product of polyU translation, poly-phenylalanine, bound to microsomal membranes while that of polyA translation, poly-lysine, did not was a factor in formulation of the model. The polyU but not the polyA results were published in 1971 (Sabatini et al. 1971). Sabatini's apparent reasoning was that poly-phenylalanine, which is hydrophobic, might direct the translating ribosome to the (hydrophobic) membrane. This may be a post facto interpretation since the 1971 model did not mention that the "X" sequence of amino acids was hydrophobic. It was mentioned in a subsequent paper several years later when some evidence in favor of the 1971 hypothesis was already available (Sabatini et al. 1975).

8. In separate interviews many years later, both Nica Borgese and Blobel used exactly this term to describe the brief 1971 paper.

9. Immunoglobulins are proteins that make up antibodies, one function of which is to block infections. Each immunoglobulin protein molecule consists of two large polypeptides (heavy chains) and two smaller polypeptides (light chains). Most immunoglobulins are secretory proteins.

22. **Left to right:** Tim Harrison (ca. 1969), Cesar Milstein (ca. 1975), and George Brownlee (ca. 1975), who in 1972 correctly interpreted their observation of a larger form of newly synthesized light chain immunoglobulin as containing an amino-terminal signal routing the protein to the ER. Not shown is Michael B. Matthews, the fourth participant in the project, who told Blobel about their results. Originally published as figure 2 in Matlin, K. S. 2011. *Nature Reviews Molecular Cell Biology* 12 (5): 333–40. © Karl S. Matlin. Original photos courtesy of the Laboratory of Molecular Biology, Cambridge University.

system) that Harrison used as starting material produced large amounts of immunoglobulins, mRNA specific for the light chain was only a small fraction of the total RNA. To overcome this obstacle, Harrison enriched for light chain mRNA by isolating rough microsomes and extracting mRNA directly from the membrane-bound polysomes instead of trying to extract mRNA from whole, unfractionated cells. Because the light chain immunoglobulin is a secretory protein, Harrison reasoned that mRNA for light chain was likely to be found almost exclusively in membrane-bound polysomes. He then fractionated extracted mRNA by size using sucrose density centrifugation.

Because there was no direct assay for specific mRNAs, Harrison used the synthesis of the light chain in an in vitro protein translation system derived from a mouse ascites tumor to determine where the light chain mRNA migrated on the sucrose gradient. Harrison incubated his various fractions in this system and then examined synthesized polypeptides by SDS-gel electrophoresis. In the first experiment, Harrison noticed that there were two translation products migrating on the SDS gel at molecular weights similar to that of the secreted light chain polypeptide; one was the same size as authentic secreted light chain, whereas the other was ~1.5 kDa larger.[10] In

10. Dalton (Da) is a measure of molecular mass. Most proteins have total masses of 10,000–100,000 Daltons (10–100 kDa).

additional experiments, Harrison linked the correctly sized product to the presence of microsomal membranes in the translation system, and he noted that the larger polypeptide was synthesized exclusively in membrane-free preparations. To determine whether both products were indeed related to the light chain, Harrison and Milstein enzymatically fragmented both the larger and smaller radioactive proteins and separated the resulting peptides into a kind of map so that they could be compared. In addition to identifying them both as light chains, the mapping showed that they differed only in the amino terminus, the first part of the protein synthesized, with the initiator methionine amino acid residue only present in the larger product. Overall, the results suggested that the larger form might be a precursor of the smaller one and that generation of a light chain that matched the size of authentic light chain depended on microsomal membranes.[11]

Drawing on Harrison's knowledge of the published literature on rough microsomes and secretion, Milstein, Brownlee, Harrison, and Mike Matthews (a fourth participant in the project) postulated that the original amino-terminus of the light chain, which seemed to be removed when the microsomal membrane was encountered, might in fact be a "signal" directing polyribosomes to the membrane. In their ensuing *Nature New Biology* paper, published in September 1972, neither Blobel nor Sabatini was mentioned because the authors were unaware of the biomembranes meeting report published a year earlier (Harrison 1972; Milstein et al. 1972). After the *Nature New Biology* paper was in press, Matthews presented the work at a summer Gordon Conference[12] in the United States that was also attended by Blobel and at that point learned of Blobel's "fantasy."[13,14]

11. Interest in purification of mRNA for immunoglobulin chains was at this time very high because it was viewed as an important step in understanding antibody diversity. Several other papers published around the same time as that of Harrison and his collaborators also reported putative immunoglobulin precursors, but none of them linked the precursors to secretion the way that they did (Swan, Aviv, and Leder 1972; Mach, Faust, and Vassalli 1973; Schechter 1973; Tonegawa and Baldi 1973).

12. The Gordon Conferences are a series of summer scientific meetings initiated in 1931 by Neil Gordon from Johns Hopkins University. At the time of the meeting attended by Blobel, most Gordon Conferences were held on the campuses of private boarding schools in New England. They were known for the presentation of new, unpublished research findings and a relaxed social atmosphere.

13. Harrison's thesis, submitted in December 1972, extensively cited Blobel and Sabatini, including the 1971 paper describing their model. The reason that Harrison's thesis cited Blobel and Sabatini and the *Nature New Biology* paper did not is because the paper was submitted before the Gordon Conference attended by Blobel and Matthews, while Harrison's thesis was submitted after the Gordon Conference.

14. Even though most of the work reported in the *Nature New Biology* paper was Harrison's,

Reconstitution

News of the results from Cambridge was greeted enthusiastically by the Blobel and Sabatini group because it suggested that their speculative model from 1971 might, in fact, be correct. While the insights in the Milstein paper were remarkable, the published results only proposed but did not demonstrate that the larger form of the immunoglobulin light chain was a precursor of the mature form. Indeed, attempts mentioned in the paper to convert the putative precursor to the mature form by adding microsomal vesicles after synthesis were unsuccessful, leading Harrison and his colleagues to conclude that removal of the amino-terminal signal might occur during synthesis in living cells (Brownlee et al. 1972; Milstein et al. 1972; Harrison, Brownlee, and Milstein 1974). Several other papers on the translation of both heavy and light immunoglobulin chains from the Milstein and Brownlee groups followed the first one, but there were no breakthroughs. Harrison graduated, and Milstein refocused on immunology, inventing monoclonal antibodies just a short time later, for which he shared the Nobel Prize (Kohler and Milstein 1975).

In 1972, the Blobel-Sabatini partnership dissolved when Sabatini left Rockefeller to become chairman of a new cell biology department at New York University School of Medicine. Within a year or so after that, Palade also left Rockefeller for Yale University Medical School, and the entire cell biology group that remained was restructured with Siekevitz at the head. Blobel, after contemplating a return to Germany, was promoted to associate professor with tenure, and several new assistant professors were hired to round out the group. For the first time, Blobel was solely in charge of his own lab and was anxious to establish his independent research program. With Palade's encouragement, he set about creating a cell-free system that would efficiently reconstitute both the synthesis of secretory proteins and their transport or *translocation* across the microsomal membrane. As stated in a short paper from a meeting published in 1974, Blobel's requirements were clear: "A system containing stripped membranes, ribosomal subunits, and either globin mRNA (in vivo translated on free ribosomes) or immunoglobulin RNA (in vivo translated on membrane-bound ribosomes) should, under protein synthesis conditions in vitro, lead to ribosome attachment

---

and Harrison was most familiar with the secretion literature, he was not the first author because he was sent off to write his PhD dissertation at the end of his fellowship, and Milstein wrote the manuscript. It was customary in the Brownlee and Milstein labs at the LMB that the person who wrote the paper would be the first author.

and to vectorial discharge of the nascent chain into a proteolysis resistant location of the membrane only in the case of immunoglobulin mRNA. Such a result would constitute unequivocal evidence for in vitro reconstitution" (Blobel 1974, 713).

Blobel decided to establish two types of in vitro systems that were both capable of protein synthesis and contained microsomal membranes. One he called a *readout* system and the other an *initiation* system. The readout system contained rough microsomes isolated in the process of protein synthesis and then extracted with detergent to dissolve the microsomal membrane, leaving behind the formerly bound polysomes and their attached nascent polypeptides. When incubated in vitro with radioactive amino acids and factors needed to stimulate further protein synthesis,[15] the released polysomes would complete (i.e., readout) the protein chains that they had started before isolation but not initiate new ones. As starting material for this system, Blobel isolated active rough microsomes that synthesized immunoglobulin light chain from a myeloma tumor similar to the one previously used by Harrison.

The initiation system, in contrast to the readout system, was more complex because it was totally reconstituted. It contained microsomal vesicles that had been chemically *stripped* of bound ribosomes and mRNA and then repurified, as well as exogenous ribosomal subunits, factors for both the initiation and continuation of protein synthesis, radioactive amino acids, and, importantly, purified mRNA. Of these components, the most difficult to obtain turned out to be microsomes that, after the stripping procedure, would not inhibit protein synthesis driven by added mRNA.

Mark Adelman, in his earlier studies with Sabatini and Blobel, developed techniques for the isolation and stripping of rat liver rough microsomes. However, in Blobel's hands, in vitro protein synthesis from mRNA was inhibited in the presence of the stripped rat liver microsomes, possibly because the enzyme RNase present in the liver preparation degraded mRNA. Searching for an alternative, Blobel isolated microsomes from a variety of animal species, but they all blocked protein synthesis. Possibly at the suggestion of George Scheele, a former postdoc of Palade's who had stayed on at Rockefeller, Blobel tried microsomes from dog pancreas. Scheele was

---

15. These factors included so-called initiation and elongation factors that facilitate peptide bond synthesis, transfer RNA molecules that carried amino acids to the ribosome during synthesis, and enzymes to activate amino acids and attach them to the tRNAs. Because this crude mix was prepared by treatment of extracts at pH 5, they were frequently referred to as pH 5 enzymes. The concentration of factors to initiate new protein synthesis was low in the pH 5 preparation that Blobel used in his readout experiments (Blobel and Dobberstein 1975a, 840).

working on the secretion of proteins from the pancreas and may have been familiar with papers cited by Colvin Redman years before that indicated that both pigeon pancreas, which Redman had used, and dog pancreas contained low concentrations of RNase (Dickman and Bruenger 1962; Redman, Siekevitz, and Palade 1966). Indeed, when Blobel prepared stripped microsomes from dog pancreas and added them to his initiation system, protein synthesis was robust (see Scheele, Dobberstein, and Blobel 1978).

Based on the Milstein results, Blobel wanted to use the immunoglobulin light chain as a model secretory protein in his experiments and needed to isolate mRNA coding for the protein. Although he could, like Harrison, extract RNA from myeloma cell polysomes, the concentration of light chain mRNA in such preparations was too low for the in vitro assay he envisaged. In 1974, Bernhard Dobberstein joined Blobel as a postdoctoral fellow and was assigned the task of purifying mRNAs from the myeloma using a new approach. While a graduate student in Germany, Dobberstein had worked with free and membrane-bound polysomes from plants and easily picked up the techniques required to work with animal cells. The procedure to isolate mRNA exploits the affinity between long sequences of the nucleotide adenosine (polyA) located at the distal end of most mRNAs and short sequences of deoxythymidine coupled to cellulose particles (oligo-dT-cellulose).[16] Because oligo-dT-cellulose can be easily recovered by centrifugation, its use enabled the isolation of mRNAs in a single step (Aviv and Leder 1972). Using this strategy, Dobberstein quickly obtained preparations enriched for the light chain messenger. As a control protein that was not secreted, Blobel decided to use globin, the core protein of hemoglobin. In this case, obtaining mRNA for globin was easy because Blobel could isolate it from polysomes derived from rabbit reticulocytes, immature red blood cells that essentially synthesized only globin and had no ER (Blobel 1971b). Now, with the key components in place, Blobel could do the crucial experiments.

As Blobel assembled his translation systems, his diverse and apparently unfocused work over the previous few years began to make sense, at least in retrospect. When asked sometime later about what Blobel had been working on in the early 1970s, his colleagues in the cell biology group at Rockefeller suggested that he had been drifting, without a clear objective. Indeed, his papers characterizing free ribosomes from reticulocytes and proteins bound to mRNAs seemed to not have anything to do with understanding

16. This affinity between adenosine and thymidine is due to the A-T association by hydrogen bonding that is the foundation of the DNA double helix structure.

the function of membrane-bound ribosomes in secretion (Blobel 1971a, 1971b, 1973; Freienstein and Blobel 1974). As late as 1974, one of Blobel's colleagues even suggested that Blobel would ultimately be best known for his discovery of proteins bound to mRNA. Now, as he began his experiments, his facility in manipulating all the components needed for in vitro protein synthesis in complex, reconstituted systems had become most useful, whether Blobel had planned it that way all along or not.

In his experiments with both the readout and initiation systems, Blobel needed to detect very small differences in molecular weight between the putative precursor and processed forms of the light chain by SDS–polyacrylamide gel electrophoresis. In early iterations of this technique, each sample was electrophoresed through a gel cast in an individual glass tube. The problem with this method was that it was difficult to align separate tube gels precisely to compare one sample to another. As an alternative, Blobel adopted a variation of this procedure using "slab" gels cast between two large glass plates, with a plastic comb placed at the top to form a series of sample loading slots.[17] With this configuration, after the acrylamide polymerizes, the comb is removed, the plates are mounted between two buffer chambers, the samples are loaded into the slots, and an electrical field is applied. After several hours, the glass plates are carefully pried apart and the released gel is fixed with methanol and acetic acid to lock separated proteins in place. The gel is then dried onto a sheet of paper. The radioactive proteins synthesized in the translation mixtures are detected by autoradiography after exposing the dried gel to a sheet of X-ray film for several days. When the film is developed, the protein samples appear as individual horizontal bands arrayed in an irregular ladder below each sample slot, with the smaller proteins nearest the bottom because they migrate the fastest through the gel matrix (see figs. 1 and 23a).

Blobel began experiments with his new in vitro systems in late 1974. At Christmas, he obtained the first successful results confirming that, in his hands, the immunoglobulin light chain was synthesized as a larger form that was processed to the correct size during protein synthesis only in the presence of microsomal membranes, thereby reproducing and improving

17. Blobel seems to have first employed slab gels in 1971, using a commercial apparatus (Blobel 1971a). By 1974, he was using custom equipment manufactured in the Rockefeller instrument shop. Slab gels were easier to use than tube gels and provided excellent protein size resolution, in addition to permitting individual samples to be directly compared on the same gel. While the latter may seem a minor advantage, it was, in fact, essential because Blobel was looking for very small differences between two polypeptides of approximately the same overall size.

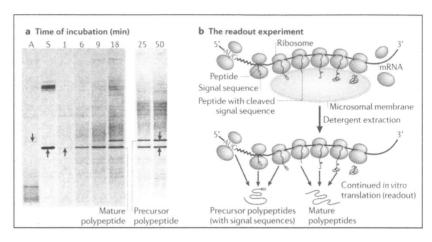

a Time of incubation (min)

A S 1 6 9 18   25 50

Mature Precursor
polypeptide polypeptide

b The readout experiment

5'   Ribosome   3'

mRNA

Peptide
Signal sequence

Peptide with cleaved
signal sequence

Microsomal membrane

Detergent extraction

5'   3'

Continued *in vitro*
translation (readout)

Precursor polypeptides   Mature
(with signal sequences)   polypeptides

23. The readout experiment that demonstrated a precursor-product relationship between the larger and smaller forms of the immunoglobulin light chain. Panel **a** is the radioautograph of the gel showing that the smaller polypeptide appears earlier than the larger (precursor) polypeptide. Panel **b** is an illustration of the experiment. In panel **a**, the bands corresponding to polypeptides detected in early time points are smaller (faster moving on the gel) than the ones appearing later (upward arrows: processed [mature] proteins; downward arrows: unprocessed precursors). The diagram in panel **b** illustrates how results such as those shown after 18 minutes of incubation are achieved. When the microsome with bound polysomes in the process of polypeptide synthesis is detergent extracted, some of the nascent polypeptides have already been processed to remove what Milstein called the extra segments (signal sequences) while others have not. With continued translation after extraction (read out) both processed and unprocessed precursor polypeptides are completed. Originally published as figure 3 in Matlin, K. S. 2011. *Nature Reviews Molecular Cell Biology* 12 (5): 333–40. © Karl S. Matlin. Panel **a**: © 1975 Blobel, G., and Dobberstein, B. Originally published as figure 8 in the *Journal of Cell Biology* 67 (3): 835–51.

on the Harrison-Milstein results. The precursor-product relationship was most dramatically demonstrated in a key experiment conducted with the readout system (Blobel and Dobberstein 1975a, 841–42) (fig. 23). Blobel extracted the rough microsomes isolated from myeloma cells with the detergent DOC to dissolve the microsomal membranes and release intact bound polysomes carrying their nascent, partially completed polypeptide chains frozen in the act of protein synthesis. He then incubated these in the readout system with radioactive amino acids so that the nascent chains were completed, taking samples at short intervals up to fifty minutes. The sizes of the completed proteins were then compared by SDS-gel electrophoresis and autoradiography. At early time points, only one major band was visible that was identical in molecular weight to an authentic light chain standard run in an adjacent slot (S in fig. 23a). After eighteen minutes of incubation, a second band that was slightly larger than the first began appearing, and both bands continued to be present at succeeding time points. The second band corresponded in size to one synthesized directly from purified light chain mRNA in a translation system that did not contain microsomes (A in fig. 23a).

To understand this experiment, it is important to remember that polysomes consist of a series of individual ribosomes positioned at various locations on a single mRNA corresponding to how far along they are in the process of protein synthesis. When the mRNA is translated by ribosomes into polypeptides, the ribosomes bind to the beginning of the mRNA (the 5' end) to initiate synthesis and then progress toward the distal (3') end of the mRNA as they elongate the nascent polypeptide chain. In polysomes, the ribosomes nearer the distal 3' end have almost completed the synthesis of the protein coded by the mRNA, while the ribosomes closer to the proximal 5' end have just begun synthesis.

Blobel's interpretation of the readout experiment was consistent with progressive polypeptide synthesis carried out by the polysomes that he had extracted from the myeloma rough microsomes (fig. 23b). He believed that the first band corresponding in size to the mature, authentic light chain resulted from nascent polypeptides that were mostly complete at the time that he extracted the microsomes bearing the polysomes with detergent. The second band resulted from polypeptides whose synthesis had just started at the time of extraction. Blobel reasoned that the first polypeptides completed in his readout system were the ones that had already crossed the microsomal membrane *before* detergent extraction. Because they were across the membrane, the extra segment at the amino terminus that Harrison and Milstein reported had already been removed, likely by a protease in the microsomal membrane. This was why the size of this protein was identical to that of the mature, secreted light chain. At later time points, the polypeptides completed in the readout system had not crossed the membrane before extraction and therefore had not been proteolytically processed. These were the precursors of the mature form and were the same size as the protein synthesized in vitro from isolated light chain mRNA under conditions in which processing could not occur.

In subsequent experiments, Blobel used the initiation system to demonstrate that the entire process, including initiation of protein synthesis, attachment of ribosomes to microsomes, membrane translocation, and proteolytic processing of the precursor, could be completely reconstituted in vitro. To demonstrate definitively that only the precursor segment of the nascent chain and not the ribosomes were directing the protein synthesis machinery to the microsomal membrane, Blobel included only free, cyto-plasmic ribosomes in the assay rather than using ribosomes that he had released from microsomal membranes. Under these conditions, Blobel observed just what he expected: light chain synthesized in the presence but not the absence of stripped dog pancreas microsomal membranes was processed to the size of the mature protein. To prove that the processed protein had actually crossed the membrane into the lumen of the microsomal vesicles, he treated the microsomes at the end of the experiment with proteolytic enzymes and showed that processed protein was protected from digestion. As a control, he digested them in the presence of detergents that disrupted the membranes; under these circumstances, the newly made light chains were degraded. As another control, he synthesized globin in the system using mRNA purified from reticulocytes, showing that its size did not change whether microsomal membranes were there or not. Furthermore, when the mixture containing microsomes and newly synthesized globin was treated with proteases, the globin was completely degraded, indicating that it had not crossed the membrane into the microsomal lumen where it would be shielded from the proteases (Blobel and Dobberstein 1975b).

Even though the experiments went smoothly after the translation systems were in place, there was a period of uncertainty. For the early experiments, a mouse myeloma tumor called MOPC 41 obtained from an investigator at the NIH was used to prepare rough microsomes for the readout system and light chain mRNA for the initiation system. Although the tumor could be maintained by passing it from mouse to mouse, at a certain point early on another isolate of the tumor was obtained from a different source (Blobel and Dobberstein 1975a, 1975b). Authentic secreted light chain or light chain synthesized from mRNA from the new tumor was slightly smaller than light chain produced from the original tumor. At this point, Blobel began to panic because he thought that the very small size difference between different forms of light chain that he had attributed to processing of a precursor might be an artifact of the extremely heterologous cell-free translation systems that he had devised. Further experimentation, however, convinced him that the second tumor isolate had a random deletion in the light chain gene that resulted in a smaller protein and that his original interpretations had been correct.

From both the readout and initiation experiments, it appeared that processing, that is, proteolytic removal of the precursor segment, occurred during synthesis of light chain on membrane-bound ribosomes, that is, *co-translationally*. Milstein and his colleagues had made a similar supposition in their paper when they were unable to remove the amino-terminal sequence that they observed on the larger form of the light chain by mixing it with microsomal membranes after synthesis was complete. Blobel repeated such experiments with the precursor made in his system, even trying detergent extracts of microsomes that he reasoned might contain the processing activity. None of these attempts yielded a processed protein and furthermore demonstrated that the precursor was unable to cross the microsomal membrane after synthesis because it remained sensitive to added proteases after the incubation with membranes. Blobel speculated that, in the absence of microsomal membranes, the completed precursor protein folded upon itself, making both proteolytic processing and membrane translocation impossible.

Based on these results, Blobel revised the model that he and Sabatini published in 1971 (see fig. 21). He now called the proposed mechanism for directing secretory proteins to and through the ER membrane the *signal hypothesis*, adopting the term "signal" from the paper by Milstein and his colleagues (Milstein et al. 1972). The signal, or *signal sequence*, referred to the short stretch of amino acids synthesized at the amino terminus of secretory proteins that directed ribosomes to membranes during synthesis and was removed during the translocation of the nascent polypeptide chain across the ER membrane into the lumen. As illustrated in a new drawing prepared by Blobel's technician Nancy Dwyer, the polypeptide crossed the membrane in an extended configuration, the signal sequence was cleaved off as it crossed the membrane, and the polypeptide folded up after it entered the lumen (Blobel and Dobberstein 1975a) (fig. 24). The new model lacked the ribosomal binding factors attached to the membrane surface proposed in 1971. Instead, ribosomes in the act of synthesis and translocation attached to a "tunnel" that assembled from subunits in the membrane in response to the signal sequence to provide an avenue through which the nascent polypeptide could cross the membrane (see fig. 24). This meant that, according to this model, the signal sequence not only directed the ribosome-mRNA-nascent chain complex to the ER membrane; it also facilitated the translocation of the nascent chain across the membrane.

Oddly enough, Sabatini, now at New York University (NYU), did not compete with Blobel in his attempt to prove the model that they had both proposed in 1971. Gert Kreibich, who accompanied Sabatini to NYU as a faculty member, continued studies he initiated at Rockefeller on the rough ER membrane (Kreibich, Debey, and Sabatini 1973; Kreibich, Hubbard, and Sabatini 1974; Kreibich and Sabatini 1974). Sabatini himself briefly entertained the alternative hypothesis that ribosomes might be targeted to microsomes through a direct interaction of mRNAs with the membrane but did not pursue such studies for long (Lande 1975; Sabatini et al. 1975; Adesnik

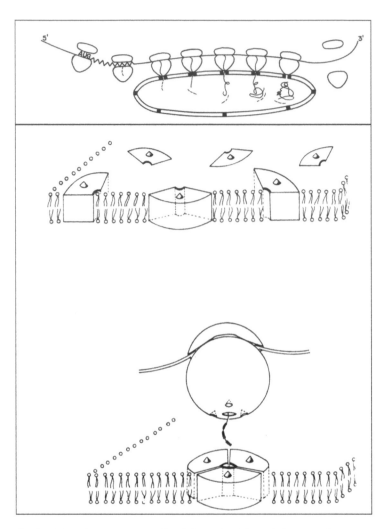

24. The signal hypothesis as proposed by Blobel and Dobberstein in 1975. The upper panel illustrates the process of ribosome binding, translocation of the nascent precursor polypeptide chain, and proteolytic processing to the mature form. The lower panel illustrates the assembly of the hypothetical tunnel through the microsomal membrane. © 1975 Blobel, G., and Dobberstein, B. Originally published as figures 15 and 16 in the *Journal of Cell Biology* 67 (3): 835–51.

et al. 1976). In the end, Sabatini was distracted by the challenges of setting up a large, new department from scratch and by his desire to investigate other biological questions.

Publication

In spring 1975, Blobel began writing two papers for submission to the *Journal of Cell Biology* describing the results and the new model. As he did so, conflicts in the laboratory that had been festering below the surface came into the open. The atmosphere was already strained by disagreements between Blobel and Christoph Freienstein, one of his postdocs, resulting in Freienstein jumping from Rockefeller to Sabatini's group at NYU. This was of some concern because it was not clear at the time if Sabatini was competing with Blobel to follow up on the Milstein results, and Freienstein had inside knowledge of the experimental system. Authorship of the papers was a second, more serious issue. Earlier, while the experimental systems were being assembled, Blobel indicated to both Dobberstein and Scheele that they would be coauthors of any resulting publications. By the time that the experiments had been completed and the significance of the results was clear, Blobel decided that he should be the sole author. Blobel had, indeed, done almost all the critical experiments with his own hands. Although Scheele's direct contributions were minimal once it was clear that dog pancreas microsomes worked and Dobberstein's role was mainly to supply purified light chain mRNA to Blobel, neither of them thought it was fair for them to be left off the papers. After some intense arguments, Scheele agreed to this arrangement on the promise that he would be coauthor on other papers. Dobberstein, on the other hand, did not. He pointed out to Blobel that the entire project was originally promised to him but that he assented to a lesser role only because he was assured by Blobel that he would be a coauthor. Faced with Dobberstein's threat to quit the laboratory, Blobel agreed to include him on the papers as second author.[18]

The two papers, "Transfer of Proteins Across Membranes I and II," were submitted to the *Journal of Cell Biology* on June 30, 1975. The reviews, received by Blobel on August 29, 1975, indicated that the "papers are considered to

18. Despite this rocky start to their relationship, Dobberstein and Blobel continued to work together productively at Rockefeller and remained friendly competitors in subsequent years (see chap. 6). According to Blobel, an amusing outcome of the authorship dispute was that he was viewed by some as Dobberstein's postdoc instead of the other way around, because the last author position is normally reserved for the senior scientist in the laboratory.

be a significant and interesting contribution" but suggested some changes.[19] At some point in the review process, Blobel learned that the two reviewers were Palade and Don Steiner from the University of Chicago. Steiner was famous for his discovery of proinsulin, a precursor of the secreted hormone. His reviews of Blobel's papers were short and positive. In addition to some stylistic suggestions, he objected to Blobel's description of his two translation systems as readout and initiation. In response, Blobel inserted two explanatory sentences in the revised manuscripts but did not change the terms.[20]

Palade's formal reviews were also very positive, stating that the results of the first paper were "very beautiful" and that the second paper "reports results of unusual interest" and will become "in time a primary reference."[21] Palade, however, departed from the usual norms by contacting Blobel directly by phone and sending him his reviews even before he submitted them to the journal,[22] apologizing that he was slow because of other responsibilities. It is conceivable that Blobel sent the manuscripts directly to Palade to solicit comments at the same time that he submitted them to the *Journal of Cell Biology* as Palade notes that he has "read and processed [the first manuscript] as suggested by your letter and phone conversation."[23] Although Palade had moved to Yale more than a year before, his letter to Blobel and comments on the manuscripts provide a sense of how Palade may have run the Laboratory of Cell Biology while he was still at Rockefeller: "I liked [the first manuscript] very much and worked on it as in the 'good old times,' putting down in the margins and back of pages whatever criticisms and suggestions I believe you should consider. In a couple of days I will send you the second ms. (now in the works) and then call you to see what we can do to discuss together the suggestions (assuming that you would like to do it)."[24]

19. Judith Tennant to Blobel, August 29, 1975, used by permission of Günter Blobel.

20. In a letter to Steiner in September 1975, Blobel thanks him for his "prompt, thorough and constructive" reviews and explains why he did not change the terms. Blobel to Don Steiner, September 9, 1975, used by permission of Don Steiner. Normally reviews in the *Journal of Cell Biology* were conducted anonymously. It is possible that Steiner notified Blobel that he had reviewed the *Journal of Cell Biology* papers because Blobel's letter provides comments on a manuscript by Steiner dealing with the possibility that proinsulin is also synthesized initially as a precursor (which Steiner calls *preproinsulin* in the manuscript), like the immunoglobulin light chain (see chap. 7).

21. Palade manuscript reviews, used by permission of Günter Blobel.

22. Palade's letter is dated prior to the official letter to Blobel from the *Journal of Cell Biology*. Palade to Blobel, August 18, 1975, used by permission of Günter Blobel.

23. Palade to Blobel, August 18, 1975.

24. Palade to Blobel, August 18, 1975.

In addition to his formal reviews submitted directly to Blobel and to the journal, Palade made dozens of comments and suggestions for rewording the manuscripts, most of which were accepted by Blobel. Many comments were intended to tighten up both Blobel's interpretations and language. Although impressed by the results, Palade urged Blobel to "remain on the 'cis' side of safety by saying that all conclusions remain to be confirmed by the further characterization of the in vitro products," adding, "You'll not lose anything—you'll show only that you are cautious."[25] He insisted that Blobel include data validating a technique that Blobel used to obtain quantitative results from his X-ray films and to clarify many methodological details. Palade did not see the need for including an explanation of the MOPC mutation, which he refers to as a "detective story." Blobel left it in, presumably because he needed to include experiments using material from both the first and second isolates of the tumor. Palade suggested that Blobel's description and diagrams of the signal hypothesis be included in the body of the manuscript instead of the appendix that Blobel had in the original version and told Blobel to back off a suggestion that ribosomes may bind directly to lysosomes and peroxisomes to synthesize proteins belonging to these organelles, stating simply, "no evidence," and reminding Blobel that lysosomal proteins are known to follow the same biosynthetic pathway as secretory proteins in leukocytes.[26] When Blobel included some discussion of the speculative tunnel in the introduction to the second paper, Palade urged him to drop it, remarking that "the important thing is transfer (assayable) not tunnel formation (not assayable)."[27]

The one area that seemed to bother Palade the most was the suggestion in the discussion and the diagram that ribosomes detach from the membranes after completing synthesis or are in polysomes that are partially attached and detached. Palade felt that this idea is inconsistent with his many observations with the electron microscope of membrane-bound ribosomes arrayed in circles and rosettes. He apparently considered the signal sequence mainly as a way to get ribosomes making secretory proteins to the membrane, ignoring Blobel's proposal that it also organizes the tunnel. Consequently, he questioned the need for the signal sequence after initial attachment has been achieved, stating, "The signal sequence could be instrumental in the initial attachment and may be 'tolerated waste' for the rest

25. Palade manuscript reviews and Palade handwritten comments on manuscripts. Used by permission of Günter Blobel.

26. Palade handwritten comments on manuscripts. Used by permission of Günter Blobel.

27. Palade handwritten comments on manuscripts. Used by permission of Günter Blobel.

of the duration of the attachment, involving many rounds of translation. There may be even devices to bypass it [the signal sequence] (eliminate [it] in conjunction with the 'circularization' of the messenger)."[28] In the end, Blobel responded by mentioning preliminary results on a ribosome detachment factor and left the diagram and discussion of ribosome detachment largely intact.[29]

Blobel resubmitted the revised manuscripts to the *Journal of Cell Biology* on September 2, 1975, only a few days after the official reviews had been received. The revised manuscripts were accepted, but Blobel was disturbed when he learned that they would not be published until the following year. This seemed unreasonable because they had been originally submitted at the end of June. Blobel asked Palade to "work his charm" on Judy Tennant, the associate executive editor handling the manuscripts, and the papers appeared in the December 1975 issue.[30]

Sequencing the Signal Sequence

By the time of Blobel's reconstitution experiments, evidence that translation of light chain immunoglobulin mRNA produced a protein that was larger than the authentic, secreted form was widespread, although Milstein and his colleagues were the only ones to clearly link the observation to the protein secretory mechanism (Milstein et al. 1972; Swan, Aviv, and Leder 1972; Mach, Faust, and Vassalli 1973; Schechter 1973; Tonegawa and Baldi 1973). Like Milstein, several of the other laboratories had determined that the extra polypeptide segment was at the amino terminus of the protein. In 1974, a research group from the Massachusetts Institute of Technology (MIT) and Harvard reported that translation of mRNA coding for proparathyroid hormone yielded a polypeptide larger than both the mature hormone and the pro-form (Kemper et al. 1974). The extra segment of this protein was, like the light chain precursor, located at the amino terminus. Because they could

28. Palade handwritten comments on manuscripts. Used by permission of Günter Blobel.

29. Blobel is referring to a manuscript in preparation about the putative detachment factor described in a footnote in the published *Journal of Cell Biology* papers. In a letter on September 25, 1975, accompanying the resubmitted manuscripts, he tells Palade that he has sent a paper on the factor to *Cell*. In the end, a short paper was published in a low-level journal, and after that the factor seems to have disappeared (Blobel 1976). Blobel to Palade letter. Used by permission of Günter Blobel.

30. Blobel was likely worried that other research groups that had reported putative light chain precursors might publish something definitive before him. See, for example (Mechler and Vassalii 1975).

not detect the larger protein in radioactively labeled whole cells, they concluded that it might be a transient precursor of proparathyroid hormone. They designated it "preproparathyroid hormone" and suggested that it was analogous in function to the putative light chain precursor found by Milstein's group.[31]

In spring 1975, Israel Schechter's laboratory reported a partial sequence of the amino-terminal part of the light chain precursor synthesized from mRNA. The extra segment of the protein was rich in the hydrophobic amino acid leucine[32] and was twenty amino acids long (Schechter et al. 1975). In Blobel's lab, the second paper of the pair that had proposed the signal hypothesis included results derived from the readout of dog pancreas rough microsomes (see Blobel and Dobberstein 1975b, fig. 6). A number of polypeptides of various sizes were synthesized. These were believed to be the digestive enzymes known to be made by the exocrine pancreas, the source of the microsomes.[33] As the other experiments were wrapping up, Blobel's lab made a concerted effort to identify and synthesize precursors of the pancreatic enzymes and sequence the amino-terminal regions. A new French postdoctoral fellow, Anne Devillers-Thiery, was placed in charge of the sequencing. Because neither she nor Blobel was an expert in protein sequencing, they collaborated with Thomas Kindt, a protein chemist at Rockefeller. Scheele also participated to help link the translated products with specific pancreatic enzymes. With the cooperation of Dobberstein, who helped them isolate mRNA for the enzymes from pancreatic polysomes, the project took off.[34]

31. The practice of designating precursor proteins that still had their signal sequence attached as *pre*-proteins stuck and was used consistently as more precursors of this type were discovered. The prefix *pro* as in proinsulin and proparathyroid hormone indicates that the protein is a precursor that will be processed in steps of secretion later than translocation of the protein into the ER.

32. As mentioned previously, the twenty most common amino acids found in proteins are classified by the chemical characteristics of their unique side chains as hydrophobic or hydrophilic. Hydrophobic amino acids would prima facia seem to be more compatible with membrane contact during translocation of a polypeptide through the lipid bilayer, although, in reality, things are not so straightforward (see chap. 8).

33. As described before, in mammals such as the dog, the pancreas consists of an exocrine part that secretes digestive enzymes into the small intestine and an endocrine part that secretes insulin and glucagon into the bloodstream. Because the two parts of the pancreas in mammals cannot be easily separated, microsomes prepared from the pancreas are derived from both the exocrine and endocrine tissues. However, microsomes isolated from the whole pancreas are mainly derived from the exocrine part because it is far larger than the endocrine part.

34. Dobberstein was not a coauthor on the eventual paper but published his procedure to isolate pancreatic mRNAs with Scheele and Blobel later (Scheele, Dobberstein, and Blobel 1978).

The method they used to sequence the amino-terminal regions of the proteins was called Edman degradation, named after its inventor, Pehr Edman (Voet and Voet 2004). Under alkaline conditions, a reagent called phenylisothiocyanate reacts with the amino-terminal amino acid to convert it into a PTC (phenylthiocarbamyl) amino acid without affecting other amino acids in the protein. The PTC amino acid is then cleaved from the polypeptide chain with a strong acid, yielding a modified and more stable PTH (phenylthiohydantoin) derivative, and isolated. This leaves the remaining part of the polypeptide (now one amino acid shorter) ready for the next reaction. The isolated PTH amino acid can be identified by comparing it to known standards. While the procedure is tedious, it could at the time be automated using a so-called spinning cup sequencer machine manufactured by Beckman Instruments. Fortunately, Kindt had such a machine, and Blobel soon acquired one as well by exceeding his research budget, much to the chagrin of the Rockefeller administration.

Even with the machine, sequencing of the precursors was a daunting task. Translation of the pancreatic mRNAs in vitro produced exceedingly small amounts of protein that were detectable only because they were radioactive. Furthermore, the concentration of radioactive amino acids (expressed as specific activity, the ratio of radioactive to total amino acid amounts) typically used in translation experiments was too low to enable individual amino acids to be detected after release by Edman degradation. Another problem was that the sequencing machine did not work efficiently with small amounts of protein. To solve the first problem, Devillers-Thiery prepared custom mixtures of concentrated radioactive amino acids. After synthesizing the pancreatic enzymes using these mixtures, she separated them by SDS-gel electrophoresis, located the bands by autoradiography, and cut the individual radioactive bands out of the dried gel using the developed X-ray film as a guide (Devillers-Thiery et al. 1975). She then mixed the isolated radioactive proteins with the unlabeled protein myoglobin prior to placing them in the sequencer; the myoglobin acted as a carrier for the radioactive proteins, solving the second problem. The radioactive PTH amino acids generated in the sequencer were then separated and identified by thin layer chromatography.

Devillers-Thiery and her colleagues sequenced the first twenty-four amino-terminal amino acids of five different bands tentatively identified as precursors of specific pancreatic enzymes by comparison of their relative mobilities on the SDS gel to known proteins. When the sequence of the trypsinogen precursor was determined, they observed that after the first eighteen amino acids the next few corresponded to the known amino-terminal sequence of mature trypsinogen from a variety of animal species. This suggested that the first eighteen amino acids were the extra segment, the signal sequence. When the sequences of the four remaining proteins were compared to that of what they now referred to as pretrypsinogen, they found that all were enriched in the hydrophobic amino acids leucine or isoleucine and were remarkably homologous, though not identical. They also were obviously similar to the partial sequence of the light chain signal sequence determined earlier by Schechter (Schechter et al. 1975). These results were clearly consistent with the idea that, regardless of specific function, secretory proteins all used a similar if not identical mechanism to attach to and traverse the microsomal membrane, as outlined in the signal hypothesis (Devillers-Thiery et al. 1975).

A Different World

The signal hypothesis formulated by Blobel and Dobberstein was strongly supported by the data provided in the two *Journal of Cell Biology* papers that appeared in December 1975 and the paper describing signal sequences published the same month. Certainly, Milstein and his colleagues provided the first evidence in 1972, and their interpretation of their results was remarkable considering that they were completely unaware of the Blobel and Sabatini model from 1971 and largely ignorant of the earlier work done by the Rockefeller group. However, they neither proved a precursor-product relationship between the larger and smaller forms of the light chain nor developed an experimental system that could be used to decipher mechanisms. Indeed, Blobel's creation of a completely reconstituted system to study the first step in the secretory process, the binding of nascent secretory proteins to and translocation through the ER membrane, was in some respects as important as the signal hypothesis itself (Blobel and Dobberstein 1975b). Because the system was composed of heterologous components isolated from mice, rabbits, rats, and dogs, the universality of the mechanism described in the signal hypothesis, at least in mammals, was demonstrated automatically.

Most significantly, by combining a mainly biochemical process, protein synthesis, with a biological event, the movement of proteins across a biological membrane, Blobel was able to make statements about a cellular process, protein secretion, even though he was using a cell-free system (see chap. 10). The reason this was possible was because the microsomal vesicles that he added to his experimental system were directly linked to the ER in whole, intact cells by the nature of the epistemic strategy developed in particular by Claude and Palade. This investigative approach coupled continuous morphological analysis of cells and cell parts to cell fractionation and biochemical characterization. As a consequence, the relationship between a cellular structure, the ER, and its function in secretion could begin to be deciphered at the molecular level using an experimental system devoid of cells. Blobel indirectly acknowledged this connection between morphology and molecular analysis by including electron microscopic images of isolated microsomes in his 1975 papers.

What was still missing was any understanding of the molecular *machinery* that carries out recognition of signal sequences and translocation of nascent secretory proteins across the ER membrane. Identification of these components would form the basis for an almost complete molecular explanation of a biological event, the initiation of protein secretion. At the same time, their discovery would mark the creation of a truly molecular cell biology. As we shall see, Blobel and his laboratory will be occupied by the search for these molecules for much of the next twenty years.

Because of the 1975 papers, Blobel suddenly went from relative scientific obscurity to prominence. In 1976, he was promoted to full professor at Rockefeller with Palade's strong endorsement and gave a keynote address at an international cell biology meeting in Boston. Blobel himself realized that he was now in a strong competitive position with his mastery of a complex reconstituted experimental system that could not be easily replicated in other laboratories. There was much work to do, but there were also some regrets. In early summer 1976, Blobel remarked to another investigator in the cell biology group that he had to greatly increase the size of his laboratory to expand the scope of his work. He added that, sadly, he would now have no time to perform experiments at the bench with his own hands.

SIX

The Strange Case of the Signal Recognition Particle

Here, There, and Everywhere

The next phase of work on the signal hypothesis began quickly after publication of the groundbreaking papers in the *Journal of Cell Biology*, and by the late 1970s, Blobel's lab had expanded significantly. Among the new members of the lab were Blobel's first graduate students. At Rockefeller, graduate students were a rare and valued commodity. Although called a university after 1955, Rockefeller in fact had only a small graduate program, an almost nonexistent curriculum, and no undergraduates. Students were expected to join laboratories as soon as they arrived and begin their thesis research almost immediately. Blobel's ability to attract them to his laboratory suggested that his rising reputation created a certain buzz about exciting new projects. Furthermore, publication of the 1975 results most certainly had an effect on recruitment of new postdocs from outside institutions, and within a short time this category of investigators also increased in number. To accommodate his new personnel, Blobel moved his laboratory to the third floor of the South Laboratory building, where he had a lot more space and a degree of independence from the rest of the Laboratory of Cell Biology.

The question was, what exactly would the new people in the lab work on? This was new territory for Blobel. In the past, he had been perhaps the most active investigator in the lab and reluctant to surrender control of projects that he considered important (see chap. 5). Now, that approach was no longer viable. There were too many things to do. Furthermore, postdocs needed to demonstrate their independence to get jobs, and graduate students were expected to conduct thesis research on their own. In any case, Blobel had no time to work in the lab because he needed to go on the speaking circuit to advertise his work both for recruitment purposes and to raise money to fund new lines of research.

Three postdocs who had entered the Blobel lab as the initial signal hypothesis experiments were under way continued their work: Bernhard Dobberstein, Anne Devillers-Thiery, and Dennis Shields. Dobberstein had been relegated to a secondary role in the work leading to the 1975 papers. Although his status as a coauthor of these groundbreaking studies likely raised his profile somewhat in the scientific community, he was anxious to begin some new research on his own that would establish his identity as an independent scientist. Devillers-Thiery, who led the effort to sequence the signal peptides, continued that work for some time before returning to France. Shields, the newest of the group, had arrived from London in late 1974, hoping to work on the mRNA-bound proteins that Blobel had discovered (Blobel 1972, 1973). Now, however, with the lab's intense focus on the signal hypothesis, he realized that that project was no longer of interest and was encouraged by Blobel to choose something completely different (see chap. 7).

Whether through planning or the force of events and personalities, research on the signal hypothesis in the Blobel lab began to fall into two categories.[1] One of these, which is described in the next chapter, was to see if the signal hypothesis also applied to proteins other than immunoglobulins and pancreatic enzymes. These included not only other secretory proteins but also integral membrane proteins that traverse but do not completely cross the lipid bilayer. Proteins that are transported from the cytoplasm into mitochondria and chloroplasts were also on the list. Blobel was, in addition, interested in seeing if prokaryotic bacterial cells use a mechanism similar to that outlined by the signal hypothesis. Although bacteria have neither nuclei nor other cytoplasmic organelles like the ER, they still secrete proteins that need to cross their plasma membrane to get outside the cell (Michaelis and Beckwith 1982). While looking at a variety of proteins and organisms might seem an unfocused and diffuse strategy, Blobel felt that this broad approach would help him establish that the signal hypothesis was both general and fundamental.

Choosing the experimental system to fit the questions under investigation was not a new strategy at Rockefeller. Earlier research by Palade and Siekevitz on the ER, microsomes, and secretion that formed the foundation

---

1. Blobel also had a second independent project in the lab on the nucleus, an interest of his since graduate school. After a postdoc working on the project left around the beginning of 1975, Blobel assigned a talented technician to the project, and then passed it along to a new graduate student, Larry Gerace, while the rest of the lab focused on the signal hypothesis. Gerace proceeded to make a major discovery, the nuclear lamins (Gerace, Blum, and Blobel 1978).

for Blobel's work focused on the mammalian pancreas (Siekevitz and Palade 1958b) (see chap. 4, above). However, this was not because they were particularly interested in the physiology or pathology of this organ. Rather, they chose the pancreas because its main function is protein secretion, the cellular process they wanted to study. Nor was there an exclusive emphasis on secretion in mammals; Redman, after all, had no problem using pigeon pancreas in his in vitro experiments because the tissue offered certain technical advantages (Redman, Siekevitz, and Palade 1966). Similarly, when Siekevitz and Palade decided to study membrane structure and biogenesis, a project not discussed here that began in the late 1960s, they initiated work on a green alga, *Chlamydomonas reinhardtii*, because its cells were filled with chloroplast membranes, not because they were suddenly interested in plants (Hoober, Siekevitz, and Palade 1969; Schor, Siekevitz, and Palade 1970). These choices were made in part because certain tissues and cells were amenable to experimental manipulation. However, there was also an underlying assumption, one central to the emerging discipline of molecular cell biology, that mechanisms fundamental to basic cellular processes are universal. That is, the molecules carrying out such processes are likely very similar if not identical regardless of organism or even cellular organelle, and their essential functions can be elucidated by comparison of multiple experimental systems (Palade 1966, 1975). As Blobel's work played out over the next few years, the truth of this assertion, at least for the signal hypothesis, was gradually established.

The second category of research pursued in the Blobel laboratory, which I focus on in this chapter (and in chap. 8), was an attempt to uncover the detailed molecular mechanisms leading to secretory protein transport into the ER as portrayed by the signal hypothesis. For all their achievements, Palade and his collaborators had only mapped the secretory pathway in cells from the ER to the zymogen granules and established some basic mechanistic parameters. They did not identify the resident molecular constituents of the ER and subsequent organelles that made everything work. Nevertheless, understanding the molecular mechanisms underlying events on the secretory pathway was always the ultimate goal of the epistemic strategy pursued by the Rockefeller group (Palade 1966, 1975). Indeed, Palade had encouraged Blobel's development of his in vitro reconstituted system for just that purpose.

While Blobel's 1975 papers established the outlines of the signal hypothesis, they did not report even one new component of the cellular apparatus needed for signal sequence–mediated *targeting* of secretory proteins to the ER and their translocation across the membrane (Blobel and Dobberstein

1975a, 1975b).[2] Among the important unanswered questions were, how is the signal sequence on secretory proteins recognized, what does the ribosome/nascent chain complex bind to on the ER membrane, how does the nascent chain cross the membrane, and what cuts off the signal sequence once membrane translocation has occurred? The answers to these questions undoubtedly involved specific molecular complexes. In the Blobel lab and elsewhere, the hunt was on. As we shall see in this chapter, serendipity and a competitive dance carried out between labs in both New York and Heidelberg, Germany, led to the discovery of two essential factors, one a molecular complex called the signal recognition particle and the other its receptor on the surface of the ER membrane.

Progress in New York

Dobberstein was the first to look for molecules involved in the targeting and translocation mechanism (fig. 25). Initially, however, he began searching for the proteolytic enzyme that removes signal sequences, an activity referred to as *signal peptidase* that converts secretory protein precursors to their mature forms. This activity was responsible for the well-known shift in the position of bands on SDS gels that had provided the first evidence in favor of the signal hypothesis (Milstein et al. 1972; Blobel and Dobberstein 1975a). Even though Dobberstein quickly obtained some promising results, he soon passed this work along to a new postdoc in the Blobel lab, Robert Jackson, enabling him to pursue other work that he thought was more challenging and important. After all, cleavage of the signal sequence seemed to occur after the most important steps of signal sequence recognition and protein translocation were either under way or completed.

Superficially, Jackson's attempts to identify the protein responsible for the signal peptidase activity appeared to be a straightforward exercise in enzyme purification. There were, however, complications. The activity was bound to ER membranes and the substrates, secretory protein precursors, were normally cleaved only during synthesis, that is, co-translationally (Blobel and Dobberstein 1975a, 1975b). Despite these challenges, Jackson was successful in demonstrating that a detergent extract of microsomes was capable of correctly cleaving signal sequences in some precursor proteins not only during their synthesis but also once the precursor proteins were complete (Jackson and Blobel 1977). He described the enzyme as "latent" since

2. The term "targeting" refers to cellular mechanisms that direct proteins to particular locations within the cell. In certain cases it is synonymous with the term "sorting" (see chapter 9).

25. Bernhard Dobberstein, photographed by the author in Heidelberg, Germany, 1999.

the microsomal membrane had to be disrupted with detergents to expose the peptidase activity and suggested that his assay worked because the signal sequence cleavage site remained exposed in some secretory proteins following their synthesis.[3] Despite the development of this useful assay, however, Jackson failed to completely purify the peptidase before he left the

3. It is possible that detergents present in the assay, including DOC and the nonionic detergent Nikkol, bound the somewhat hydrophobic signal sequences and facilitated their exposure and cleavage.

Blobel laboratory in 1979 (Jackson and Blobel 1980). Others completed the purification several years later (Evans, Gilmore, and Blobel 1986).[4]

In the meantime, Dobberstein began looking for factors associated with the microsomal membrane that might be important for signal sequence recognition and translocation. One of the first things he did was to simplify the complicated reconstituted system that Blobel devised for the 1975 papers. As described previously (see chap. 5), Blobel's "initiation" system consisted of components isolated from rats, mice, rabbits, and dogs and was, consequently, both finicky and, to Dobberstein at least, unnecessarily laborious to assemble (Blobel and Dobberstein 1975b). A former student colleague of Dobberstein's, now in Philadelphia, was working with a much easier system for in vitro protein synthesis derived from, of all things, wheat germ (Roman et al. 1976). Dobberstein soon traveled there to learn how to prepare the system so that he could transfer the technology to Rockefeller. In addition to being very inexpensive, one advantage of wheat germ, which is essentially the plant embryo, is that it contains ribosomes and enzymes needed for protein synthesis but almost no endogenous mRNA. This was in contrast to another common translation system made from rabbit reticulocytes, the precursor cells of red blood cells, that was also beginning to be used in the lab. The reticulocyte lysate, as it was called, is also simpler than Blobel's original system but is loaded with mRNA for globin, the protein constituent of hemoglobin. Translation of globin mRNA by the lysate reduces its ability to translate other added mRNAs, making use of the reticulocyte system somewhat undesirable.[5] Another advantage of the wheat germ system is that it is exceedingly easy to prepare. All that is required is to grind frozen wheat

4. Reid Gilmore, another postdoc, took over the signal peptidase project when he arrived in 1980 just after Jackson departed. He made progress purifying the enzyme but thought that his preparation might be contaminated with other proteins because of its complex composition. After Gilmore moved on to another project in the Blobel lab, a graduate student, Emily Adams, picked up the work with Gilmore's input and succeeded in completing the complicated purification (Evans, Gilmore, and Blobel 1986). As it turned out, Gilmore's preparation was almost pure; the preparations from both Gilmore and Evans consisted of a complex of six polypeptides. It was later established that the signal peptidase resided on the inside of the ER membrane, explaining Jackson's observation of "latency."

5. To prepare the reticulocyte system from rabbits, the rabbits are first made anemic by several injections of phenylhydrazine (Shields and Blobel 1978). This causes the bone marrow to begin making large amounts of new red blood cells from their precursor cell, reticulocytes. Reticulocytes but not mature red blood cells are active in protein synthesis. Normally reticulocytes never appear in the blood stream in significant quantity. In response to anemia, however, they flood the blood stream and can be isolated from blood drawn from the anemic rabbits. After that, they are simply washed and then broken open or lysed in an appropriate buffer solution. The translation system is essentially the clarified lysate.

germ with a mortar and pestle, extract the resulting powder with a buffered salt solution, and pass the extract over a small chromatography column to get rid of amino acids and other small molecules.[6]

Once he set up the wheat germ system for protein synthesis in New York, Dobberstein tested whether it could be used to study protein translocation by incubating it with dog pancreas microsomes and purified light chain mRNA. He quickly demonstrated that the immunoglobulin precursor was synthesized by the wheat germ extract, translocated across the microsomal membrane, and correctly processed by removal of the signal sequence (Dobberstein and Blobel 1977). The results not only showed that the new in vitro translation system was useful for his experiments but also underscored the universality of the mechanism proposed in the signal hypothesis because a mammalian mRNA from mouse cells was translated into protein using plant ribosomes provided by the wheat germ extract, and the plant ribosomes were capable of correctly interacting with dog pancreas microsomal membranes to mediate translocation and processing.

Dobberstein soon began to use the new translation mix to dissect the translocation machinery. The focus of his work was very clearly on the microsomes, the vesiculated remnants of the ER membrane. He was almost certain that recognition of the signal sequence and translocation was carried out by components of the microsomal membrane and not by ribosomes because years of previous work had failed to detect significant differences between ribosomes derived from free cytoplasmic pools and ribosomes that were bound to the ER membrane.

In the development of the original reconstituted translocation system by Blobel, microsomes had always been "stripped" of endogenous ribosomes and mRNA before addition to the mix (Blobel and Dobberstein 1975b). This stripping procedure was needed to eliminate any background protein synthesis of dog pancreatic enzymes driven by bound ribosomes and mRNA attached to and copurified with microsomal membranes. Too much background synthesis might swamp out the synthesis of proteins from added, exogenous mRNA such as the one coding for the light chain, as well as confuse the overall results. In the lab, various stripping methods were used, including extraction of microsomal membranes with EDTA, which binds magnesium and disrupts ribosomes, or with high concentrations of the salt

---

6. A minor complication was that it was sometimes difficult to locally obtain wheat germ that had not been "killed" by toasting, a way of preparing it for consumption as a health food. Eventually, a contact at the General Mills Company in Minnesota sent bags of raw wheat germ to Dobberstein for free.

potassium chloride. In some cases these procedures were combined with treatment with the protein synthesis inhibitor puromycin, which releases the nascent polypeptide chain from any bound ribosomes (see chap. 4), facilitating removal of both ribosomes and mRNA from the membrane surface (Blobel and Dobberstein 1975b).

When Dobberstein demonstrated that wheat germ extract worked in the translocation assay, he used pancreatic microsomes stripped with EDTA alone (Dobberstein and Blobel 1977). With other stripping methods, however, the amount of translocation was highly variable, suggesting that certain stripping procedures might remove a component that was essential for the translocation activity. After refining his methods, he discovered that translocation was consistently inhibited when he used membranes stripped by treatment with puromycin *and* extraction with high salt concentrations. Strikingly, readdition of the salt extracted material to the membranes after lowering the salt concentration reconstituted translocation activity. Dobberstein hypothesized that some part of the translocation machinery was attached to the microsomal membrane surface by a salt-sensitive linkage. He showed his results to Blobel, but Blobel was not that impressed, considering them simply a reflection of the variability in the translocation assay that was commonly observed in the lab.

During this period, Dobberstein's work in the lab was somewhat unfocused. He was working on several projects at once, including a collaborative one involving the synthesis of chloroplast proteins (see chap. 7). He was also looking for a job back in Germany and had applied for positions throughout the country. One of his applications was sent to the European Molecular Biology Laboratory (EMBL) in Heidelberg. The EMBL was both new and, from a European perspective, a completely different type of institution (Kendrew 1980). It was common at the time for German and many other European graduate students who had completed their initial research training to go to the United States for two to three years of postdoctoral work. However, when they returned to Europe most available positions required them to work as a junior scientist under the direction of a senior professor rather than pursue independent work of their own choosing. By the 1970s, this way of organizing research laboratories was considered counterproductive because it suppressed the creativity of young scientists. The EMBL was created precisely to circumvent this system. Scientists who had completed postdoctoral training both abroad and in their home countries were able to obtain well-funded and independent group leader positions at the EMBL. Although these positions were generally not permanent, they enabled young scientists to develop their own research programs for up to (originally) nine years before moving to a professorship at a university or research institute in their home country. While Dobberstein awaited the outcome of his applications, Blobel gave a talk in Heidelberg and learned about the EMBL. Back in New York, he urged Dobberstein to seriously consider a position there. When a job offer at the EMBL came through, Dobberstein immediately accepted, shelving his salt wash experiments for the moment and preparing to move.

Translocation to Heidelberg

When Dobberstein arrived in Heidelberg in 1977, the EMBL was still more a concept than a well-developed institution. Although a laboratory building was planned, at the time of Dobberstein's appointment the EMBL shared cramped space with the German Cancer Research Center. For Dobberstein, the shortage of space turned out to be a blessing in disguise. Graham Warren, another new group leader, soon arrived from Cambridge University. Lacking a laboratory of his own and still unsettled about his eventual direction in research, he camped out in Dobberstein's small laboratory. Anxious to get started, the two of them began collaborating and quickly repeated Dobberstein's reconstitution of translocation by salt extracts of microsomes using the wheat germ system. The combination of Dobberstein and Warren, who shared interests in membranes and mechanisms, turned out to be synergistic.

Lacking the patience to wait for overnight runs of SDS gels and autoradiography to determine the outcome of an experiment, the standard approach of Blobel's lab, Warren developed a rapid quantitative assay for translocation that did not involve SDS-gel electrophoresis at all. The assay was based on the use of proteases to determine if translocation into microsomal vesicles had occurred. Secretory proteins synthesized in the wheat germ system that are translocated across added microsomal membranes are "protected" from degradation by proteases added after synthesis is complete because they are trapped on the inside of the microsomal vesicles where the membrane shields them from the protease (Sabatini and Blobel 1970; Blobel and Dobberstein 1975a; Warren and Dobberstein 1978a). Proteins not translocated, on the other hand, remain outside of the vesicles and are completely degraded by added proteases. For Warren's assay, pairs of in vitro protein synthesis reactions were conducted in the presence of microsomal membranes. One sample of each pair was then treated with proteases and the total amount of radioactivity incorporated into protein in each sample measured. The ratio of these two measurements constitutes the approximate

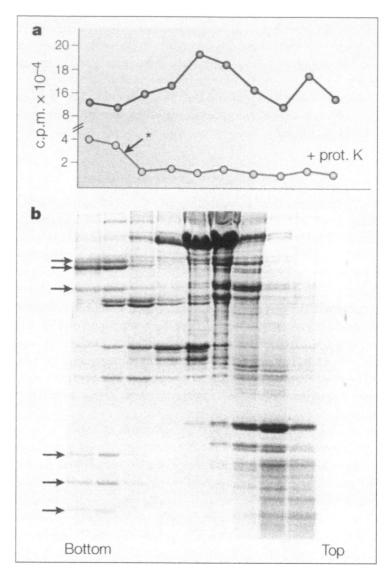

26. The fractionation of Dobberstein's high salt extract of pancreatic microsomal membranes. These unpublished data were presented to the German Biochemical Society in March 1978. Panel **a** shows analysis of sucrose-gradient fractions in the wheat germ–based translocation assay developed by Graham Warren. The lower trace represents radioactivity in translocated polypeptides, detected by proteinase K treatment. Active fractions are indicated by the arrow and asterisk. Panel **b** is the SDS gel of the same sucrose-gradient fractions, with the active fractions near the bottom of the gradient (left). The six polypeptides that were later identified as the signal recognition particle subunits by Peter Walter are indicated with arrows. The labeling of the lower panel is by the author. Original data courtesy of Bernhard Dobberstein. Originally published as figure 2 in Matlin, K. S. 2002. *Nature Reviews Molecular Cell Biology* 3 (7): 538–42. © Karl S. Matlin.

amount of translocation. With this new assay, the work progressed quickly. Within a few months, a paper was accepted in *Nature* (Warren and Dobberstein 1978a). Simultaneously, Warren and Dobberstein submitted an abstract for a meeting of the German Biochemical Society, to be held in Frankfurt in March 1978 (Warren and Dobberstein 1978b).[7]

By the time of the meeting, Dobberstein had more data. The high-salt extract, which could reconstitute translocation, sometimes precipitated out of solution when Dobberstein lowered the amount of salt by dialysis. This concentrated the unknown factor and made it easier to analyze. When the material was separated by centrifugation on a sucrose gradient and the fractions were assayed for translocation activity, the activity was found near the bottom of the gradient, suggesting that it was a large molecular complex.[8] On SDS gels, active fractions consisted of six main polypeptide bands (fig. 26). Dobberstein presented these results at the meeting, but they were not included in the published abstract because it had been submitted before the new findings were obtained.[9]

The New York Story Resumes

Back in New York, a new graduate student, Peter Walter, had joined the Blobel laboratory (fig. 27). His mass of curly hair and full beard gave Walter, who was a German from Berlin, an impish appearance that was reinforced by his love of practical jokes. Walter had arrived at Rockefeller from Nashville, Tennessee, where he had spent a year at Vanderbilt University in an exchange program getting a master's degree in chemistry. Although he had originally planned to return to Berlin for his doctorate, his mind was changed by a visit from Stanford Moore, a Rockefeller Nobel laureate and Vanderbilt board member, who convinced him to apply to Rockefeller.

Like Dobberstein before him, Walter was attracted by the challenge of understanding the translocation machinery and began to treat the microsomal membranes in various ways to inactivate them. By this time, Blobel's lab had become less centralized, with various technical approaches floating

7. The meeting was formally known as the Frühjahrstagung der Gesselschaft für biologische Chemie, held on March 12–15, 1978, in Frankfurt am Main.

8. Sucrose density gradients, which had been introduced earlier into the Rockefeller group by Siekevitz (Britten and Roberts 1960; Tashiro and Siekevitz 1965a), were commonly used to separate a variety of very large molecules (macromolecules), including mRNAs, polyribosomes, and large protein complexes.

9. Dobberstein never published the results showing the six polypeptides, something he may have regretted.

27. Peter Walter, photographed by the author in Dresden, Germany, in 2001.

around. Instead of using the wheat germ translation system, Walter picked up the one that was based on rabbit reticulocytes. Blobel's lab had previously used the rabbit reticulocyte lysate as a source for ribosomes and protein synthesis factors (Freienstein and Blobel 1974; Blobel and Dobberstein 1975b). As mentioned previously, it had not been used much before as a complete translation system for translocation assays because of interference from the huge amount of globin mRNA included in the lysate and from the resulting globin protein band that appeared on SDS gels. By the time of Walter's experiments, however, a technique to degrade the globin message with a specific nuclease had been developed by Hugh Pelham and Richard Jackson at Cambridge University (Pelham and Jackson 1976; Shields and Blobel 1978). This made the reticulocyte system more acceptable because of the improvement in the signal-to-noise ratio with exogenously added mRNAs.

To inactivate the translocation activity of microsomal membranes, Walter based his approach on methods that had been used to analyze membrane-bound proteins in other experimental systems. These included extractions of various types and even protease treatments that would shave off parts of proteins exposed on the membrane surface (see, e.g., Bennett 1978). During this time, the Warren and Dobberstein *Nature* paper appeared describing reconstitution with a high-salt extract. However, in Walter's hands, salt extraction did not block the ability of the membranes to translocate proteins. After multiple trials, Walter finally settled on an unlikely solution. One set of microsomal membranes was treated extensively with the proteolytic enzyme trypsin to inactivate the membranes for translocation. A second set was treated more gently with trypsin and high salt concentrations and the resulting extract collected. After neutralizing the trypsin and lowering the salt concentration in the extract, Walter added it back to inactivated membranes. Remarkably, translocation activity reappeared.

Although somewhat concerned about the differences between his findings and those of Warren and Dobberstein, Walter reasoned that the pancreatic microsomal membranes that were used by Warren and Dobberstein for high salt extraction might have been slightly proteolyzed during isolation, which made them equivalent to his trypsin-treated membranes (Walter et al. 1979). After all, the pancreas synthesizes many proteolytic enzymes, including trypsin, and, if one is not careful, preparations of pancreatic microsomes might be contaminated with proteases. In December 1978, Walter and Blobel submitted a paper, calling their approach "tryptic dissection" of translocation, with publication occurring shortly thereafter (Walter et al. 1979).

Heidelberg Looks to New York

Back in Heidelberg, things were not going well. The rapid progress on the characterization of the salt wash extract had ground to a halt. Warren had left the project to pursue his own interests, and a new Swedish postdoctoral researcher in Dobberstein's lab had lasted only a few weeks, unable to adjust to life in Germany. On top of these setbacks, the EMBL had finally completed its new building, and in winter 1978–79 all the laboratories relocated.

In March 1979, Dobberstein welcomed a new postdoc. David Meyer had just finished a first fellowship in Basel, Switzerland. With his neat moustache and fluent knowledge of the Basel Swiss-German dialect, he superficially

28. David Meyer, early 1980s, in Heidelberg, Germany.
Photo courtesy of Evi and David Meyer.

resembled a young Swiss scientist (fig. 28). Meyer was, however, an American from Los Angeles, although, ironically as it turned out, his parents had immigrated to the United States from Berlin just before World War II. Meyer's first attempts to replicate the Warren and Dobberstein results were somewhat successful, but in his hands the reconstitution of translocation activity was very inefficient. Without knowing whether his problems were due to inexperience with translations or some flaw in the assay, he set about optimizing the conditions.

In the meantime, Dobberstein had begun a completely new project in the lab looking at the synthesis of histocompatibility antigens (see Kvist et al. 1982). These are membrane proteins that are transported after synthesis from the ER to the plasma membrane of cells, where they play essential roles in the immune system. Dobberstein recognized this project as an opportunity to show that he could start something on his own while answering significant questions not only about the histocompatibility antigens but also about membrane protein biogenesis in general. Although Blobel's lab had by then studied the synthesis of a viral membrane protein and its insertion into the ER membrane, a variation on the signal hypothesis (see chap. 7), no naturally occurring membrane proteins like the histocompatibility antigens had been investigated.

Soon after he began the project, Dobberstein discovered that he could not translate mRNA for histocompatibility antigens in the wheat germ system but found that the new, improved reticulocyte lysate worked well. A second postdoc soon arrived in his lab to work on the new project. His initial goal was to clone and sequence the mRNA for one histocompatibility antigen, and his approach required large numbers of translation reactions using the reticulocyte system. Taking advantage of the abundance of research funds at the EMBL and the availability of the rabbit reticulocyte translation system from a commercial supplier, he stopped preparing the reticulocyte lysate himself and began ordering it in large quantities (Kvist et al. 1981).

Meyer also switched to the reticulocyte system for his translocation experiments. Although the wheat germ system was easy to prepare, Meyer decided that it was even easier to order the commercial reticulocyte lysate. He also thought that the new system might solve some of his problems replicating the Warren and Dobberstein results. With the reticulocyte lysate, however, the situation was worse, not better. Both Meyer and Dobberstein began to wonder if the original salt wash results were an artifact.

After Walter's paper describing the use of trypsin was published in April 1979, Meyer decided to shift gears. In initial experiments, Walter's procedure seemed fickle; too much trypsin or slightly too long an incubation, and translocation activity could not be recovered. Nevertheless, by carefully controlling parameters, the procedure worked, and the original Warren and Dobberstein result was apparently reproduced using Walter's combination of trypsin and high salt extraction. The one problem was the gradient. When the tryptic extract was centrifuged in a sucrose gradient, the activity reconstituting translocation appeared near the top of the gradient and not at the bottom where Dobberstein had originally seen it (see fig. 26).

Armed now with a functioning assay, Meyer began to optimize the procedure. Reasoning that trypsin might be causing problems because it was too aggressive an enzyme, he tried a range of proteases, finally settling on elastase to both inactivate the membranes and prepare the protease–high salt extract. After scaling up this approach and applying conventional protein purification techniques to the extract, Meyer was able to identify a 60 kDa polypeptide that reconstituted translocation when added back to proteolyzed microsomal membranes. Meyer believed that the 60 kDa polypeptide was a fragment of a microsomal membrane protein that was released from the membrane by both cleavage with elastase and extraction with high concentrations of salt. When this fragment was added back to the inactive microsomal membrane after reduction in the salt concentration, it apparently rebound to the membrane in its correct location and orientation and was still able to perform its function in translocation even though its polypeptide chain had been cleaved. Meyer's conclusions were similar to those expressed by Walter in his 1979 paper, but now Meyer had an actual protein fragment to work with. With this finding, Meyer and Dobberstein published their results in two back-to-back papers in the *Journal of Cell Biology*, appearing in November 1980 (Meyer and Dobberstein 1980a, 1980b).

Meanwhile, Back in New York . . .

In New York, things had also changed for Walter. Lacking the financial resources of the EMBL that enabled purchase of the commercial reticulocyte lysate and unhappy with the trauma of bleeding anemic rabbits himself, Walter had switched to the wheat germ translation system. In addition to its simplicity and low cost, the ready availability of an RNase inhibitor purified in a neighboring lab at Rockefeller made the wheat germ extract as efficient at translation as the improved reticulocyte lysate (Blackburn, Wilson, and Moore 1977). Now, however, high salt extraction alone without trypsin treatment inactivated the pancreatic microsomes, and readdition of the high salt extract to the inactivated membranes reconstituted translocation, just as Dobberstein had observed more than two years earlier.

On sucrose gradients, the activity migrated at 11S, consistent with a large protein complex, but it was incredibly unstable.[10] After many frustrating months during which Walter tried adding everything he could think of to his preparations that might stabilize the translocation activity, he finally figured out that inclusion of a small amount of nonionic detergent during purification was key.[11] Now he was able to isolate the activity. On SDS gels,

10. "S" stands for Svedberg units, named after the pioneer in centrifuge design (see chap. 3). The Svedberg unit is a measure of sedimentation coefficient related to both the size and the shape of the object undergoing centrifugation.

11. Many years later, Walter's factor was crystallized in the presence of detergent. When its structure was determined by X-ray crystallography, a detergent molecule was seen binding in the site where the signal sequence is thought to bind in the functional molecule (Keenan et al. 1998). All that was needed for stability was, as Walter later stated, a "greasy" (i.e., hydrophobic) molecule like the included detergent.

it consisted of six polypeptides. After quickly writing up these results, Walter and Ibraham Ibrahimi, a fellow in Blobel's lab, immediately characterized the activity further and found that the complex bound to polysomes that were actively synthesizing secretory proteins, where it arrested translation. Then, on association of the polysome-bound complex with microsomal membranes, translation resumed, followed by translocation of the nascent chain across the membrane. The paper that described the purification of the salt wash activity was published in December 1980, only one month after Meyer's results (Walter and Blobel 1980). The characterization of the complex—now called signal recognition protein, or SRP—appeared in three separate Rapid Communications in the *Journal of Cell Biology* in November 1981 (Walter and Blobel 1981a, 1981b; Walter, Ibrahimi, and Blobel 1981).[12]

But what was the relationship between SRP and Meyer's 60 kDa protein fragment? Walter's SRP, which seemed similar to Dobberstein's original factor, was released from membranes by high salt extraction alone, while Meyer's protein required both elastase digestion *and* high salt extraction. The question was, how did these results fit together?

Mystery Solved in Heidelberg

The initial reaction of Meyer and Dobberstein to Walter's first results was incredulity. As far as salt washing alone was concerned, they had been there before without success. Furthermore, Meyer now had an antibody against his 60 kDa factor and used the antibody to determine that the original uncleaved protein was a 72 kDa polypeptide incorporated into the ER membrane (Meyer, Louvard, and Dobberstein 1982). With the antibody, he was also able to demonstrate that the protein was found only in the ER of intact cells using the light microscopic technique of immunofluorescence, a result expected if the protein was functioning as part of the translocation apparatus (Meyer, Louvard, and Dobberstein 1982).[13]

---

12. At the suggestion of David Anderson, another graduate student in the Blobel lab, Walter originally named the complex the Translocation Initiation Complex, abbreviated TRIC. Anderson chose the name because of the clever tricks that Walter used to purify the complex. Some in the lab felt that the name should be changed because of its bawdy connotations. Ultimately, George Palade, who was on Walter's thesis committee, told Walter to change the name because the discovery was "too important to be funny."

13. At the time, the use of antibodies as experimental reagents was becoming more and more common. With a protein such as Meyer's, the usual approach is to inject the purified or partially purified protein either into rabbits or mice and wait for the animal to recognize the

Meyer's results were subsequently confirmed by Reid Gilmore, working with Walter in Blobel's lab. Gilmore used procedures similar to both Meyer's elastase proteolysis and Walter's early trypsin "dissection" to release a factor from the microsomal membrane that reversed SRP's inhibition of translation. Reasoning that this factor must bind SRP, he then took advantage of Walter's supply of purified SRP to use it as an affinity substrate to bind and pick out the factor from a complex mixture of other ER proteins. Gilmore observed that his factor had the same 72 kDa molecular mass as Meyer's protein, and, after Meyer gave him some antibody, he confirmed that both his protein and Meyer's were identical (Gilmore, Blobel, and Walter 1982; Gilmore, Walter, and Blobel 1982).

In June 1982, Meyer put both his findings and those of Walter and Gilmore together in a paper published in *Nature* (Meyer, Krause, and Dobberstein 1982). It had been known for many years that when ribosomes synthesize proteins, the first part of the nascent polypeptide to emerge from the large subunit is the N-terminus, corresponding to the first amino acid synthesized. In secretory proteins, the signal sequence that will eventually direct the ribosome/nascent chain complex to the ER membrane and facilitate translocation is located at the N-terminus. Like Walter, Meyer showed that SRP bound to both the emerging signal sequence and ribosome, temporarily inhibiting translation. When the SRP/ribosome/nascent chain complex reached the ER membrane, it bound to Meyer's 72 kDa protein, which he now called *docking protein*. Docking protein then released the SRP-mediated translation arrest, permitting completion of the secretory protein and, in some as yet undetermined manner, its movement across the ER membrane (Walter and Blobel 1981a, 1981b; Walter, Ibrahimi, and Blobel 1981; Meyer, Krause, and Dobberstein 1982).

Meyer also solved the mystery that had bedeviled both the New York and Heidelberg groups for the previous three years (Meyer, Krause, and

---

protein as a foreign body (or antigen) and mount an immune response against it. Specific antibody molecules are then typically purified from the animal's blood using different forms of chromatography or certain affinity binding procedures (Garvey, Cremer, and Sussdorf 1977). The purified antibodies can then be used to either pull the protein out of cell extracts and separate the *immunoprecipitate* on an SDS gel (the approach used by Meyer) or identify the protein by applying the antibody directly to an SDS gel so that it binds to a specific band, a procedure called Western blotting (Burnette 1981; Moritz 2020). One can also, like Meyer, localize the protein in whole cells by microscopy using immunofluorescence. For the latter, the specific antibody is bound to fixed cells grown on microscope slides, and this so-called primary, or first, antibody is then visualized using a secondary antibody that binds to the primary antibody. The secondary antibody is coupled to a dye that glows with fluorescence when illuminated in the microscope with the appropriate wavelength of light (Ploem 1989).

Dobberstein 1982). The key to explaining why it had been difficult to reproduce Dobberstein's early results resided in the choice of translation system. In New York and in initial experiments with Warren in Heidelberg, Dobberstein had used the wheat germ translation system and observed that he could block translocation by salt washing microsomes and then reconstitute translocation by adding back the salt extract (Warren and Dobberstein 1978a). When Meyer took over the project, he switched to the reticulocyte lysate translation system and could not repeat the original results. Working in New York, Walter initially used the reticulocyte system and had no luck with salt washing alone, finally resorting to trypsinization to inhibit translocation and showing that something released by trypsinization could then restore translocation (Walter et al. 1979). Because this approach was tricky, Walter once again tried salt washing by itself. By the time he did this, he had switched to the wheat germ system because it was easier than bleeding rabbits to obtain reticulocytes. Now he was able to reproduce Dobberstein's original findings, including, but unbeknownst to him, the existence of a complex of six polypeptides that turned out to be SRP (Walter and Blobel 1980). Back in Heidelberg, Meyer, frustrated with salt washing and still using the reticulocyte system, refined Walter's trypsinization approach and identified docking protein, the receptor for SRP on the ER membrane (Meyer and Dobberstein 1980a, 1980b).[14]

Why was it that anyone working with the reticulocyte system was unable to identify SRP? Reticulocytes are immature red blood cells that synthesize only the cytoplasmic protein globin and lack any recognizable ER or other secretory organelles (Fawcett 1986, 248–49). Investigators using the reticulocyte system assumed, therefore, that the reticulocyte lysate lacked proteins involved in translocation such as SRP. After all, a cell like the reticulocyte that had no ER and did not make secretory proteins had no use for SRP. Meyer discovered, however, that the reticulocyte lysate in fact contained sufficient soluble SRP to reconstitute translocation to salt washed microsomes (Meyer, Krause, and Dobberstein 1982). This meant that it was impossible for anyone to test for the absence of SRP on salt washed microsomes using the reticulocyte system because they added reticulocyte SRP back to the microsomes during the experiment.[15] Presumably, reticulocytes,

14. Possibly because of Palade's advice to avoid cute names in favor of names describing established functions, Gilmore and Walter called Meyer's "docking protein" the "SRP receptor," a name that eventually stuck.

15. As described previously, investigators working on translocation in Blobel's lab worked with microsomes that had been chemically stripped of ribosomes and mRNA using some combination of EDTA extraction, high salt extraction, and puromycin treatment developed by Mark

which develop from other red cell precursors that do secrete proteins, retained some SRP from these earlier cells.

A Final Surprise

Even after first purifying SRP, Walter continued to use sucrose density gradients for his SRP preparations. The technique was also commonly used in
Blobel's lab for the isolation of mRNAs and polyribosomes (Britten and
Roberts 1960). Typically, sucrose density gradients are constructed in thin
plastic centrifuge tubes such that there is a continuous, linear change in sucrose concentration from more dense sucrose on the bottom to less dense
sucrose on the top. To separate molecules, concentrated samples are placed
on the top of the gradient, and the tube is centrifuged for several hours
at very high speed. Macromolecular components of the sample distribute
throughout the gradient according to their relative velocity, that is, how fast
they move down the tube, a measure that is related to their overall size and
shape.

After samples are centrifuged, the tubes are removed and then divided
into small samples or fractions based on their position in the gradient. In the
Blobel lab, this fractionation was carried out by carefully pushing the open
top of each centrifuge tube into a device called a flow cell and then inserting a
needle into the very bottom of the centrifuge tube. The needle was connected
through fine plastic tubing to a large syringe filled with a very dense sucrose
solution that was then pumped slowly from the syringe into the bottom of
the tube, gradually pushing the contents of the centrifuge tube out the top
and through the flow cell. As the contents of the gradient passed through the
flow cell, proteins and nucleic acids were detected and quantitated by absorption of ultraviolet (UV) light and the gradient fractions serially collected.

Adelman in Sabatini's lab and subsequently modified by Blobel and Dobberstein and Dennis
Shields (Adelman, Sabatini, and Blobel 1973; Blobel and Dobberstein 1975b; Shields and Blobel 1978). Given that at first Dobberstein and later Walter used stripped membranes in their
translocation assays, it is reasonable to ask why they worked in the wheat germ system when the
stripping procedures could have removed SRP? In Dobberstein's early, unpublished experiments
in New York, he did indeed find that wheat germ had variable translocation activity, possibly
due to partial depletion of SRP from the stripped membranes. He was most successful when he
stripped the membranes with EDTA alone. Later, in Heidelberg, the stripping procedures used
by Warren and Dobberstein avoided high salt extraction completely (Warren and Dobberstein
1978a). In his experiments leading to the discovery of SRP, Walter also avoided salt extraction
to remove endogenous ribosomes and mRNA from the microsomal membranes (Walter and
Blobel 1980).

The instrument for quantitating the gradient fractions that contained the flow cell was a simple UV light spectrophotometer with two wavelength settings, 260 nm and 280 nm. Nucleic acids absorb UV maximally at about 260 nm, while proteins absorb maximally at 280 nm.

The bench with the fractionation apparatus was a crossroads in the Blobel lab because so many students and postdocs ran sucrose gradients. Many gradients leaked slightly when punctured by the needle that delivered heavy sucrose, and occasionally the tubing came off of the large sucrose-containing syringe, causing the entire area around the instruments used for fractionation to be sticky and encrusted with sucrose and dried up salt solutions. To the casual onlooker, it seemed impossible that refined and quantitative data could emerge from such a mess.

In his first paper describing SRP, Walter had demonstrated that it moved to the middle of a sucrose gradient ranging from 5% to 20% sucrose after a centrifuge run of 20 hours in a rotor spinning at 40,000 rpm (Walter and Blobel 1980). Sometime later, Walter was fractionating a routine gradient containing SRP and was surprised to see an unusually large peak of absorbance at a point in the gradient where he expected to find the complex. Because he had run such gradients many times before, he was puzzled because the absorbance was two times more than he had seen when earlier centrifuging similar amounts of SRP. Walter soon realized that someone (rumored to be Keith Mostov, another student in the lab) had set the spectrophotometer to 260 nm instead of the usual 280 nm. Because proteins absorb some UV light at 260 nm, Walter then compared the ratio of SRP absorbance at 260 nm to the absorbance at 280 nm and found that the ratio was not characteristic of proteins alone. He began to sense that the SRP complex might contain a nucleic acid molecule in addition to its six protein chains. If true, it was a stunning discovery.

Within a short time, Walter had extracted the SRP complex and identified a small RNA molecule (Walter and Blobel 1982). When he digested SRP with a nuclease capable of cleaving RNA, the complex lost its ability to either arrest in vitro translation or facilitate the translocation of secretory proteins across the microsomal membrane. Even though the polypeptides making up SRP appeared to remain together as a complex after digestion, the migration of digested SRP on sucrose gradients was altered, suggesting that the RNA had, at the least, some structural role. Walter characterized the RNA molecule further and prepared a partial sequence, finding that it was about 260 nucleotides long. The sequence, however, was not immediately recognizable. To account for the presence of RNA, he revised the name signal recognition protein to *signal recognition particle*.

Walter presented his new results in a seminar at Yale Medical School. When he showed the sequence of the RNA, Elisabetta Ullu, a scientist sitting in the audience, immediately recognized it as "her" 7S RNA, a cytoplasmic molecule of previously unknown function that she had studied for some time (Ullu and Melli 1982; Ullu, Murphy, and Melli 1982). Ullu had just arrived at Yale from the EMBL in Heidelberg.

Chance Favors the Experimental System

As all scientists know, the sequence of discoveries documented in research publications is often fabricated to allow logical presentation of results. Nobody really cares that the experiment presented first was really done last, or that some controls were only added later at the behest of a reviewer, long after the original manuscript was complete. But the process of writing scientific papers also expunges the anxieties and doubts, missteps and mistakes that punctuate the path to completion of the work.

The story of SRP and its receptor on the ER membrane exemplifies the underlying vagaries of scientific research as it is really practiced. If not for changing geographic locations, a new laboratory building, and flush financial circumstances, Dobberstein and Meyer might have discovered SRP: the complex was, in fact, in Dobberstein's hands in 1978. Initially unfamiliar with Dobberstein's results and then unable to reproduce them, Walter turned to techniques used to study other membranes and tried proteases to inactivate and then reconstitute translocation. That worked for a time but was eventually dropped. Then Meyer, frustrated with his own experiments, picked up and refined Walter's discarded approach, discovering docking protein, the SRP receptor. By then, Walter had shifted, managing to finally repeat Dobberstein's first results and discover SRP. Then, in a final twist, the wrong setting on an instrument led Walter to find that SRP was not just a protein complex, but instead a "particle" containing proteins and RNA, a result that was at the time astounding.

In subsequent years, detailed knowledge about the role of SRP and its receptor in targeting of secretory proteins to the ER and their co-translational translocation across the ER membrane expanded tremendously in both breadth and depth. SRP apparently moves in a cycle between the cytoplasm, where it is soluble, and the microsomal membrane, where it attaches to the SRP receptor (Walter, Gilmore, and Blobel 1984; Shan and Walter 2004; Rapoport 2007) (see fig. 32). That is why it was found both bound through a salt-sensitive linkage to the microsomal membrane and free in the cytoplasm of reticulocyte lysates. The SRP receptor was later shown to consist of not one but two polypeptide chains and to possess activities that enabled it to release SRP from the ribosome–nascent chain complex (Walter, Gilmore, and Blobel 1984; Shan and Walter 2004; Rapoport 2007). Eventually, some form of SRP was shown to be universally present in eukaryotes and bacteria, and even in some cellular organelles (Akopian et al. 2013). The mechanisms by which SRP recognizes signal sequences and then targets the nascent chain to the ER membrane for translocation are now almost completely understood in great molecular detail. Even the three-dimensional molecular structure of SRP as it binds to both a signal sequence and the large ribosomal subunit has been determined at high resolution (Halic et al. 2004). Altogether, these findings as well as others that occurred later (see chaps. 7 and 8) constitute an explanation at the molecular level of a biological phenomenon, the initiation of secretion in cells.

Why is the work described here cell biology rather than biochemistry or molecular biology? An argument about the definition of molecular biology as a discipline is beyond the scope of this chapter. At the moment, suffice it to paraphrase Erwin Chargaff, a biochemist whose work provided a key clue to the structure of DNA, and say that molecular biology is practicing biochemistry without a license (Rose 2011).[16] Describing the discovery and characterization of SRP and its receptor as biochemistry makes some sense. Most of the methods used in this work were biochemical techniques, including protein and RNA isolation, purification, and sequencing, as well as enzyme activity measurements, protein fragmentation with proteases, and so on. These were all of critical importance. What was crucial, however, was the reconstituted translocation assay because it measured biological events, the targeting and transfer of a protein across the microsomal (i.e., ER) membrane (Blobel and Dobberstein 1975b). The inclusion of a cell fragment that replicated functions of the living cell in the assay meant that the application of biochemical procedures yielded molecular understanding of a cell biological function.[17]

While circuitous, the discovery of SRP and its receptor were dependent on an exquisite understanding of the reconstituted experimental system developed by Blobel. Through the epistemic strategy originally devised by Claude and then refined by Palade and Siekevitz, it became clear that microsomes were derived from the ER (see chap. 4). Because isolated microsomes retained key ER functions, namely, the translocation and sequestration of

16. See the epilogue for further discussion of molecular versus cell biology.
17. Note that at this point no so-called molecular biology techniques, such as the use of recombinant DNA procedures, played significant roles in any of the outlined discoveries.

newly synthesized secretory proteins, the molecular basis of these functions could be studied in Blobel's cell-free system. This then enabled the biochemical dissection of microsomes by extraction and proteolysis that led to SRP and its receptor while at the same time linking these molecules to specific biological events in the ER.

The epistemic strategy employed by the Rockefeller group depended on morphological analysis with the microscope to guide cell fractionation and biochemical characterization. Even though Blobel rarely used microscopy in his experiments after 1975, his in vitro assay nevertheless explored the morphology of the cell because it dealt with the form of the ER as manifested in isolated microsomes and the relationship between the cell cytoplasm and the ER. Blobel's microsomal vesicles no longer resided within the cell, but they retained a topological organization that was identical to that of the ER in the intact cell. That is, the outside of the vesicles corresponded to the cytoplasmic side of the ER membrane while the inside corresponded to the ER lumen. While the methods associated with Blobel's assay were biochemical, his results enabled him to come to morphological conclusions about relationships between cytoplasmic and lumenal events as they occurred in the intact ER. When a secretory protein traversed the microsomal membrane in his assay, it was equivalent to crossing the boundary separating the cellular cytoplasm from the ER lumen.

We now turn to the efforts by Blobel and his laboratory colleagues to generalize the signal hypothesis to not only secretory proteins but also membrane proteins and proteins from mitochondria and chloroplasts, using everything from fish to fungi to bacteria. As we shall see, Blobel's large footprint begins to step on many toes.

SEVEN

Enemies, Real and Imagined

A Broadening Strategy

Blobel described the kind of work that led to the discovery of SRP and its receptor as a vertical investigation. Once the phenomena of translocation and signal sequence removal were established in his reconstituted system, his lab began to drill down, in a sense vertically, to find new molecular components underlying these processes. The alternative was what he called a horizontal investigation. This was intended to help establish the generality of the signal hypothesis as well as expand it to different situations by exploring a variety of proteins beyond the secretory proteins already studied.

He knew both from his results and from those of Milstein and others that immunoglobulins had signal sequences that facilitated their translocation across microsomal membranes and that these signals were very similar to those performing an identical function for the enzymes secreted from the exocrine pancreas (Milstein et al. 1972; Devillers-Thiery et al. 1975; Schechter et al. 1975). But the pancreas also has an endocrine part that secretes insulin and glucagon into the bloodstream. Did their secretion use the same mechanism? What about integral membrane proteins that penetrated the lipid bilayer? There was some evidence that they were synthesized on ribosomes bound to the ER (Grubman, Ehrenfeld, and Summers 1974; Grubman et al. 1975). Were they directed there by signal sequences? Further afield, did the hydrolases resident in the lysosome, the degradative organelle of the cell, begin their journey to the lysosome through a signal sequence–mediated process? Even further afield, bacteria, though lacking any organelles, still secreted proteins. Was it possible that the mechanism proposed

The title of this chapter comes from a Nixon-era cartoon that Blobel reportedly had in his office.

in the signal hypothesis was so conserved evolutionarily that it worked in bacteria? Blobel was driven to answer these questions. Conveniently, this horizontal approach also helped him deal with the wave of postdocs and students who descended on his now well-known lab: he would just assign them different proteins to study.

The approach to most of these questions was straightforward. Find cells or tissues that primarily synthesize one or a few major proteins, isolate mRNAs, and translate them in the reconstituted in vitro system containing dog pancreas microsomes. Compare the size of the translation products synthesized in either the presence or absence of microsomes by SDS-gel electrophoresis to see if the membranes reduced the size of the synthesized proteins by an amount equivalent to the size of a signal sequence. And finally, check that the smaller proteins synthesized in the presence of membranes are protected from proteases added to the outside of the microsomal membrane vesicles, an indication of translocation. Certainly, extending this approach to bacteria would require development of an equivalent translation system. In general, however, the existing reconstituted system or simplified versions using wheat germ extracts or reticulocyte lysates should work. Even better, no other lab had such a system and the technology was not easy to set up. Other labs interested in asking similar questions about their favorite protein would either be forced to collaborate with Blobel's lab or lose valuable time trying to develop reconstituted systems on their own.

In this chapter, I describe several such horizontal investigations. Included are not only some of the examples mentioned above but also unexpected explorations into the biogenesis of mitochondria and chloroplasts. While these two membrane-bounded organelles responsible for energy production in animal and plant cells are not like the ER, they still need to get proteins synthesized in the cytoplasm across their membranes. Some of this work was carried out mainly by the Blobel lab; in other instances, collaborations with other research groups were necessary. As we shall see, Blobel began these studies with a sort of naive exuberance but was brought to earth when he realized that not everyone was invested in his success, particularly when this success required an invasion of their territories. The result was the erection of a mental defensive barrier around the Blobel lab that convinced those inside that they were engaged in a zero-sum game, with other labs viewed as competitors. Ultimately, these horizontal studies demonstrated the broad applicability of the ideas embedded in the signal hypothesis but left scars on participants on both sides of the wall.

Insulin

Blobel's postdoc Dennis Shields needed a new project. He had arrived in the lab in late 1974 with the intention of working on mRNA-bound proteins, but that subject was passé after completion of the 1975 papers announcing and largely demonstrating the signal hypothesis. As a new project, Blobel suggested that Shields look at the biosynthesis of insulin by the pancreas. Blobel may have thought that such a project would be straightforward after the lab's success in isolating mRNAs for digestive enzymes from dog pancreas preparations. There was a problem, however, as Shields soon discovered. The endocrine part of the pancreas, which secretes insulin and other hormones into the bloodstream, consists in mammals of small groups of cells called the Islets of Langerhans scattered like an archipelago in the exocrine pancreas, the part secreting digestive enzymes. In humans, there are more than a million islets, but these are not easily visible to the naked eye and, more significantly, constitute only 1% to 2% of pancreatic volume (Fawcett 1986, 721). If Shields used the entire dog pancreas, both exocrine and endocrine, for mRNA preparation, then the fraction of mRNAs coding for insulin would likely be exceedingly small, with mRNAs for exocrine proteins dominating. To get around this, Shields attempted to isolate islets from the pancreas using enzymatic digestion of the extracellular matrix that supported the cellular structures but was unsuccessful.

Insulin, whose name refers to its origin in the Islets of Langerhans, was the first peptide hormone discovered and among the most important because it controls the amount of sugar in the blood. Failure of the pancreas to secrete insulin leads to diabetes, a disease that is only controlled by daily or even hourly injections of insulin purified from animals or, more recently, recombinant sources. Secreted insulin is a small protein, consisting of only fifty-one amino acids distributed in two polypeptide chains, called A and B, that are bonded together. Because of the protein's size and significance and the availability of purified insulin, it was the first protein for which the entire polypeptide sequence was determined (Sanger and Tuppy 1951a, 1951b; Sanger and Thompson 1953a, 1953b). In 1967, Donald Steiner and Philip Oyer of the University of Chicago reported that insulin is synthesized as a precursor referred to as proinsulin (Steiner and Oyer 1967). Proinsulin is a single polypeptide chain consisting of the two chains present in the secreted form of the protein linked into a single polypeptide by an intervening C peptide. By the 1970s it was known that proinsulin is synthesized by ribosomes bound to the ER and transported along the secretory pathway prior to release at the cell surface (Steiner 2011). At a late stage of its transport through the cell, proinsulin is proteolytically processed to cut out the C chain and produce the mature protein (Steiner et al. 1974).

Blobel's interest in seeing if a proinsulin precursor was translocated and processed by pancreatic microsomes in his reconstituted system may have arisen in conversations with Steiner, the discoverer of proinsulin. Sometime in 1975, an East German fellow from Steiner's lab met with Blobel at Rockefeller and learned about the signal hypothesis. He returned to Chicago very excited about the model, which he described to Steiner. Shortly thereafter, Steiner went to Rockefeller himself and heard the story directly from Blobel. Even before this, Steiner was very aware of the proposal made by the Milstein laboratory that there was a precursor of the immunoglobulin light chain that might be involved in targeting the molecule to the ER. Indeed, one of the coauthors of the Milstein paper, Mike Matthews, had given a seminar on the results at the University of Chicago prior to its publication in 1972.[1]

As described previously (chap. 5), Steiner ended up as one of the reviewers of the Blobel and Dobberstein 1975 papers, something that Blobel was aware of. In fact, Blobel wrote to Steiner in September 1975, just as the papers were accepted for publication in the *Journal of Cell Biology*, to both thank him for his quick review and provide comments on a review article manuscript that Steiner had written for a CIBA[2] Foundation volume.[3] In the manuscript, which eventually appeared in 1976, Steiner discusses the Milstein results and the possibility that there is a similar precursor of proinsulin, even mentioning that a slightly larger form of proinsulin that might contain a signal sequence had been observed (Lernmark et al. 1976). Blobel critiqued some of Steiner's conclusions about the Milstein results, pointing to his own papers that Steiner had reviewed as providing the evidence missing from the Milstein work. Despite these interactions, Blobel seemed to naively ignore Steiner's clear interest in determining in his own lab if the signal hypothesis applied to insulin. This should have been obvious because

1. As described in chapter 5, Blobel first learned of the Milstein results when Matthews gave a talk at a Gordon Conference. It is likely that this was part of the same lecture tour that led Matthews to Chicago.

2. CIBA, or Ciba, was an acronym for Company for Chemical Industries Basel, in Switzerland. Through a series of mergers it is now part of the Novartis AG, a pharmaceutical and chemical holding company. Primarily in the latter half of the twentieth century, the Ciba Foundation held symposia in London, and the proceedings were published in a series of volumes.

3. Letter, Blobel to Steiner, September 9, 1975. Used by permission of Don Steiner.

Steiner's entire career had been focused on the biosynthesis of insulin, and he considered anything to do with this subject his professional territory.

After Shield's failure with the dog pancreas, he searched the library for another strategy. He discovered that in fish species the endocrine portion of the pancreas is a separate organ rather than scattered islets in the exocrine part. This meant that the tissue secreting insulin could be easily removed without contamination from the exocrine pancreas, and the chance of preparing mRNA coding for insulin was consequently greatly enhanced. Shields managed to locate sources of two species of fish at the MBL in Woods Hole and the New England Aquarium in Boston and arrange for the live fish to be shipped to him at Rockefeller. The first shipment of about twenty fish arrived on a Friday afternoon, a rather inopportune time, and Shields had to enlist several members of the lab who happened to still be around to dissect the endocrine pancreas.

At this point the project began moving very quickly. When Shields translated mRNAs from the fish endocrine pancreas in the wheat germ system in the presence or absence of dog pancreas microsomes, he discovered that a polypeptide was synthesized in the absence of membranes that was slightly larger than the presumed size of fish proinsulin. With membranes, some of this polypeptide appeared to shift to the molecular mass of authentic proinsulin and was protected from digestion with proteolytic enzymes. From this it seemed clear that a precursor of proinsulin, referred to as preproinsulin, was the primary translation product of insulin mRNA and was translocated across the microsomal membrane and processed to the size of authentic proinsulin in the in vitro reconstituted protein synthesis system. To demonstrate the existence of a signal sequence, both putative preproinsulin and proinsulin were isolated from SDS gels by excising their bands and partial sequences of the N-terminal regions determined. Although the sequences were incomplete, it was evident that there was a signal sequence on preproinsulin that approximated the size and some sequence characteristics of signal sequences from other proteins. Furthermore, it appeared that the dog pancreas microsomes cleaved the signal sequence at the correct location to yield the proper N-terminus of fish proinsulin.

Shields and Blobel wrote up their results with the intention of submitting them to the *PNAS*. At the time, it was not possible to send a manuscript directly to *PNAS* for consideration. Instead, a member of the Academy had to "communicate" the paper to the journal before it could be reviewed, a step that indicated a positive endorsement by the member. Blobel asked Steiner, a member of the Academy, to communicate their paper, seemingly blind to the idea that Steiner was a competitor who might consider this an intrusion on his turf. Steiner agreed to consider it but was very slow to do so. Ultimately, he sharply criticized the results and rejected the paper, leading to a delay in its publication. By this time, Steiner had told Blobel that he had a student looking for preproinsulin in his lab, suggesting, to Blobel at least, that Steiner sat on Blobel's paper so that he could publish his results first. Indeed, Steiner published a paper in *PNAS* in June 1976 that he communicated himself (Chan, Keim, and Steiner 1976). In it, he reported in vitro synthesis of rat preproinsulin and the sequence of an N-terminal signal peptide. Possibly because he was unsuccessful in setting up a reconstituted system with microsomes or was concerned that Blobel would scoop him, the paper did not attempt to demonstrate translocation and processing of the rat protein. Shields and Blobel soon asked an Academy member at Rockefeller to communicate their paper to *PNAS*, and it was quickly accepted, appearing in May 1977 (Shields and Blobel 1977). At the time it was communicated, the Steiner paper was out, and Shields and Blobel were forced to include Steiner's findings.

The results of Shields and Blobel were a major step in demonstrating the generality of the signal hypothesis. They showed not only that proinsulin possessed a signal sequence that facilitated its translocation and entry into the secretory pathway but also that the signal hypothesis applied to the secretion of peptide hormones, a class of proteins not previously examined in the Blobel lab. Inadvertently, Shields's choice of fish as a source of proinsulin mRNA, a decision based mainly on technical expediency, showed that the mechanism depicted by the signal hypothesis applied to organisms other than mammals. Furthermore, the mechanism seemed to be highly conserved since dog pancreas microsomes correctly processed fish preproinsulin into proinsulin. Intentionally or not, Blobel's horizontal strategy had begun to work.

Blobel was stung by the conflict with Steiner but considered it a learning experience about the territoriality of science, something that he had not properly considered before. Nevertheless, his reaction was to stake out his own broad territory—anything to do with the signal hypothesis. The concepts embedded in his model, including among others co-translational translocation and processing, needed to be broadly verified but also defended.

The Origins of Membrane Protein Asymmetry

At the same time that the insulin work was under way, Blobel initiated another project to determine if proteins inserted into membranes used a signal-mediated mechanism similar to that followed by secretory proteins.

As described previously (chap. 2), advances in the study of biological membranes in the early 1970s had established that certain proteins associated with membranes traverse the lipid bilayer. The orientation of these transmembrane proteins is fixed and asymmetric such that different specific parts of the proteins are exposed on either one side of the membrane or the other. In particular, it was known from studies of the red cell that many plasma membrane proteins were glycoproteins with attached sugar chains and were oriented in the membrane such that the sugars, or *oligosaccharides*, faced the outside (Bretscher 1971b; Steck, Fairbanks, and Wallach 1971; Singer and Nicolson 1972; Steck 1974). Other work showed that the lipids making up the bilayer were also asymmetrically arranged in the membrane (Bretscher 1972). How the asymmetry of both membrane proteins and lipids came about was unknown.

Blobel was not particularly interested in membranes per se but did want to extend the signal hypothesis to as many types of proteins as possible. As with secretory proteins, the challenge was to isolate an mRNA fraction that coded mainly for one or a few membrane proteins. This would enable the translation products from these messages to be easily identified by SDS-gel electrophoresis. Unfortunately, most cells and tissues synthesize a large variety of membrane proteins, and the fraction of mRNA devoted to any individual protein is small. Exceptions, however, are cells infected with animal viruses.

In the mid-1970s, when Blobel's project began, biosynthetic studies of membrane proteins were limited and confined mainly to investigation of viral spike glycoproteins. Certain classes of viruses that infect animal cells are known as enveloped or membrane viruses because the virus particle consists of a tiny membrane vesicle derived from the plasma membrane of cells that the virus infects. The proteins on such virus particles that bind to host cells and facilitate infection are transmembrane glycoproteins usually referred to as spike proteins because they project from the virus membrane like porcupine quills. Structurally and biochemically, spike proteins are identical to normal cellular plasma membrane proteins. When some viruses, such as the one causing influenza, infect cells, they take over the host cell's protein synthesis machinery, forcing it to synthesize only viral proteins (Matlin and Simons 1983). Because the spike proteins constitute a large fraction of the proteins made by the virus, a substantial amount of the mRNA isolated from infected cells codes for the viral membrane proteins.

Blobel initially sought a collaboration with a large virology laboratory at Rockefeller led by Purnell Choppin. Choppin's group investigated the synthesis of proteins from a variety of viruses in infected mammalian cells grown in culture, including those of Sendai virus and several varieties of influenza virus (Lazarowitz and Choppin 1975; Lamb and Choppin 1977). Negotiations about how to get the project off the ground did not, however, progress, and Blobel began to look for a new partner.

He next contacted Harvey Lodish at MIT in Boston. Lodish, who had been a student at Rockefeller, likely knew Blobel either through his connections to the institution or through their mutual interest in protein synthesis. A few years earlier, Lodish had begun working with cells infected with vesicular stomatitis virus (VSV) to learn about how the virus controls protein synthesis (Morrison et al. 1974). VSV is a cousin of the virus that causes rabies and, like rabies, produces oddly shaped virus particles that look like bullets. VSV is an enveloped virus, and infected cells produce only a small number of viral proteins, including a single membrane glycoprotein called G protein. Most important, Lodish's lab and others had shown that in infected cells mRNA for G protein was associated with membranes and, when isolated, could be translated in an in vitro protein synthesis system (Grubman, Ehrenfeld, and Summers 1974; Morrison et al. 1974; Grubman et al. 1975; Knipe, Rose, and Lodish 1975). Although he was not a typical cell biologist, Lodish was interested in the intracellular events involved in the transport of G protein from its site of synthesis in the cell to the plasma membrane, where it was incorporated into new virus particles when they budded from the infected cell. Indeed, Lodish's postdoctoral fellow David Knipe was in the process of doing groundbreaking studies of how fast the G protein traversed the cell following its synthesis (Knipe, Baltimore, and Lodish 1977b; Knipe, Lodish, and Baltimore 1977). Overall, Lodish's orientation and interests, as well as his familiarity with Blobel's work on the signal hypothesis, led him to quickly agree to collaborate with Blobel.

In 1976, Lodish sent a new graduate student, Flora Katz, to New York with mRNA isolated from VSV-infected cells. Blobel assigned Vishwanath Lingappa from his lab to work with Katz on the project. Lingappa was a medical student from Cornell Medical College across the street from Rockefeller. He had started working in Blobel's lab as an unaffiliated visitor and had quickly become a valuable member of the group. Lingappa had earlier met Blobel when he interviewed for a position in the MD/PhD program run jointly by Rockefeller and Cornell Medical College. When he was not selected for the double degree program but was admitted to the MD part, he entered medical school but convinced Blobel to allow him to simultaneously conduct a research project in the lab. He found time to do this by frequently skipping his medical school classes and spending most of his days and many nights conducting experiments while surviving on a loaf of supermarket bread and a jar of peanut butter. He first followed up on Shield's work with insulin to demonstrate that endocrine hormones synthesized by the pituitary gland also followed the signal hypothesis (Lingappa, Devillers-Thiery, and Blobel 1977). His work ethic and apparent mastery of the in vitro reconstituted translation system so impressed Blobel that he thought Lingappa was the right person to collaborate with Katz on the VSV project.

In concert with Blobel and Lodish, Katz and Lingappa hypothesized that the VSV G spike protein is initially synthesized as a larger precursor with a signal sequence that directs it to the ER membrane where it is inserted. This would place the protein at the beginning of the intracellular transport pathway leading to the plasma membrane. Like secretory proteins, they predicted that the signal sequence would be cleaved off during this process. Unlike secretory proteins, they believed that G protein would not completely translocate through the ER membrane but stop partway, leaving it permanently associated with the lipid bilayer. Thus, in a variation of the signal hypothesis, the signal sequence would both target the G protein to the ER and help determine its asymmetric transmembrane orientation.

Their initial experiments translating the VSV mRNAs progressed quickly but with a brief period of confusion. When translated in a wheat germ system in the absence of microsomal membranes, several polypeptides were synthesized. All but the G protein were the same molecular weight as polypeptides seen in VSV-infected cells, as shown by SDS-gel electrophoresis, and were determined to be internal structural proteins of the virus. The G protein synthesized in vitro, on the other hand, was smaller than its counterpart in infected cells. This difference in molecular weight was attributed to the absence of attached sugars since it was known that the G protein is glycosylated during its synthesis in cells but presumably not during its synthesis in vitro. When they translated mRNAs in vitro in the presence of membranes, however, they were baffled by the results. On SDS gels, they expected the band corresponding to G protein to shift to a lower molecular weight in the presence of membranes due to cleavage of a signal sequence. Instead, it shifted to a higher molecular weight (Katz et al. 1977). Disheartened, they began to believe that their ideas about the biosynthesis of G protein might be wrong. While mulling over the results, Katz suddenly figured out a possible interpretation: when nascent G protein began crossing the microsomal membrane, its signal sequence was cleaved, reducing the size of the polypeptide. Simultaneously, however, the microsomes added a mass of sugars to the protein that more than compensated for the loss of the signal sequence. This resulted in a band with a higher molecular weight (fig. 29).

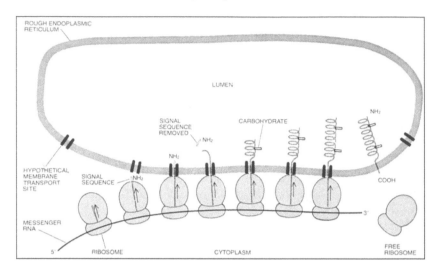

29. Co-translational translocation and glycosylation of the VSV G membrane protein, as illustrated in *Scientific American* (Lodish and Rothman 1979). As the G protein crosses the microsomal membrane, its signal sequence is removed, but oligosaccharides (called "carbohydrate" in the figure) are added. The end result is that the completed G protein has a higher molecular weight, as detected by SDS-gel electrophoresis, than the original precursor protein that includes the signal sequence. Original illustration © George Kelvin. Used with permission.

If true, these results showed that Blobel's reconstituted system was capable of not only recapitulating the functions of the ER in protein targeting and translocation, but also the glycosylation of transmembrane and possibly other types of proteins.

At the time of the VSV work, how proteins are glycosylated was just beginning to be understood. Detailed studies carried out by Rosalind Kornfeld and Stuart Kornfeld suggested that the structures of sugars attached to glycoproteins are very intricate, consisting of treelike *core oligosaccharides* containing mainly the sugar mannose and an outer section of more diverse and unusual *complex oligosaccharides* made up of sugars such as galactose and sialic acid (Kornfeld and Kornfeld 1976). Synthesis of these had been linked by biochemists to cellular membranes, including microsomes, with an initial step in the process carried out by a specialized lipid called dolichol (Waechter and Lennarz 1976). Hence, the idea that VSV G was glycosylated during synthesis by microsomes was plausible.

At this point, a new player entered the project. James Rothman was nominally a new postdoc with Lodish, but, like Athena springing from Zeus's head, appeared in Lodish's lab as a fully formed scientist. Rothman had attended college at Yale, hoping to become a theoretical physicist, but a basic biology course changed his mind. After graduation, he entered Harvard Medical School as a compromise that satisfied both his family's hope that he would become a physician and his increasing interest in biology. While taking basic science courses during his first year he gained a reputation among fellow students as possibly a more reliable source of information than the professors and became enamored with the cell biology that he learned in his histology class. After a period of indecision, Rothman's drive to understand biological systems more deeply overwhelmed any thoughts of clinical medicine, and he switched from medical school to the graduate program in the Department of Biological Chemistry, where he had already begun research with Eugene Kennedy, a noted lipid biochemist.

Rothman had become interested in membranes during his last year at Yale when he conducted research on the biophysics of lipids and published several papers (Engelman and Rothman 1972; Rothman and Engelman 1972; Rothman 1973). In Kennedy's lab, he started characterizing lipid asymmetry in the membrane of influenza virus to determine to what extent it resembled that seen already in the red cell and other membranes (Lenard and Rothman 1976; Rothman et al. 1976). Although Rothman's approach was biochemical, he considered the questions that he was asking biological because asymmetry in membrane organization was a consequence of biological processes in the living cell. In fact, Rothman believed that the biochemical mechanisms that maintained and propagated spatial asymmetry in cells were what distinguished cell biology from pure biochemistry. As he finished his graduate degree, he published an influential article in *Science* with John Lenard, one of his collaborators, reviewing all aspects of membrane asymmetry, including not only lipids but also proteins and oligosaccharides attached to lipids and proteins (Rothman and Lenard 1977).

By the time that Rothman entered the Lodish lab, he was very much aware of Blobel's work on the signal hypothesis and planned to use the VSV system to study the origins of membrane protein asymmetry. One obstacle, however, was that Katz was working on essentially the same problem and had already obtained key results in her experiments with Lingappa. Determined to get involved, Rothman pushed himself onto the project, soon joining Katz when she visited Rockefeller the next time. After some tense moments, Katz and Rothman figured out how to collaborate, and the work proceeded to their mutual benefit. Katz valued Rothman's input and help with experiments and managed to avoid being crushed by the bulldozer bearing down on her.

Rothman's insinuation into the project was viewed by the Blobel lab as an aggressive attempt to seize control. It was undeniable, however, that Rothman's ideas were helpful, and his direct contributions justified his inclusion on the first paper describing the translocation of the VSV G protein, even though it meant that Lingappa was pushed from second author to third (Katz et al. 1977). In addition to demonstrating that G protein was synthesized as a precursor and glycosylated as it crossed the microsomal membrane, they used trypsin digestion of microsomes to show that it spanned the membrane asymmetrically with the N-terminus inside the microsomal lumen and the C-terminus of the protein facing the outside, the side equivalent to the cytoplasm in intact cells. In a subsequent paper, still in collaboration with the Lodish lab, Lingappa sequenced the G protein signal peptide, demonstrating its similarity to signal sequences for secretory proteins (Lingappa et al. 1978). Furthermore, through a clever competition experiment, he was able to suggest that G protein and a secretory protein used the same putative translocation sites on the microsomal membrane, intimating that the same translocation machinery might be employed by both types of proteins (Lingappa et al. 1978).

Shortly thereafter, Rothman and Lodish proceeded with the project to the exclusion of the Blobel lab. They published a paper in *Nature* and then an overview in the lay journal *Scientific American* that barely mentioned the Blobel lab contributions (Rothman and Lodish 1977; Lodish and Rothman 1979). The exclusion of Blobel at this point was completely justifiable because Lodish and his lab had provided the key VSV experimental system and independently produced other key insights into the biosynthesis and transport of VSV G protein (Knipe, Baltimore, and Lodish 1977a, 1977b; Knipe, Lodish, and Baltimore 1977). Nevertheless, some in the Blobel lab considered it somewhat of an affront. Eventually, Blobel realized that he and Rothman shared many attributes. Both were highly ambitious, but their ambition was motivated more by devotion to research and a drive to figure out how cells worked than a hunger for fame and fortune. Rothman soon established his own lab at Stanford University, assimilating Blobel's cell-free approach and running with it (Fries and Rothman 1980).[4] Blobel continued both his horizontal and vertical strategies, entering avenues of research far afield from his previous efforts.

---

4. Rothman went on to win the 2013 Nobel Prize in Physiology or Medicine with Randy Schekman and Thomas Südhof. Rothman's work was initiated using a cell-free reconstituted system that reproduced vesicular transport on the secretory pathway. His approach is described in chapter 10. Some of Schekman's work is discussed in chapter 8.

Off the Secretory Path

While Dobberstein was still working with Blobel in New York, he frequently conversed with Nam-Hai Chua, another faculty member at Rockefeller whose lab was just across the hall from Blobel's. Chua was a plant biologist who had worked on the single-celled alga *Chlamydomonas reinhardtii* since graduate school at Harvard University. Prior to Palade's departure for Yale, Chua came to Rockefeller for postdoctoral research with Siekevitz, attracted by previous work there on *Chlamydomonas* (Hoober and Blobel 1969; Hoober, Siekevitz, and Palade 1969; Schor, Siekevitz, and Palade 1970). Soon after he arrived, he realized, like Blobel before him, that Siekevitz's interests had gone in a different direction, and he began to gravitate to Blobel for advice. No doubt with Blobel's encouragement, Chua began characterizing the two classes of ribosomes in *Chlamydomonas*, larger 80S ribosomes that synthesize proteins in the cytoplasm and smaller 70S ribosomes found inside the chloroplast (Chua, Blobel, and Siekevitz 1973).

The chloroplast is the organelle in plants that uses sunlight to produce energy and sugar through the process of photosynthesis. Chloroplasts contain stacks of so-called thylakoid membranes incorporating two protein-chlorophyll complexes called reaction centers that give green plants their color. When light is absorbed by the reaction centers, the captured energy is used to generate energy-rich molecules such as ATP and to convert or "fix" carbon dioxide into carbohydrates while releasing oxygen into the atmosphere (Voet and Voet 2004). Like mitochondria (see below), chloroplasts are believed to have originated as bacteria-like symbiotes that invaded early eukaryotic cells and gradually evolved into fully integrated organelles. As a vestige of their original prokaryotic existence, they retain a DNA genome. Despite this, over the millions of years of evolution after chloroplasts and mitochondria took up residence in eukaryotic cells, most chloroplast and mitochondrial genes were somehow transferred from the organelles to the nucleus. Consequently, the proteins produced from these genes are synthesized in the cytoplasm and then transferred to the organelles. The residual DNA genomes of the organelles continue to produce mRNAs coding for some essential chloroplast and mitochondrial proteins. These are synthesized using protein synthesis machinery resembling that of prokaryotes, including smaller 70S ribosomes instead of the larger 80S ribosomes used to make cytoplasmic proteins. Thus, the full complement of proteins that carry out chloroplast and mitochondrial functions are a mixture of those made in the cytoplasm and others synthesized inside the organelles (Voet and Voet 2004).

Chua's work on *Chlamydomonas* ribosomes recapitulated Blobel's research as a graduate student and at Rockefeller characterizing cytoplasmic ribosomes, something not surprising given Blobel's association with the project (Chua, Blobel, and Siekevitz 1973). Chua's most important discovery was that chloroplast ribosomes bound to the thylakoid membranes and that this binding could be synchronized with the cell cycle (Chua et al. 1973, 1976). This observation was of obvious interest to Blobel, and the two of them published three papers on the phenomena together with Siekevitz and Palade (Chua, Blobel, and Siekevitz 1973; Chua et al. 1973, 1976). Chua believed that the thylakoid-bound ribosomes were likely synthesizing chloroplast membrane proteins. When Palade left Rockefeller as this work was being completed, Chua was offered an independent faculty position in the Laboratory of Cell Biology and decided to focus his research on proteins of the *Chlamydomonas* thylakoid membranes, about which little was known.

When Dobberstein began speaking with Chua, they discussed the question of how chloroplast proteins that are synthesized in the cytoplasm are transported into the chloroplast. The basic problem is analogous to the one addressed by the signal hypothesis: somehow proteins made in the cytoplasm have to find and cross a membrane boundary. In the case of chloroplasts (and mitochondria) the problem is even more complicated than in the ER because proteins must cross two membranes to reach the interior of the organelle. Dobberstein, who had worked on plant protein synthesis as a graduate student and in Blobel's lab had become adept at the isolation of mRNA, suggested that he and Chua look at proteins synthesized in vitro from *Chlamydomonas* cytoplasmic mRNA to see if some kind of precursor protein was involved in transport from the cytoplasm into the chloroplast. Chua thought that they should focus on the small subunit of the enzyme ribulose diphosphate carboxylase. Carboxylase is the enzyme in chloroplasts that fixes carbon dioxide by adding its carbon to the complex sugar ribulose diphosphate, converting it into the metabolic intermediate 3-phosophoglycerate (Voet and Voet 2004).[5] What is interesting about the biosynthesis of carboxylase is not only that the small subunit is made in the cytoplasm, but that a second, large subunit is synthesized inside the chloroplast using the chloroplast protein synthesis machinery. This means that the two subunits must assemble after the small subunit reaches the chloroplast interior.

---

5. Since this reaction is so widespread in plants, some feel that carboxylase is the most abundant protein in the world, if not the most important.

To begin their experiments, they isolated RNA from *Chlamydomonas* cells and enriched for cytoplasmic mRNAs using affinity chromatography to select out messages with polyA tails, a modification not believed to occur on chloroplast mRNAs. In addition, Chua used his previous experience with *Chlamydomonas* ribosomes to isolate free cytoplasmic polysomes from broken cells. When Dobberstein incubated both isolated mRNAs and the free polysomes in the wheat germ translation system, a variety of proteins were synthesized.

To pick out the carboxylase small subunit from the mixture of translation products, they immunoprecipitated it using a specific antibody against the small subunit made by Chua. The technique of immunoprecipitation was becoming more common at the time because it enabled the detection and isolation of proteins that are present in complex mixtures at low concentrations. For experiments using in vitro protein synthesis, immunoprecipitation eliminated the requirement that cells or tissues have large amounts of mRNAs coding for particular proteins of interest. This technical advance made it possible to examine the biosynthesis of practically any protein.[6]

When Dobberstein and Chua immunoprecipitated the small subunit from the wheat germ system and examined it by SDS-gel electrophoresis, they observed that a polypeptide synthesized from isolated cytoplasmic mRNA was larger than the authentic small subunit by 3,500 Da. They also observed that the small subunit was synthesized by isolated free polysomes. When they treated the larger form of the small subunit with a soluble *Chlamydomonas* extract, it was converted to the authentic size. A small peptide fragment likely corresponding to the extra segment seen in the larger form was also detected. From these experiments, they proposed that the larger form of the small subunit was a precursor of the authentic small subunit, and possibly related to transport of the small subunit into the chloroplast (Dobberstein, Blobel, and Chua 1977). While there were obvious parallels to the protein precursors involved in translocation across the ER membrane, this seemed distinctly different. First of all, cytoplasmic ribosomes were not known to bind to the outer chloroplast membrane like they bound to the

---

6. The immunoprecipitation procedure that Dobberstein and Chua used employed a primary antibody from rabbits specific for the carboxylase small subunit and a secondary antibody that binds the primary antibody. At a certain ratio of primary and secondary antibodies, the antibody complex precipitates out of solution, carrying the protein of interest with it (Dobberstein, Blobel, and Chua 1977). Shortly thereafter, an innovation using killed bacteria carrying an antibody-binding protein on its surface (called protein A) became commonplace. This made it much easier to perform immunoprecipitations because the bacteria replaced the secondary antibody and, because of their size, could be easily recovered by centrifugation (Kessler 1975).

ER. Indeed, the small subunit was made by free, not bound, polysomes. (The ribosomes that Chua had earlier found bound to the thylakoid membranes were clearly not involved since they were inside the chloroplast, while the small subunit was synthesized outside on cytoplasmic ribosomes.) Second, putative "processing" of the precursor occurred after translation was complete instead of co-translationally, as was observed during translocation of proteins into the ER. They were unable to say whether the enzyme that processed the precursor was originally found in the cytoplasm or inside the chloroplast because the chloroplast membranes ruptured whenever *Chlamydomonas* was broken open.

Blobel was taken with the results from Chua and Dobberstein. Although they did not strictly conform to the signal hypothesis, they seemed consistent with the general concept that precursor segments of proteins were involved in determining the proteins' final destinations in the cell. Dobberstein and Chua wrote up the results for publication with Chua as the senior (last) author and Blobel as the second author. The paper was communicated to the *PNAS* by Siekevitz in December 1976 and was published the following March (Dobberstein, Blobel, and Chua 1977). In the meantime, Blobel had been invited to present his work on the signal hypothesis in a plenary session at the annual meeting of the American Society for Cell Biology. The meeting, held in Boston in late 1976, was billed as an international conference because it was organized in conjunction with societies from other countries, and attendance was larger than usual. When Blobel gave his talk, he included the results on the putative precursor found in *Chlamydomonas*. It is not clear if Chua, who was not at the meeting, had given Blobel permission to present his unpublished results. In any case, when he learned about Blobel's talk he was livid. Although Chua's reputation was beginning to be established within the plant cell biology community, the discovery of a possible clue to the biogenesis of chloroplasts was a big deal, and Chua did not want it associated with Blobel.

The dispute with Blobel resulted in the end of Chua's previously friendly relationship with him. As with Steiner, Blobel had either naively intruded into territory that Chua considered his own or simply could not control his enthusiasm about the new results. After a number of arguments, Blobel backed off. Although Chua soon lost out on the race to prove that the small subunit precursor actually mediated transport into the chloroplast, his lab quickly caught up (Chua and Schmidt 1978; Highfield and Ellis 1978). Gregory Schmidt in Chua's lab published the sequence of the small subunit precursor segment, demonstrating that it was distinct from Blobel's signal sequences (Schmidt et al. 1979). Schmidt and Chua were, however, forced to do so with Blobel because his lab provided both the postdoc to do the sequencing and the sequencing machine. Despite this last collaboration, Chua's relationship with Blobel was permanently damaged. Within two years, Chua managed to obtain funding from the Monsanto Corporation to study gene expression[7] in plants and essentially left cell biology to establish his own independent plant molecular biology laboratory at Rockefeller. Blobel, relentless as ever, moved into the chloroplast transport field just as Chua left, with mixed success.

At about the same time that this was happening, Blobel got involved in a project analogous to the one on chloroplast biogenesis, the transport of cytoplasmically synthesized proteins into mitochondria. While attending a meeting in Europe in spring 1976, Blobel met and had dinner with Gottfried Schatz, a biochemist working at the new Biocenter associated with the University of Basel in Switzerland, and Schatz's graduate student, Maria Luisa Maccecchini. Schatz, an Austrian whose prolific research on mitochondria was well known, had recently returned to Europe after a number of years at Cornell University working in association with Efraim Racker, a noted mitochondrial expert (Schatz 2000).

Maccecchini had become interested in how proteins might cross the mitochondrial double membrane after reading Blobel's work on the signal hypothesis and proposed a PhD dissertation project on the subject to Schatz. She began by preparing antibodies directed against major subunits of the mitochondrial ATPase, the enzyme that produces much of the ATP in cells, that were known to be synthesized in the cytoplasm. Once she had the antibodies, she went to Cambridge University to learn the improved reticulocyte lysate translation system. Back in Basel, she translated yeast mRNA in the system but was unsuccessful in her attempts to immunoprecipitate the newly synthesized ATPase subunits with her antibodies. The main problem was that the antibodies precipitated multiple polypeptides, making it impossible to determine conclusively which ones were related to the ATPase subunits and which ones were nonspecific contaminants.

Before Maccecchini and Schatz met with Blobel, they had heard a rumor that Blobel was trying to set up a mitochondrial import project in his lab.[8] Blobel, who may have also heard about Maccechini's efforts, was no doubt

7. Gene *expression* refers to the transcription of DNA genes in the nucleus into mRNA and (usually) translation of that mRNA into a protein. As recombinant DNA procedures developed, the terminology was carried over when scientists began expressing genetically engineered DNA in cells. This is accomplished by artificially introducing the engineered DNA into cells, inducing the cells to transcribe mRNA from the engineered DNA and then synthesize the encoded protein.

8. In fact, Blobel had a new graduate student who was beginning to work on mitochondrial import, but his project never got off the ground and the student eventually left Rockefeller.

interested in interrogating her about their work. During the dinner, they learned that Blobel, whose lab had begun using immunoprecipitation as it expanded its search for different precursor proteins, had managed to greatly improve the technique.[9] He suggested that Maccecchini and Schatz send him the antibodies to see if he could get better results. Maccecchini insisted instead that she should come to New York with the antibodies and do the experiments in Blobel's lab herself. By summer 1976, she was there.

Maccecchini found both Blobel's busy lab and New York City invigorating, much different from relatively provincial Basel. Furthermore, once she tried Blobel's immunoprecipitation procedure, everything seemed to work, and she quickly demonstrated that three ATPase subunits were synthesized as putative precursors in vitro. Most significantly, she was able to use isolated, intact yeast mitochondria to show that the precursors were transported after synthesis into the mitochondria and were processed to their authentic mature size during transport. Schatz was delighted with the results and soon began urging Maccecchini to return to Basel to complete the work. Maccecchini resisted, and Schatz began to pressure her by intimating that her funding would be withdrawn. In the end, her visit, which was supposed to last three to six months, went on for a year, and she returned to Basel in fall 1977.

The paper reporting the results was communicated to *PNAS* on October 28, 1978, and published in January 1979 (Maccecchini et al. 1979). In May 1979, Maccecchini and Schatz submitted another, similar paper to the *Journal of Biological Chemistry* on the import of the mitochondrial enzyme cytochrome c peroxidase that was published that August (Maccecchini, Rudin, and Schatz 1979). Blobel was not a coauthor, and his previous assistance was not acknowledged, which he considered somewhat of an affront. Schatz, no doubt, believed that the mitochondrial import work was his, and he did not want to give Blobel a chance to get credit for the discovery of mitochondrial protein precursors or an opportunity to invade his territory.

Once again, Blobel did not back off, soon assigning a postdoc, Katsuyoshi Mihara, to a project investigating the import of mitochondrial cytochrome oxidase subunits.[10] In the 1980 paper reporting their first results,

---

9. Blobel's innovation was to treat the translated proteins with the strong, ionic detergent SDS and then boil them for three minutes before adding an excess of a nonionic detergent, Triton X100. The latter neutralized the SDS, permitting addition of the primary antibody without fear of denaturation. Addition of the SDS step often eliminated nonspecific binding of proteins during the immunoprecipitation.

10. In a well-known anecdote, Blobel is said to have told Mihara, who was competing alone against not only Schatz but also another large lab: "You are one and they are many. You must work day and night or you will be crushed like a cockroach."

Blobel referred to the extra segments of the mitochondria precursors as signal sequences, subtly looping the work with organelles together with his findings on ER translocation (Mihara and Blobel 1980). While he may have done this to indicate conceptual linkages between the two mechanisms, both of which used protein precursors to target proteins made in the cytoplasm to their cellular destinations, his terminology was interpreted by some other labs working on organelle biogenesis as an aggressive attempt to take credit for their discoveries. In Maccecchini and Schatz's second paper that did not include Blobel, they refer to the extra segment as a "leader sequence," and Chua, in his paper reporting the sequence of the carboxylase precursor segment, calls it a "transit sequence" (Maccecchini, Rudin, and Schatz 1979; Schmidt et al. 1979). By using these terms, both Schatz and Chua were likely staking out their own territories while resisting Blobel's attempt at occupation. Nevertheless, as with chloroplasts, Blobel's efforts to stay in the mitochondrial transport field eventually ended but not before his relationship with Schatz had been damaged.

Despite the conflicts resulting from the work on chloroplast and mitochondrial protein import, the discoveries of different types of precursor proteins and a distinct, post-translational mechanism to translocate proteins across organelle membranes were major advances. One could argue that Blobel's involvement in both projects was incidental and that he himself does not deserve much credit for the discoveries. Closer examination belies such an interpretation. In addition to committing his expertise, laboratory space, and, in the case of Dobberstein, a postdoc to the projects, it is clear that his work on the signal hypothesis motivated the search for chloroplast and mitochondrial precursors that mediated targeting to the organelles and facilitated their transport across the membranes. Both Blobel and Dobberstein contributed their thinking to Chua, and Maccecchini was stimulated to embark on her project by Blobel's papers. As for Blobel, what he took away from the experiences was a conceptual broadening of his ideas about how proteins are targeted or sorted to different parts of the cell, ideas that would lead to an important theoretical paper in 1980 (Blobel 1980) (see chap. 9). Unfortunately, however, what Blobel still failed to appreciate was the effect that his aggressiveness was slowly having on his reputation.

Conceptual relationships aside, extensive research over the next few years proved that transport of proteins into chloroplasts and mitochondria occurred by mechanisms distinct from those used to get proteins across the ER membrane. Not only was transport post-translational, but the components of the transport machinery were different, reflecting the evolution of a distinct system to accomplish the same goal of crossing membranous boundaries (Neupert and Herrmann 2007; Bölter and Soll 2016). In an unexpected development a few years later, the post-translational aspect of organelle transport turned out not to be unique. In some cases, transloca-tion across the ER membrane and across bacterial membranes occurs post-translationally (Rapoport 2007). While there are distinct features, in these cases the mechanisms turned out to be just a variation on Blobel's signal hy-pothesis and did not ultimately challenge his original findings (see chap. 8).

Under the Spotlight and into the Wilderness

By 1981, Blobel's horizontal research strategy had accomplished the goals of not only generalizing the signal hypothesis but also extending it concep-tually. In addition to hormones, membrane proteins, and components of mitochondria and chloroplasts, he and his collaborators reported studies on lysosomal hydrolases (which use the signal mechanism), peroxisomal en-zymes (which do not), the complex membrane protein opsin, a key retinal component of the visual system, and bacterial proteins (Goldman and Blobel 1978; Chang, Model, and Blobel 1979; Erickson and Blobel 1979; Erickson, Conner, and Blobel 1981; Goldman and Blobel 1981). In this period, a visit to the Blobel lab resembled a visit to the zoo, with multiple species and tis-sues on display. Not only could you encounter fish in the lab, not to speak of parts of rats, mice, and dogs, but removal of the lid from an ice bucket might disconcertingly reveal a dozen or two bovine eyeballs staring back at you.

It is easy to dismiss Blobel's efforts to generalize the signal hypothesis as so much dabbling. Excluding his involvement in studies of protein import into the chloroplast and mitochondria, it is certainly the case that his lab's work on a diverse group of proteins might seem diffuse. However, looked at from a different perspective, it is apparent that his in vitro reconstituted system to study protein translocation into the ER had reset expectations of what could be accomplished when studying the biosynthesis of proteins in the cell. Ironically, the ease with which his lab generated novel results with a variety of proteins somehow trivialized his accomplishments. Simply trans-lating different mRNAs in his reconstituted system rewrote what was known about the biosynthesis of whole classes of proteins, including hormones, lysosomal hydrolases, and membrane proteins. With Blobel's reconstituted system, generating novel findings about the biosynthesis of a wide array of proteins was relatively easy, and the impact of these results were anything but insignificant.

Blobel's success and aggressiveness in pursuing his research on the signal hypothesis had made him famous but had also inspired resentment and jealousy. In late May 1979, he and Sabatini organized a meeting of more than one hundred scientists at the Cold Spring Harbor Laboratory on Long Island on the subject of membrane biogenesis. A wide range of topics was included under this rubric, giving Blobel and his laboratory the opportunity to show off the breadth of their accomplishments. Over the five days of the conference, tensions began to build among participants who were unhappy that Blobel, one of the organizers, had arranged for his lab to present ten papers, many more than almost anyone else. This broke out in the midst of a session when Lodish vociferously complained about this evident bias in the program.[11] The situation was defused by Palade, who directly addressed Lodish, stating that Lodish would have done the same if he had enough presentable results. Nevertheless, it was clear that among some scientists in the fields represented at the meeting, an anti-Blobel cohort had emerged.

Blobel seemed to relish and even be motivated by the division into two camps: his lab and the rest of the scientific world. He viewed members of his lab and a few associates as part of his nuclear family; anyone else working on similar topics was treated with initial distrust (see fig. 31). The conflict had real and immediate consequences. In late 1979 or early 1980, Blobel was nominated for membership in the National Academy of Sciences of the United States, a singular honor and recognition of substantial accomplishments. Initial support for Blobel was high.[12] However, Bernard Davis, a prominent Harvard microbiologist, circulated a letter to a group of "personal acquaintances" who, like Davis, were members of the Academy. In the letter, Davis strongly objected to Blobel's nomination and indicated his intention to oppose it on the floor at the next meeting. Davis, who had presented a paper at the Cold Spring Harbor meeting and had just completed an extensive review of protein secretion across membranes for *Nature* (Davis and Tai 1980), decided that Blobel was getting too much credit for a discovery more properly attributed to Milstein on the basis of his 1972 paper in

11. Sabatini, the other organizer, also presented a number of papers from his lab at New York University, but the ire was mainly directed at Blobel. Participants may have viewed independent papers from scientists who had recently left his lab or collaborated with him as Blobel lab papers as well.

12. In an undated note apparently sent to Bernard Davis and perhaps others by Siekevitz, an Academy member and chairman of the Academy's Section of Cell and Developmental Biology, Siekevitz states that Blobel received thirty-one of thirty-seven votes in favor of his nomination out of a total of forty-seven eligible members of the section. While stating that he should remain neutral, given that he is at the same institution as Blobel (Rockefeller), Siekevitz does indicate some sympathy with arguments that Blobel's election to the Academy might be premature (Undated letter from Philip Siekevitz to Bernard Davis, Bernard Davis Papers, Folder 6, Box 9, Center for the History of Medicine, Harvard University. Used with permission.)

Nature New Biology. In the letter that he circulated he claimed that Milstein's work had been overlooked.

> In 1975 Blobel began an extensive series of publications, first employing Milstein's experimental system (immunoglobulin synthesis) and then extending it to other proteins. Unfortunately Milstein's key contribution was almost entirely ignored in the literature, and the mechanism of co-translational secretion, employing a signal sequence, had become almost universally cited as Blobel's model. . . . Blobel has added a good deal to our knowledge of the subject, though his contributions to a more detailed model have been largely speculative. . . . I am not questioning that the quality and the energy of Blobel's work will make him a promising future candidate for the Academy. But I am questioning whether he would have become a leading candidate so very rapidly if the literature on the subject were not dominated by a defective version of its history.[13]

Given that Blobel's nomination to the Academy preceded the discovery of SRP, a major novel contribution to understanding the mechanism of protein translocation across the ER, Davis's reservations about Blobel might seem to have merit. However, Davis ignored Blobel's development of the in vitro reconstituted system that enabled the Milstein suppositions to be definitively proven and his efforts to demonstrate that the signal hypothesis was a general mechanism. Also, if Milstein was, as Davis claimed, "ignored in the literature," it was not Blobel's fault. In the first of his 1975 papers, Blobel acknowledges Milstein's independent contribution, as well as another study by Schecter that provided a partial sequence of the N-terminal extra segment of the immunoglobulin light chain.

In follow-up correspondence with one of his "acquaintances" who had received the letter staking out his position against the Blobel nomination, Davis indicates that the Milstein issue is not all that is going on.

> [I] might not have sent the letter if there were not more. The inadequate recognition of the work of Milstein and of Schecter was part of a style that is continuing to the present day. As a specific example of an unusual imbalance in relating his own work to others, I would note a symposium on membranes that Blobel and Sabatini organized at Cold Spring Harbor last year. By now many people are working in this field, and there were a couple of hundred of

---

13. Bernard Davis to multiple individuals, "Lincoln's birthday 1980" (Bernard Davis Papers, Folder 6, Box 9, Center for the History of Medicine, Harvard University. Used with permission).

these present from all over the world. There was open expression of intense resentment that Blobel, and those associated with him[,] had a total of twenty 10-minute papers, while every other laboratory group, except for the other organizer[,] was allowed only one 10-minute slot.[14]

Davis's campaign to block Blobel's nomination in 1980 was successful, although a few years later he was elected to the Academy. Palade, who was certainly aware of Davis's campaign against Blobel and may have witnessed his challenge in person, was apparently very displeased with Davis's actions, and Davis later felt it necessary to mend bridges with both Palade and Blobel. It is not known if Blobel accepted Davis's apology. What is clear is that others continued to resent and oppose Blobel, and Blobel continued to tilt at the windmills of his perceived enemies.

14. Bernard Davis to Sanford Palay, February 28, 1980 (Bernard Davis Papers, Folder 6, Box 9, Center for the History of Medicine, Harvard University. Used with permission).

EIGHT

The Light at the End of the Tunnel

The Speculative and "Non-Assayable" Component

In the 1975 model of the signal hypothesis, Blobel proposed that nascent polypeptide chains traverse the ER membrane through a "tunnel" that also provides binding sites for ribosomes in the process of synthesizing secretory proteins on the cytoplasmic side of the membrane (Blobel and Dobberstein 1975a) (see fig. 24). He did not conceive of the tunnel as a permanent structure in the membrane. Instead, Blobel proposed that the tunnel assembled from membrane protein subunits in response to the appearance of the signal sequence in the neighborhood of the membrane and was stabilized by ribosome binding and subunit cross-linking.[1] Although Redman and Sabatini had earlier suggested that nascent chains penetrated the membrane through a "discontinuity," Blobel's proposal was much more concrete (Redman and Sabatini 1966).[2] Nevertheless, the existence of the tunnel was

---

1. As stated in the paper, "The signal sequence of the nascent chain emerging from within a tunnel in the large ribosomal subunit may dissociate one or several proteins which have been found to be associated with the large ribosomal subunit of free ribosomes. . . . Dissociation of these proteins may in turn uncover binding sites on the large ribosomal subunit. At the same time the emerging signal sequence also recruits two or more membrane receptor proteins and causes their loose association so as to form a tunnel in the membrane. . . . This association is stabilized by each of these membrane receptor proteins interacting with the exposed sites on the large ribosomal subunit, with the latter playing a role of a cross-linking agent. Binding of the ribosome would link the tunnel in the large ribosomal subunit with the newly formed tunnel in the membrane in continuity with the transmembrane space" (Blobel and Dobberstein 1975a, 848–49).

2. In a paper with Blobel in 1970, Sabatini mentions this again, stating this time, "In this tentative model, the space within the large ribosomal subunit (through which the attachment to the membrane occurs) communicates—through a permanent or intermittent discontinuity of the membrane—with the cisternal space or the microsomal cavity" (Sabatini and Blobel 1970, 156). In an accompanying figure, the discontinuity is illustrated as a simple hole through the membrane. A similar illustration appears in the written version of a symposium presentation (Blobel 1977).

completely speculative, and the absence of evidence for it or even a way to biochemically detect such a tunnel caused Palade to encourage Blobel to deemphasize it when he reviewed Blobel's 1975 papers (see chap. 5).

Blobel wrote that the tunnel proposal was informed by a number of "theoretical considerations," although he did not state explicitly what these are (Blobel and Dobberstein 1975a). Presumably, some are related to the problem of getting largely hydrophilic secretory proteins through the hydrophobic interior of the lipid bilayer, something that prima facia appears to be energetically unfavorable (see chap. 2; and see below, this chapter). Existence of a tunnel would circumvent this issue because the opening through the membrane created by the assembled subunits would be hydrophilic, resembling structures through which charged ions cross membranes, although these too were poorly defined at the time (Hille 1992, 236). While Blobel understood the problem posed by membrane hydrophobicity, he was anything but an expert on biological membranes.[3] At roughly the same time that the papers appeared, Blobel described his idea to the physical biochemist Jacqueline Reynolds after she presented a seminar at Rockefeller, asking her if it seemed plausible. Reynolds and Charles Tanford, another physical biochemist, had recently published a series of papers on the interactions of detergents with lipids, proteins, and membranes and were experts on the so-called hydrophobic effect responsible for membrane formation and stability (Makino, Reynolds, and Tanford 1973; Tanford 1974). Reynolds endorsed Blobel's idea, although it is likely that she did not give it detailed consideration, given that the conversation occurred at a bar before dinner.

Blobel was likely aware at the time that the signal sequences postulated to cause tunnel assembly were somewhat hydrophobic (Devillers-Thiery et al. 1975; Schechter et al. 1975). However, in his 1975 paper reporting homology between the signal sequences from several pancreatic secretory proteins, he does not invoke their hydrophobicity as a means to facilitate penetration of the lipid bilayer or, if he even considers this, is very vague about how it works. Instead, he emphasizes that the sequence homology

3. In her paper on visual representations of the signal hypothesis, Michelle LaBonte points out that inclusion of the tunnel in the proposal may have been influenced by the 1972 Singer and Nicolson fluid mosaic model of membranes, noting that Singer himself had visited Rockefeller in about 1971 (Singer and Nicolson 1972; LaBonte 2017). Many years later, in the midst of the controversy about the tunnel, Singer and his colleagues published a pair of papers supporting the idea of an aqueous tunnel for protein translocation (Singer, Maher, and Yaffe 1987a, 1987b) (see discussion below).

is consistent with a conserved mechanism for getting ribosomes with their nascent chains to bind to the ER membrane.

> The striking homology of the amino terminal sequence of pancreatic precursor proteins satisfies the most important criterion for the hypothetical signal sequence, which in the signal hypothesis . . . was postulated to constitute an essential but metabolically short-lived amino-terminal extension of all proteins to be segregated in membrane-bounded compartments. The preponderance of hydrophobic residues in this sequence may represent an essential feature for its proposed function, namely to provide the topological conditions for its transfer across the membrane by establishing a functional ribosome membrane junction. (Devillers-Thiery et al. 1975, 5020)

In this chapter, I describe the twenty-year search for the elusive tunnel. This effort was initially characterized by challenges not only to Blobel's tunnel idea but also to the signal hypothesis itself. As with other disputes occurring at the same time, Blobel viewed some of these as attacks on him from the forces arrayed around his familial laboratory compound, feelings that may have been, on occasion, justified. While Blobel's lab contributed to the ultimate discovery and characterization of the tunnel, the key findings were generated by others. In particular, the novel application of first bacterial and then yeast genetics to cell biological problems opened the door to discovery of the components of the tunnel. This was followed by sophisticated molecular and structural characterization of the tunnel that ultimately validated the signal hypothesis and provided an incredibly detailed look at the molecular mechanisms by which proteins cross the membrane boundary.

Who Needs a Tunnel Anyway?

Bill Wickner, who became well known for his counterproposal to the signal hypothesis, blazed a pathway at the beginning of his career that resembled in many respects the one that was later followed by Jim Rothman (see chap. 7). Like Rothman, Wickner was an undergraduate at Yale and began his professional education at Harvard Medical School. While there, he was, again like Rothman, attracted to lipid and membrane research through stints in Gene Kennedy's laboratory, experiences that ultimately caused both of them to lose interest in clinical medicine. While Rothman never received his Harvard medical degree, Wickner did, although his story is a bit complicated. Wickner's entry into medical school in 1967 coincided with the height of the Vietnam War, and if he, like Rothman, had left medical school for graduate studies, he likely would have been drafted into the military. To avoid this while still pursuing research, he convinced the medical school to grant him an extended leave to work with the famous biochemist Arthur Kornberg at Stanford University, finally receiving his MD degree in 1973 as the Vietnam War was winding down.

Kornberg had been awarded the Nobel Prize for his discovery of DNA polymerase. When Wickner entered Kornberg's lab, he began working on the bacteriophage M13, a virus that infects *Escherichia coli*, as a way of studying various aspects of DNA and RNA synthesis, a project consistent with Kornberg's general interests. M13 is a filamentous phage with a structure resembling a bottlebrush with bristles formed in part by a coat protein synthesized when the virus infects cells (Wickner et al. 1980). Wickner became intrigued with the coat protein because in infected *E. coli* it was found in the bacterial plasma membrane. Even though it was very small, only 5.2 kDa, its structure resembled that of a typical transmembrane protein with an acidic amino terminus, a basic carboxy terminus, and a central hydrophobic region. When the phage infects *E. coli*, coat proteins move from the virus particle into the inner bacterial membrane, and new copies synthesized from the phage genome during infection are also inserted in the bacterial membrane before assembling into new virus particles as they are released from the infected cell.

When Wickner moved to the University of California, Los Angeles, in 1976 for his first faculty position, he decided to focus on how the phage coat protein intercalates into the bacterial membrane. Although bacteria cells are very different from eukaryotic cells, the mechanistic challenges of getting proteins into lipid bilayer membranes are identical. Wickner believed that by studying the process in bacteria, he might also provide insights into what was happening in eukaryotes. At about this time, the Lodish and Blobel papers were published describing the insertion of the VSV G membrane protein into the ER membrane through a process consistent with Blobel's co-translational signal sequence–mediated mechanism (Katz et al. 1977). It was known that the M13 coat protein was initially synthesized as a precursor called procoat with an amino-terminal extension of twenty-three amino acids (Sugimoto et al. 1977). Although this resembled a signal sequence, there was a problem. When Wickner began to study the biosynthesis of procoat, he found that it was made on free ribosomes and only entered the membrane after synthesis, that is, post-translationally, with removal of the amino-terminal extension occurring shortly thereafter (Wickner et al. 1978; Ito, Mandel, and Wickner 1979). Even more surprising was his finding that procoat spontaneously inserted itself into lipid vesicles in the absence of other proteins (Wickner et al. 1978). Based on these observations, Wickner proposed an alternative mechanism to the signal hypothesis for insertion of membrane proteins that he called the "membrane trigger hypothesis" (Wickner 1979).

Wickner's idea was that assembly of proteins into membranes occurs when the lipid bilayer "triggers" a conformational change in the soluble protein precursor such that the protein inserts itself into the membrane after translation without what he called the need for "catalysis," referring to the enzymes driving protein synthesis during co-translational insertion: "The essence of the membrane trigger hypothesis is that the thermodynamics of protein folding governs membrane assembly with little intervention of catalysis. The leader peptide [the amino-terminal extension] is proposed to function by modifying the folding pathway where necessary. It may be proteolytically removed when assembly is complete to permit the protein to function or to drive the assembly reaction. In the latter function, the leader peptide may be viewed as a means of activating a protein for assembly" (Wickner 1979, 37).

Surprisingly, Wickner's proposal was published as the lead article in the 1979 *Annual Review of Biochemistry*. Normally, articles in this prestigious series critically review the literature relevant to well-established areas of research rather than introduce new and largely speculative models. Paul Boyer, a biochemist, future Nobel laureate, and member of the *Annual Review of Biochemistry* board, assembled the volume and wrote a brief introduction. Perhaps he saw an opportunity to promote Wickner, a junior faculty member in Boyer's Institute of Molecular Biology at UCLA, and invited or permitted him to publish his hypothesis in this high-profile venue. It certainly caused the article to receive considerable attention.

Although Wickner's proposal is designed to account for his observations with M13 procoat, he uses the opportunity afforded to him to call into question the relevance of the signal hypothesis to the biosynthesis of membrane proteins in general and even, to some extent, its validity with secretory proteins. He particularly challenges the notion expressed in Blobel's 1975 papers that a proteinaceous tunnel through the membrane is required for protein translocation. Wickner discusses several membrane proteins other than procoat, including complicated ones that span the lipid bilayer multiple times and even the VSV G protein, the protein that the Lodish and Blobel labs had already shown to follow the signal hypothesis (Wickner 1979). Other than G protein and to some extent procoat, biosynthesis of the other examples that Wickner mentions had not been rigorously examined in reconstituted in vitro systems, and his argument that they might conform to his membrane trigger idea seems to be that their biosynthesis might be difficult to explain by the signal hypothesis. He also states that certain proteins exhibit "conformation flexibility" because their solubility and ability to bind detergent changes after denaturation. These behaviors, however, are explainable by the known characteristics of protein folding and detergent binding and are not necessarily supportive of the membrane trigger hypothesis. In the case of the *E. coli* lipoprotein, one of his examples, Wickner points out that a mutation in the signal sequence reported by another lab does not prevent the protein from inserting in the membrane (Wickner 1979). This observation, though, is not inconsistent with either the signal hypothesis or the membrane trigger mechanism. For VSV G, he describes results from Rothman and Lodish indicating that membranes must be added early in the synthesis of G protein precursor to get proteolytic processing, glycosylation, and membrane "sequestration" (Rothman and Lodish 1977). On this basis, Wickner argues that "this striking result indicates that the nascent chain will fold into a conformation that does not enter the assembly pathway unless it interacts with membranes at an early stage of its synthesis" (Wickner 1979, 32). However, this finding would seem to support a requirement for co-translational membrane insertion, and it is not evident that membrane triggering would be a better alternative. Overall, Wickner's arguments are provocative but not particularly convincing.

Early in his article, Wickner drops the use of "signal sequence" in favor of "leader peptide." While it is possible that he believes that leader peptide is less prescriptive than signal sequence, he seems to forget that the signaling by the signal sequence has less to do with any particular membrane insertion mechanism than with the specificity of targeting to a particular membrane, an informational function. He mentions the targeting issue in an early section, asking, "How do membrane proteins make the correct membrane choice?," but then never explains how membrane triggering resolves this fundamental issue (Wickner 1979, 24). Perhaps Wickner's choice of terms has more to do with distinguishing his idea from that of Blobel's than any mechanistic consideration. Alternatively, he may have wanted to place himself in the emerging contra-Blobel camp of scientists. Wickner's lab had presented a paper at the May 1979 Cold Spring Harbor meeting dominated by Blobel's lab that had offended both Lodish and Davis (see chap. 7). Although Wickner's abstract for the meeting reports both in vivo and in vitro results on the biosynthesis of the M13 coat protein and promises a discussion "in light of current knowledge and speculation as to the mechanisms of membrane protein assembly," only the in vivo results were presented, and it is unclear if membrane triggering was mentioned. Perhaps he was hesitant to fully expose his ideas in such a setting.[4]

At the same time that Wickner published his hypothesis, Gunnar von Heijne in Stockholm provided a purely thermodynamic argument that generally supports Wickner's statement that a tunnel is unnecessary. Von Heijne, who worked in the Department of Physics at the Royal Institute of Technology, proposed what he called a "direct transfer mechanism" in which a hydrophobic signal sequence inserts into the lipid bilayer and the bound ribosome pushes subsequent hydrophilic residues through the membrane in the absence of a tunnel (von Heijne and Blomberg 1979). He buttressed his model with calculations of the free energy of transfer of protein segments into the membrane, concluding that direct transfer is energetically favorable. Although von Heijne did not suggest that transfer is post-translational, he did think that Wickner's finding that procoat inserts in pure lipid vesicles supported his proposal (von Heijne and Blomberg 1979). Two years later, in 1981, the structural biochemists Donald Engelman and Thomas Steitz at Yale also made a theoretical thermodynamic argument in favor of spontaneous insertion of proteins into membranes that they call the helical hairpin hypothesis (Engelman and Steitz 1981). As with von Heijne, their model did not invoke a post-translational mechanism but still dispensed with the necessity for an aqueous tunnel through the membrane, and they cited Wickner's results as consistent with their ideas.

The membrane trigger hypothesis continued to be discussed for many years after these early arguments in its favor. Along the way, Wickner discovered that procoat insertion required a membrane potential and isolated and characterized "leader peptidase," the enzyme responsible for removing the precursor segment, before the similar eukaryotic signal peptidase had been fully purified (Date, Goodman, and Wickner 1980; Zwizinski and Wickner 1980). Overall, Wickner's ideas opened up the discussion of protein

---

4. At roughly the same time, others were producing results challenging the signal hypothesis. According to Blobel, prior to a trip to MIT to present a seminar, his Rockefeller collaborator Peter Model burst into his office stating that "the shit has hit the fan." Model had just reviewed and accepted a paper from David Botstein, Blobel's host at MIT, indicating that the amino-terminal signal sequence was insufficient for secretion of the bacterial protein β-lactamase and that carboxy-terminal sequences were also essential. Botstein suggested that his findings might be consistent with Wickner's membrane trigger hypothesis (Koshland and Botstein 1980). Blobel believed that Botstein was going to spring the results on him prior to his seminar to orchestrate an attack on the signal hypothesis. After Blobel convinced Model to show him the manuscript, he quickly recognized a technical problem with the paper and was able to parry Botstein's challenge during his visit.

translocation, providing a foundation for interpretation of later results. Eventually, however, it became clear that the insertion mechanism of tiny procoat was an exceptional case (Stiegler, Dalbey, and Kuhn 2011), and, to paraphrase something that Blobel was famous for reciting to members of his lab, a beautiful hypothesis was killed by ugly facts when evidence for the channel began emerging. Ironically, Wickner later made major contributions to studies of the bacterial tunnel, or channel, as it came to be known, a structure that he had claimed was unnecessary (see below).

Genetic Strategies

The road to discovery of the channel begins with Jonathan Beckwith (fig. 30). Beckwith was a bacterial geneticist and a molecular biologist in the classical sense, although his work as a political activist eventually put him at odds with reductionistic and prescriptive aspects of molecular biology (Beckwith 1996, 2002). After undergraduate and graduate studies at Harvard University, Beckwith successively did postdoctoral research with Arthur Pardee at Berkeley and Princeton, Sydney Brenner at Cambridge, and François Jacob at the Institute Pasteur in Paris, all influential molecular biologists known for their work on the genetic code and the regulation of gene expression. While in Paris he developed early gene cloning techniques and back at Harvard as a junior faculty member used this approach in 1969 to isolate the DNA segment containing the E. coli lacZ gene, part of the lac operon made famous by Jacob and Jacques Monod. This was the first gene that had been cloned, and the result made Beckwith famous and then infamous as he used the attention his discovery garnered to warn about the dangers of genetic engineering (Beckwith 2002).

Although Beckwith's work on gene regulation brought him success, during a short sabbatical in Naples in 1970, he began to consider other research directions. His former mentor Jacob had begun advising young scientists that work on bacterial systems was old-fashioned. Beckwith, however, believed that genetics could be used to probe biological processes in E. coli other than gene expression. He had developed techniques to delete and recombine parts of genes as a way of identifying regulatory regions. Now, he wondered if the same approach could be used to study bacterial secretion (Beckwith 2002). Although bacteria do not have secretory organelles like the ER in eukaryotes, they do secrete proteins, and it was possible that mechanisms of secretion and even elements of the translocation apparatus might be conserved between prokaryotes and eukaryotes. Bacteria such as E. coli have both an inner membrane that is structurally identical to the plasma

30. Jonathan Beckwith, 2003. Photo courtesy of Jonathan Beckwith.
Original photo by Mark Ostow.

membrane in other cells and an outer membrane containing lipopolysac-
charides. Secretion occurs directly through the plasma membrane. Many
proteins secreted by bacteria end up in the periplasmic space between the
inner and outer membranes, whereas others are components of either the
inner plasma membrane or the outer membrane and cell wall.

Beckwith initially began studying the periplasmic enzyme alkaline phosphatase. He synthesized the protein in an in vitro translation system, observing that the primary translation product was a larger precursor that could be
processed to mature size by an enzyme apparently associated with the outer

membrane (Inouye and Beckwith 1977). The precursor protein was enzymatically active but seemed more hydrophobic than the processed form, a property possibly related to the precursor segment. At about the same time, Linda Randall at the Wallenberg Lab in Sweden reported that three other *E. coli* proteins synthesized on ribosomes bound to the cytoplasmic surface of the bacterial plasma membrane were also made as precursors in vitro, and other observations of bacterial secretory protein precursors also began to appear (Randall, Hardy, and Josefsson 1978; Randall, Josefsson, and Hardy 1978). The parallels between these findings and Blobel's signal hypothesis seemed apparent and were noted.

At the same time, Beckwith and a talented postdoctoral fellow, Thomas Silhavy, began using genetic techniques to fuse portions of the *lacZ* gene to segments of the *malF* gene (Silhavy et al. 1976; Silhavy et al. 1977). *LacZ* codes for β-galactosidase, a cytoplasmic enzyme required for degradation of the sugar lactose, and *malF* codes for a maltose (another sugar) transport protein that is localized to the inner membrane (Silhavy et al. 1976). The objective of these experiments was to determine which part of the membrane protein contained the information necessary for directing it to the inner membrane after its synthesis in the cytoplasm. They reasoned that if they fused the part of the membrane protein gene with the localization information to the β-galactosidase gene, then the resulting hybrid protein would be redirected from the cytoplasm to the membrane. Using this approach, they discovered that hybrid proteins produced from fusions composed of the amino-terminal part of the maltose transport protein and a carboxy-terminal portion of β-galactosidase ended up stuck in the inner bacterial membrane. This suggested to them that the localization information found in the maltose transport protein fragment that directed the hybrid protein to the membrane was in its amino-terminus, consistent with the location of signal sequences observed in eukaryotic secretory proteins (Silhavy et al. 1976; Silhavy et al. 1977).

Several years later, Beckwith came up with a strategy using bacteria expressing a similar gene fusion to try to identify mutants in the actual secretory apparatus, about which nothing was known. To do this he used a bacterial strain expressing a gene fusion between the amino-terminal part of a periplasmic secretory protein and β-galactosidase. These bacteria had very low β-galactosidase activity, presumably because the hybrid protein produced from the fused genes was partially secreted and the β-galactosidase activity of the fusion protein was trapped in an inactive state in the membrane (Oliver and Beckwith 1981). Because cytoplasmic β-galactosidase is required for the metabolism of lactose, the strain was unable to grow on lactose-containing medium. Beckwith reasoned that he could use this metabolic abnormality to select for mutants that retained the β-galactosidase activity of the hybrid protein in the cytoplasm. Such mutants might have defects in the translocation machinery affecting multiple secretory proteins. Taking advantage of the inability of his strain to metabolize lactose, Beckwith mutagenized the bacteria expressing the hybrid protein and selected for variants that could grow on lactose. He presumed that in such mutants the hybrid protein had failed to insert in the membrane, leaving the β-galactosidase activity in the cytoplasm. When he analyzed these mutants further, he found that they were unable to secrete several other periplasmic and outer membrane proteins, suggesting that the mutation was indeed not in the hybrid protein itself but in the secretory apparatus. Beckwith mapped this mutation to the *E. coli* genome and called the new gene that he found *secA* (Oliver and Beckwith 1981, 1982).

At about the same time, Scott Emr and Silhavy, who by now had his own laboratory, used an analogous approach to identify another gene that might code for a component of the bacterial secretory apparatus. In their case, they started with several mutants in the signal sequence of the λ receptor protein, an outer membrane protein, that failed to be exported from the cytoplasm. They then isolated mutants that corrected or *suppressed* the original signal sequence mutations such that the λ receptor was now secreted (Emr, HanleyWay, and Silhavy 1981). The suppressor mutations also restored the export of other proteins with similar signal sequence mutations. The mutations were in a gene that they called *prlA* because of its involvement in protein localization (Emr, Hanley-Way, and Silhavy 1981). Two years later, Koreaki Ito in Japan identified an *E. coli* mutant that failed to secrete and proteolytically process the precursors of several secretory proteins. When he mapped the mutation, which he called *secY*, to the bacterial genome, it appeared to be identical to *prlA* found by Emr and Silhavy (Ito et al. 1983). Overall, the genetic studies of bacterial secretion provided some of the first evidence that proteins are involved in the translocation process itself.

While the work on secretion in bacteria was under way, Randy Schekman at Berkeley was pioneering a novel genetic approach to the study of secretion in eukaryotes using the microorganism yeast. Schekman was, like Wickner, a product of the Kornberg lab at Stanford, and the two of them were close friends. Both left DNA-focused molecular biology to pursue questions in cell biology using unconventional organisms.

Schekman was motivated by the realization that for all of Palade's accomplishments in deciphering the secretory pathway in the pancreas, not a single molecule involved in the mechanism of secretion had been identified at the time he began his work in the mid-1970s (Schekman 2013). Schekman believed that he could find such molecules by isolating yeast mutants defective in various steps in the secretory process. The biology of yeast, chiefly baker's yeast, *Saccharomyces cerevisiae*, had been studied extensively since at least the nineteenth century because of its economic importance, and when Schekman began his work, the ability to manipulate yeast genetically was highly developed. Because yeast is a microorganism, many techniques for mutagenesis and selective growth on different combinations of metabolites had been adapted to yeast from bacterial procedures. Furthermore, yeast offered, from a genetic perspective, certain unique advantages such as the ability to grow as either haploid or diploid cells (Strathern, Jones, and Broach 1982). Even though yeast is a eukaryote with a nucleus and organelles such as mitochondria, some wondered if the process of secretion in yeast would be similar to that of the higher eukaryotes that had been studied by Palade and others, and Schekman's strategy was initially treated with some skepticism by funding agencies (Schekman 2013). Schekman, nevertheless, persevered.

With a graduate student, Peter Novick, Schekman set about producing a series of yeast mutants that would be impaired in secretion. While it was straightforward to introduce mutations in the yeast genome by treating cells with chemicals, the problem was selecting mutants that were specifically defective in secretion. Novick realized that mutants in steps of the secretory pathway after the actual synthesis of secretory proteins would not only accumulate concentrated secretory proteins inside their cells, but also would likely not divide, rendering the mutant cells denser than normal cells. On this basis he set up a way to enrich for denser cells by centrifugation (Novick, Field, and Schekman 1980). Using this approach, he and Schekman identified secretion mutants in twenty-three different genes. When they examined the mutant cells by electron microscopy, they observed distended secretory organelles, in patterns dependent on the particular mutation.[5]

Over the next few years, Schekman's lab began characterizing their mutants, trying to identify both mutated genes and the precise steps in secretion where the abnormal proteins functioned. Even after isolation of additional mutants, however, they realized that they had not obtained any that were defective in the initial steps in secretion, the targeting and translocation of proteins into the ER. As with the initial cohort, selection for such mutants was challenging.

---

5. In his Nobel Lecture in December 2013, Schekman notes that the suggestion to examine secretory mutants in the electron microscope came from Palade during a visit to Berkeley. He states that observing the enlarged vesicles in the first mutant examined was one of the most exciting experiences in his life (Schekman 2013).

In the mid-1980s, Raymond Deshaies joined Schekman's lab as a graduate student and came up with a most clever way to find translocation mutants. The logic of his approach resembled the fusion protein strategy originally used by the Beckwith lab for *E. coli*. Deshaies constructed a yeast strain in which the gene for an enzyme required for the synthesis of the amino acid histidine was fused to a signal sequence such that, when synthesized, the enzyme was misdirected from the cytoplasm and translocated into the ER lumen. Once there, the sequestered enzyme could not participate in histidine synthesis. Under these conditions, the yeast strain did not grow in medium lacking supplemental histidine (Deshaies and Schekman 1987). Deshaies reasoned, however, that in mutants unable to translocate proteins into the ER, the essential enzyme would remain in the cytoplasm and the mutant cells would grow without added histidine. Using this strategy, Deshaies soon isolated a mutant called *sec61* that was unable to translocate or process several different precursor proteins destined for the secretory pathway (Deshaies and Schekman 1987). Using the same approach, two other translocation mutants were soon isolated, *sec62* and *sec63*. When the SEC61 gene was eventually cloned and sequenced, it was found to be homologous to the bacterial *secY* gene (Stirling et al. 1992).[6]

Before proceeding to a description of the biochemical and structural characterization of the genes and gene products discovered by Beckwith, Silhavy, Emr, Ito, Deshaies, and Schekman, it is worth commenting on a particular characteristic of the genetic strategy. Even though mutations interfere with particular cellular functions, they are always expressed in fully formed cells. This means that the mutations are automatically viewed in a biological context. As stated by Beckwith in his 2002 memoir:

> We geneticists did not break open cells, extract proteins, and study reactions in the test tube to solve fundamental problems. Instead, we studied biological phenomena directly in the living bacteria. . . . Our path to understanding biological phenomena began with the isolation of mutations of an organism that were altered in the expression of some trait. We then "mapped" these mutations to locate them to a particular gene on a chromosome. We compared how the mutant and wild-type (nonmutant parent) organisms behaved with regard to this trait. These comparisons often led to new biological

---

6. One should keep in mind that recombinant DNA techniques and DNA sequencing, while advancing continuously in the 1980s, were still laborious and time consuming. It was not easy to go from the identification of a mutation to the cloning and sequencing of its gene, as illustrated by the gap between the discovery and sequencing of *SEC61*.

insights. . . . The important point for us geneticists is that we are looking at an in vivo phenomena—the behavior of the organism while it grows. (Beckwith 2002, 78–79)

If we reflect on the epistemic strategy developed by cell biologists such as Claude and Palade and carried to the molecular level by Blobel, an essential aspect is that the cellular context is never lost, even after cells are disrupted and functional analysis is reduced to the molecular level, because the whole cell is always a touchstone in the analysis. This is reminiscent of the way that the context of the bacterial cell is never lost with Beckwith's mutations because they are always studied in the living organism. In both instances, this preservation of context ensures that insights are automatically biological and not just molecular, making it possible to provide molecular explanations of *biological* phenomena. In this sense the work of geneticists in identifying mutations in cellular processes is akin to that of cell biologists. As we shall see, however, even geneticists had to resort to the use of a version of the reconstituted system assembled by Blobel, a product of the cell biologists' epistemic strategy, to figure out what their mutations ultimately explained. While Blobel did "break open cells, extract proteins, and study reactions in the test tube," an approach eschewed by Beckwith, microsomes, pieces of the cell, were always there, and that made all the difference (see chaps. 10 and 11).

Clues

Although Blobel had proposed the existence of an aqueous channel through the ER membrane in 1975, by the early 1980s, he not only had not provided any evidence for it, but really had not seriously looked for it at all unless you count the experiments leading to the discovery of SRP. According to the 1975 model, the channel not only recognized the signal sequence but also provided a ribosome binding site when assembled, although the details were vague. It is conceivable that when Dobberstein, Meyer, and Walter began extracting and proteolyzing microsomes to block translocation of nascent chains across the microsomal membrane, they might have found evidence for the channel. Instead, they found SRP and its receptor, elements of the translocation apparatus that were completely unexpected.[7]

7. That is, unless you want to go back to the original 1971 Blobel and Sabatini model that proposed a "binding factor" that circulated between the cytoplasm and the surface of the ER membrane, just like SRP (Blobel and Sabatini 1971b) (see fig. 21). Nobody, however, seemed to take the details of Blobel's 1971 "fantasy" that literally.

Despite the lack of evidence for the channel, Blobel adhered to the attitude that if there was no evidence against it, then there was no reason to jettison that element from the signal hypothesis. That went for other aspects of the model, including the requirement that translocation across the ER membrane is co-translational. These factors, perhaps, were reasons that Blobel bristled at the challenge presented by Wickner's membrane trigger hypothesis, which not only dispensed with the need for any translocation apparatus in the membrane but also suggested that co-translational translocation might be optional at best. It did not help that Blobel was competing directly with Wickner by studying the membrane insertion of a similar phage coat protein in collaboration with Peter Model, an established phage investigator at Rockefeller (see fig. 31). In contrast to Wickner's findings, their experiments, carried out by a postdoc, Chung Nan Chang, found membrane insertion of the coat protein to be co-translational, although it later turned out that a flaw in their experimental design led them to an incorrect conclusion (Wickner 1988). It was not that Blobel was rigidly dogmatic about the translocation mechanism, it was just that he needed convincing data to change his mind. Indeed, when experiments in his lab and those of Walter and Meyer demonstrated post-translational translocation in yeast, he apparently had no difficulty accepting the findings (see below). In Wickner's case, however, Blobel just did not believe his results.[8]

Before Reid Gilmore, who had worked on the SRP receptor, left Blobel's lab for a faculty position at the University of Massachusetts in 1985, he carried out experiments to chemically probe the interaction of the signal sequence and nascent chain with the ER membrane once translocation was initiated. His previous results suggested that the signal sequence is transferred to the membrane from SRP once SRP docks with its receptor (Gilmore and Blobel 1983). Using a wheat germ in vitro system containing the mRNA for preprolactin, SRP, and microsomes, Gilmore arrested translation on membrane-bound ribosomes after either the first 70 amino acids or the first 158 amino acids of the nascent chain had been synthesized. In the former case, this was sufficiently long that the entire signal sequence was exposed to the membrane, while in the latter case, enough of the polypeptide had been

8. For reasons that have been difficult to decipher, Wickner and Blobel continued to have a very contentious relationship, even after Wickner's early hypothesis was no longer an issue. Blobel resorted to insulting Wickner openly in meetings and blamed Wickner for rejection of one of his grants by the NIH. On his part, Wickner was most gracious to people from Blobel's lab when they met at meetings, making the dispute even harder to understand. Indeed, Wickner is credited with providing an antibody to Chang for the studies that contradicted his own findings (Chang, Model, and Blobel 1979).

synthesized to permit translocation and even signal sequence cleavage to commence. When he stripped the ribosomes, both short and long nascent chains remained attached to the membrane, and Gilmore could only release them by extracting with strong and hydrophilic protein perturbants. From this he concluded that the nascent chains were likely interacting with proteins and not only lipids during translocation (Gilmore and Blobel 1985). Although Gilmore's evidence was indirect because he had not identified any presumptive channel protein, his experiments were not consistent with the lipid-only models of Wickner, von Heijne, or Engelman and Steitz. Two years later, Jon Singer, who had developed the fluid mosaic model of biological membranes informed by thermodynamic considerations (Singer and Nicolson 1972) (see chap. 2), published a pair of theoretical papers arguing that an aqueous channel through the membrane made more thermodynamic sense than a lipid-mediated mechanism, further undermining the earlier proposals (Singer, Maher, and Yaffe 1987a, 1987b).

In the late 1980s Sandy Simon began a novel set of experiments in the Blobel lab designed to provide additional evidence for a channel through the ER membrane. Simon was a neurophysiologist who had studied calcium transport in synapses and believed that techniques used to detect ion channels could be adapted to microsomal and bacterial plasma membrane vesicles. The experiments he proposed to Blobel centered on the idea that a membrane channel large enough for the passage of a polypeptide chain might, under the proper circumstances, also carry ions and be detectable as an electrical conductance. The approach was controversial because one of the arguments *against* the existence of the protein channel was that it would collapse transmembrane electrical gradients by allowing ions to freely cross the membrane, something incompatible with living cells. Although Blobel was unfamiliar with electrophysiology, he agreed to let Simon go ahead with the experiments.

The experimental system set up by Simon was a planar lipid bilayer, a fairly standard technique among transport physiologists (Hille 1992, 223). To create the planar bilayer, Simon used a two-compartment chamber filled with an aqueous buffer and ion solution, with the two compartments separated by a teflon divider containing a tiny hole. When he injected lipids into one compartment, they spontaneously formed a single lipid bilayer across the hole, creating an electrically tight barrier. He then added membrane vesicles with presumptive translocation channels to one side of the bilayer, and they fused with the existing lipids such that any membrane proteins in the vesicles now spanned the artificial membrane separating the two chambers. Fusion was monitored by using electrodes on either side of the bilayer to detect increases in ion conductance that signaled the insertion of channels (Simon, Blobel, and Zimmerberg 1989).

In initial experiments, Simon fused pancreatic rough microsomes and inverted plasma membrane vesicles from *E. coli* to the bilayer and observed changes in conductance consistent with a 115-picosiemens (a measure of conductance) channel. While this work, which was published in 1989, demonstrated that Simon's experimental approach to look for a translocation channel was feasible, in reality it provided little to no evidence for such a channel (Simon, Blobel, and Zimmerberg 1989).

Subsequent experiments were different. After fusing microsomes with attached polysomes and their nascent polypeptide chains to the bilayer, Simon added puromycin, which causes the premature termination of protein synthesis (Yarmolinsky and De La Haba 1959). His idea was that this might permit translocation and release of the nascent chains through the microsomal membrane fused to the planar bilayer and open up the postulated channels for the movement of ions. If this happened, he expected to observe the appearance of an electrical conductance. If, on the other hand, translocation occurred directly through the hydrophobic core of the membrane, as proposed by Wickner and others, then no change in conductance would be expected. In fact, Simon observed that puromycin induced a large and sustained increase in conductance, consistent with the existence of an aqueous channel (Simon and Blobel 1991). When he extracted ribosomes from the membrane, the conductance disappeared. Simon believed that his results supported the idea of a translocation channel composed of proteins that was unable to conduct ions when filled with the nascent chain but could conduct ions when the nascent chain was released from the ribosome by puromycin and had passed through the membrane. As long as the ribosome remained attached to the membrane, ion flow was still possible, but the channel completely closed when the ribosome was removed. In a following paper, Simon went on to show that in *E. coli* membranes, addition of synthetic signal peptides in the absence of ongoing protein synthesis opened ion conducting channels. This suggested to him that the putative translocation channel was regulated by the signal sequence (Simon and Blobel 1992).

When Simon's findings were published, many investigators were willing to admit that the results were consistent with a translocation channel but considered the evidence less than completely convincing, because, like Gilmore's earlier findings, no channel protein had been identified. What Simon had observed were the predictable consequences of a channel, not the channel itself. Another factor may have been that the cell biologists and biochemists interested in the translocation problem were unable to appreciate or understand electrophysiological arguments. In any case, indirect findings like Simon's were about to be overtaken by other events.

Comrade in Arms

Earlier, in 1986, results appeared in *Nature* that signaled the arrival of Tom Rapoport as a significant scientific competitor in the search for the channel. Rapoport was a biochemist from East Germany (Deutsche Demokratische Republik [DDR]) who in the late 1970s had become interested in the mechanisms underlying the signal hypothesis. The paper from Rapoport's lab described a technique for chemically cross-linking the nascent chain to proteins encountered during its synthesis. The cross-linking reagent was in the form of a modified amino acid that could be incorporated into the growing polypeptide chain. When instantaneously activated by ultraviolet light during protein synthesis, the reagent covalently coupled the nascent polypeptide to any other nearby proteins. Rapoport added it to a translation mix containing SRP and activated it when translation was arrested, demonstrating that the signal sequence was cross-linked to the SRP 54 kDa subunit (Kurzchalia et al. 1986). While this was an interesting result, Rapoport clearly had another goal in mind, stating at the end of the paper, "This method may serve to identify the environment of a protein during its passage through a membrane."

Rapoport was by this time well known to the Blobel lab. His background was complicated. His father, Sam, who was also a biochemist, came from a Russian Jewish family that had moved to Vienna during the Russian Revolution, where he studied chemistry and medicine. While on a fellowship in Cincinnati in the late 1930s, the Nazis took over Austria, and Sam remained in the United States, where he married and Tom was born. Sam had become a member of the Communist Party in Vienna, and when the McCarthy anti-Communist campaign began in the United States in the 1950s, the Rapoports left for Austria and became citizens there. Unfortunately, because of the influence of the United States after the war, Sam was unable to work in Vienna or elsewhere in Western Europe. With the help of the Austrian Communist Party, he and his family moved to the DDR, where he joined Humboldt University in East Berlin (Rapoport 2010; Hoffmann 2017).

After receiving his PhD in 1972 in enzymology, Tom's initial work focused on mathematical modeling of metabolism. Shortly thereafter he moved from Humboldt University to the Institute of Molecular Biology of the DDR Academy of Sciences, also in East Berlin. His research director had been charged with the development of molecular biology in East Germany, and he asked Rapoport to isolate mRNA from carp with the goal of cloning the insulin gene.[9] When he translated polysomes containing insulin mRNA, he observed that the protein produced in vitro was larger than proinsulin. Rapoport soon met Blobel at a meeting and learned about the signal hypothesis. He realized that the larger form of proinsulin was likely a precursor responsible for guiding the protein through the ER membrane. In his paper submitted in July 1976 and published the following October, he referred to the larger form as preproinsulin, citing the brief Blobel and Sabatini symposium paper from 1971 (Rapoport et al. 1976). Because current journals only reached the DDR after a long delay, he had presumably not yet seen the 1975 issue of the *Journal of Cell Biology* announcing the signal hypothesis.

Conducting high-level research in the DDR was challenging. Research budgets were very small, and many reagents were unavailable and had to be ordered a year in advance from the West. To compensate, research institutes throughout the country agreed to prepare certain key reagents to share, and Rapoport made ^{35}S-methionine for the consortium every few months (Rapoport 2010). As an Austrian citizen, he had been treated with a certain amount of suspicion by the authorities, leading him to change his citizenship to that of the DDR in 1968. His motivation was partly political. Once he was a DDR citizen he joined the Socialist Unity Party of Germany (Sozialistische Einheitspartei Deutschlands [SED]), eventually becoming a party secretary of the institute. By all accounts, he was a true believer in the socialist state. Ten years after the switch, Rapoport was allowed to travel to the West, and he set about developing new contacts and taking advantage of his visits (Rapoport 2010; von Plato et al. 2013, 55; Hoffmann 2017).

Rapoport soon got to know Dobberstein in Heidelberg, and in 1982, he traveled to the United States. Blobel had invited him to New York, and Rapoport decided to use Blobel's lab as a kind of base of operations while he went around the country giving seminars (fig. 31). The routine he developed for his trips was to take the honoraria he received for his lectures and use them to purchase lab supplies, collecting everything in Blobel's lab until he could take it with him on the plane back to Europe. His first trip to New

9. Rapoport chose carp for the same reason that Dennis Shields in Blobel's lab used fish: the endocrine pancreas is separated from the exocrine pancreas, making isolation of the insulin-producing tissue and insulin mRNA much easier. Although both Rapoport and Shields worked on their respective projects at roughly the same time, there is no evidence that either was aware of the other, although Shields did cite the 1976 Rapoport paper when his paper on the insulin precursor was published.

31. "Blobelites" at a 1985 Gordon Conference. Left to right: Peter Walter, Keith Mostov, Peter Model, the author, Tom Rapoport, and Reid Gilmore. Seated on the floor: Larry Gerace. Walter, Mostov, and Gerace are Blobel's former graduate students, and Gilmore is a former postdoc. Model, a faculty member at Rockefeller, collaborated with Blobel on his studies of bacterial translocation. Rapoport was, at the time, a frequent visitor to the Blobel lab from East Germany. Photo by the author.

York coincided with Walter's discovery of the 7S RNA in SRP, and he was assigned space in the lab adjacent to Reid Gilmore. Although he did not carry out any significant experiments, he freely participated in discussions. As fellow Germans, both Blobel and Dobberstein considered scientists from the DDR poor cousins who were trying hard to do quality research. Rapoport certainly took advantage of this condescending attitude but viewed himself as an equal who was able to compete with the best of the West. He soon proved this to be the case.

Translocation: During or After?

In the mid-1980s, the question of whether translocation of secretory and membrane proteins occurs co-translationally or post-translationally in both prokaryotes and eukaryotes became a confused mess. As the search for the channel and other potential components of the translocation mechanism proceeded, clarification of this issue became more and more important.

The 1975 signal hypothesis was constructed around the co-translational mechanism, and strong evidence in favor of this had not only been provided by Blobel, but also earlier by Milstein (Milstein et al. 1972; Blobel and Dobberstein 1975a, 1975b). In the meantime, Blobel's lab had discovered SRP, which clearly played a role in co-translational translocation. Wickner, of course, had proposed a post-translational mechanism for the insertion of membrane proteins based mainly on his experiments with procoat. Other investigators of bacteria leaned initially toward a co-translational mechanism but gradually accepted the possibility that both may be occurring. Beckwith, for example, accepted Blobel's signal hypothesis as a starting point when he began observing precursors of bacterial secretory proteins, but as a geneticist, he was less committed to a particular mechanism, preferring to let secretory mutants lead the way (Inouye and Beckwith 1977). Linda Randall, one of the first biochemists to study bacterial secretion, reported in an early paper that exported proteins were synthesized on membrane-bound ribosomes, consistent with a co-translational mechanism (Randall and Hardy 1977). Soon afterward, she signaled a willingness to accept post-translational translocation after observing that the precursor of periplasmic arabinose-binding protein could be detected in vivo inside bacterial cells, something that could only occur if synthesis of the polypeptide was completed before secretion (Randall, Hardy, and Josefsson 1978). Part of the problem was that experimental systems capable of synthesizing bacterial proteins in vitro and translocating them into vesicles derived from the bacterial plasma membrane were underdeveloped relative to eukaryotic systems, and obtaining clear results was not easy (Muller and Blobel 1984).

At the same time, Schekman was busily collecting mutants in the secretory pathway of yeast, a eukaryote, and others had taken notice of his approach. Peter Walter accepted a job at the University of California, San Francisco, in part because the noted yeast geneticist Ira Herskowitz had recently joined its Department of Biochemistry and Biophysics. In Heidelberg, David Meyer realized that if yeast was going to be used to study translocation, then an in vitro system similar to that available for higher eukaryotes was needed and set about developing it with a postdoc, Jonathan Rothblatt (Rothblatt and Meyer 1986a). Blobel was also working on yeast in parallel. In November 1985, Walter submitted a paper reporting that a precursor of the yeast secretory protein α-factor is translocated post-translationally across the yeast microsomal membrane (Hansen, Garcia, and Walter 1986). The paper was revised in February 1986 and published in May in *Cell*. Walter was in communication with Blobel and notified him of their results. Within just five days in February, Blobel managed to push through the acceptance of a paper in the *Journal of Cell Biology* with similar findings, and it appeared the same month as Walter's (Waters and Blobel 1986). When Walter visited Meyer at the EMBL in winter 1985, Meyer learned of Walter's results and quickly reexamined his lab's findings on the translocation of the α-factor precursor, realizing that they supported a post-translational mechanism as well. Meyer then scrambled to put together a paper, managing to get a manuscript accepted by *EMBO Journal* within a week in February 1986 that also appeared in May (Rothblatt and Meyer 1986b). Walter was not pleased.

The yeast results were surprising and unexpected but in themselves provided little direct insight into the mechanisms of either co-translational or post-translational translocation. There were even more questions than before. Did co- and post-translational translocation occur in all organisms, or was post-translational translocation only a characteristic of the lower eukaryote yeast and bacteria? Did either require a channel? If so, was the type of channel different for the different processes? Resolution of these issues would depend on the biochemical characterization of the various secretory mutants that had accumulated in both prokaryotes and eukaryotes.

The Channel: Hiding in Plain Sight

By 1984, the Wickner laboratory had begun to study the translocation of proteins other than procoat in *E. coli*. Wickner's earlier results had shown that procoat could insert in lipid vesicles in the absence of other proteins with the exception of leader peptidase, consistent with his membrane trigger hypothesis (Wickner 1988). Leader peptidase, which Wickner originally discovered, is the enzyme in bacteria responsible for cleaving the precursor "leader sequence" from procoat and other exported proteins. Leader peptidase itself is an integral membrane protein without a cleavable leader (or signal) sequence, and its mechanism of insertion into the bacterial plasma membrane was of interest (Zwizinski and Wickner 1980; Wickner 1988). Despite Wickner's apparent preference for the membrane triggering mechanism, the export mutants discovered by Beckwith, Silhavy, Emr, and Ito were hard to ignore. After obtaining bacterial strains carrying the *secY* and *secA* mutations from Ito and Beckwith, he set about reproducing their earlier results. He quickly confirmed that the export of both the periplasmic maltose binding protein and the outer membrane protein OmpA was dependent on both SecA and SecY, as shown earlier by others (Wolfe, Rice, and Wickner 1985). When he examined the insertion of leader peptidase into the membrane, he determined that it too required both gene products (Wolfe, Rice, and Wickner 1985). Based on these results, Wickner began to realize that the bacterial export mechanism was more complex than he had imagined. He also noticed that there was an opportunity to investigate this mechanism biochemically since most previous work had been done using a genetic approach.

Over the next several years, Wickner and others made considerable progress in characterizing both the SecA and SecY proteins. SecA was determined to be an ATPase, consistent with a requirement for ATP hydrolysis during export, and a peripheral membrane protein loosely attached to the cytoplasmic face of the bacterial plasma membrane (Lill et al. 1989). SecY was found to be an integral membrane protein, with involvement in at least post-translational translocation (Bacallao et al. 1986). During the same period, Blobel continued to study bacterial secretion, managing to set up an improved in vitro system to study translocation (Muller and Blobel 1984). However, in contrast to the rest of the field, he claimed that SecY was not necessary for translocation (Watanabe, Nicchitta, and Blobel 1990; Watanabe and Blobel 1993). In 1990, Wickner succeeded in obtaining SecA-dependent translocation of the OmpA precursor into lipid vesicles that incorporated a complex of SecY and SecE (SecY/E), the latter a membrane protein that was a product of a gene discovered by Emr and Silhavy when they screened for suppressor mutants (Emr, Hanley-Way, and Silhavy 1981; Brundage et al. 1990; Driessen and Wickner 1990). This was clearly a major advance, but, to Wickner at least, it did not explain the exact role of SecY/E in the translocation process. At the end of his second paper, he summarized the outstanding questions.

> First, it might function with SecA as part of the proOmpA/SecB receptor; second, . . . the SecY/E protein might directly bind the precursor proteins after their release from SecA; third, the SecY/E protein might conduct protons and couple the proton flux to either the action of SecA itself or to work performed on the proOmpA molecule; and fourth, a major question will be whether the SecY/E protein serves as a pore [i.e., channel] to conduct proOmpA and other precursor proteins across the bilayer, or whether these proteins cross through the lipid phase per se. (Brundage et al. 1990, 654–55)

Clearly, Wickner was not quite ready to give up on all aspects of his membrane trigger hypothesis.

At about the same time, Christopher Nicchitta, a postdoc in the Blobel lab, initiated a strategy to identify the putative channel protein and other components of the translocation machinery in dog pancreatic rough microsomes. His approach was to dissolve translocation competent microsomes in the detergent sodium cholate, fractionate the detergent extract to separate the dissolved proteins, and then incorporate different combinations of proteins into lipid vesicles (Racker et al. 1979; Nicchitta and Blobel 1990). He accomplished this by mixing the detergent-protein complexes with purified lipids and then reducing the concentration of detergent by dialysis such that the hydrophobic membrane proteins were forced to associate with lipid vesicles instead of detergent, a process known as reconstitution.[10] He believed that if he could find a membrane protein or protein complex that was capable of translocating proteins into the lumen of the protein/lipid vesicles, then this would strongly suggest that one of the incorporated protein(s) was the elusive channel.[11]

Nicchitta's approach was unbiased in the sense that he did not initially try to purify any specific candidate proteins for reconstitution. For Blobel, just showing that a protein was required may have been enough at first because he was still obsessed with disproving Wickner's membrane trigger hypothesis that lipids alone are sufficient.[12] Although Schekman's yeast translocation mutants had suggested possible components of the translocation machinery, their genes had not yet been sequenced, and similar mammalian homologues were unknown. Furthermore, in yeast the predominant translocation mechanism appeared to be post-translational, and it is possible that the Blobel lab did not at the time consider the yeast translocation machinery that relevant to co-translational translocation in mammals.[13] In

10. Detergents dissolved in aqueous solutions exist as monomeric detergent molecules in equilibrium with aggregates called micelles. Micelle formation is driven when the monomeric detergent concentration becomes too high to be stable in aqueous solution because detergents have a hydrophilic "head" and a hydrophobic "tail." In micelles, the heads all face the aqueous solution and the tails are sequestered in the hydrophobic micelle interior. The concentration at which micelle formation occurs is characteristic of the detergent and is referred to as the critical micelle concentration, or CMC (Helenius and Simons 1975). Detergents such as octylglucoside are ideal for reconstition because their CMC is rather high, and the high monomeric detergent concentration makes them easier to remove by dialysis. Sodium cholate, the detergent used by Nicchitta, also has a high CMC. Detergent monomers can pass through the polymeric material used for dialysis, but micelles are too large.

11. Early on, Blobel's lab had competition on this project from both Peter Walter and Gert Kreibich (Sabatini's former postdoc at NYU), who were trying to do the same thing, although their efforts turned out to be limited (Yu et al. 1989; Zimmerman and Walter 1990).

12. In two of Nicchitta's three papers dealing with components of the translocation machinery and reconstitution, Wickner's original membrane trigger hypothesis paper and the theoretical papers in support of Wickner by von Heijne and Engleman and Steitz are cited, even though Wickner was at the same time beginning to have doubts that his proposal was valid in most instances (Nicchitta and Blobel 1989; Nicchitta, Migliaccio, and Blobel 1991).

13. The Walter lab had by this time identified yeast homologues of SRP, but co-translational translocation in yeast had not yet been demonstrated (Poritz et al. 1988; Hann, Poritz, and Walter 1989; Hann and Walter 1991).

any case, Nicchitta was making progress, reporting in 1991 that binding of precursors to the membrane and translocation could be uncoupled in vesicles reconstituted with partially purified protein preparations, thus narrowing the list of proteins that might be involved in the translocation process (Nicchitta, Migliaccio, and Blobel 1991).[14]

In the meantime, Rapoport pursued his cross-linking strategy, identifying other microsomal proteins by their proximity to the nascent chain that were candidate members of the translocation apparatus (Wiedmann et al. 1987; Wiedmann et al. 1989). His progress was occurring in the midst of huge changes in his professional circumstances as the Berlin wall was breached and the DDR became a thing of the past. By the early 1990s, his institute in Berlin had been rebranded by the unified German government and he was able to obtain vastly larger amounts of funding through the Deutsche Forschungsgemeinschaft and other Western European sources (Rapoport 2010).[15] As these changes were under way, however, Rapoport continued to exploit his connections to both Dobberstein and Blobel. In particular, one of his students, Dirk Görlich, visited Nicchitta in New York and questioned him about the details of his reconstitution strategy, something that intimated that Rapoport was considering a similar approach.

In 1992, Rapoport reported that, in yeast, Sec61 and Sec62, the latter another gene product implicated in translocation by Schekman, were cross-linked to nascent polypeptides as they moved through the membrane, suggesting a direct role in the process (Musch, Wiedmann, and Rapoport 1992). Simultaneously, Schekman's lab, using a different cross-linking approach, also showed that Sec61 was intimately involved in translocation (Sanders et al. 1992). Shortly thereafter, Görlich and Rapoport, taking advantage of the newly available sequence of yeast Sec61 (Stirling et al. 1992), demonstrated that a mammalian form of the protein was associated with nascent chains and the ribosome during translocation (Görlich, Prehn, et al. 1992). The

14. Nicchita's use of Efraim Racker's reconstitution strategy highlights Racker's reconstitution of mitochondrial ATP synthesis in lipid vesicles in the 1970s. Racker's work, combined with morphological studies of mitochondria by George Palade and Peter Mitchell's chemiosmotic hypothesis, demonstrates an early example of how the epistemic strategy of cell biology was applied to oxidative phosphorylation to yield molecular mechanisms, although, in this case, the work was distributed among different labs (Racker et al. 1979; Nicchitta and Blobel 1990; Weber 2005, 106–8; Matlin 2016; see also Grote 2019).

15. After the end of the DDR, Rapoport was criticized because of his involvement with the SED, the DDR Communist Party, and it became clear that his prospects for advancement in the newly reunified Berlin were limited. In 1995, he was recruited to Harvard Medical School. German institutions quickly had regrets and attempted to keep him, but to no avail (von Plato et al. 2013, 55; Hoffmann 2017).

sequence of Sec61 indicated that it was similar to SecY, a protein required for translocation in bacteria. Suddenly, the results from bacteria, yeast, and mammals began to converge.

Soon Görlich and Rapoport announced that they had obtained translocation of the secretory protein preprolactin into proteoliposomes consisting of only mammalian Sec61, in a complex with one other protein, the SRP receptor, and purified lipids (Görlich and Rapoport 1993). This was a stunning result, not only demonstrating the simplicity of the basic translocation machinery, but also strongly suggesting that the Sec61 complex was the channel protein. Other proteins, including the VSV G membrane protein, required a third component, a membrane protein called TRAM that had been previously identified by cross-linking (Görlich, Hartmann, et al. 1992).

Unlike Nicchitta, Görlich's starting point had been purification of the Sec61 complex, the SRP receptor, and TRAM (Görlich and Rapoport 1993). Görlich and Rapoport then reconstituted combinations of these into lipid vesicles until they were able to demonstrate the minimal requirements for translocation. Their reconstitution method appeared to be based on Nicchitta's, but they referred to it as "improved," mainly, so it seems, by the use of different but related detergents (Görlich, Hartmann, et al. 1992; Görlich and Rapoport 1993). These included a synthetic detergent called BigCHAP and digitonin, a plant derivative. Both are steroidlike detergents, as is the sodium cholate that Nicchitta used in his experiments. They note in their paper that cholate can be used instead of the others but fail to acknowledge that Nicchitta and Blobel had shared their reconstitution protocols using cholate with them (Görlich, Hartmann, et al. 1992; Görlich and Rapoport 1993). It is not clear whether they informed Nicchitta and Blobel of their results before publication. Blobel, in particular, was annoyed with this because of his long and substantial support of Rapoport over the years, and his previous friendship with him was affected negatively.

The reconstitution results were buttressed by several ancillary findings. Using a modified cross-linking strategy, Rapoport's group showed that the nascent chain is in contact with Sec61 continuously during its translocation across the membrane (Mothes, Prehn, and Rapoport 1994). Dobberstein, in collaboration with Rapoport, then demonstrated that two types of integral membrane proteins also contact Sec61 during their insertion, supporting the idea that both secretory and membrane proteins use the same translocation machinery (High et al. 1993). Finally, Arthur Johnson, using fluorescent probes incorporated into the polypeptide chain, provided evidence that the growing chain remains in an aqueous environment during its translocation (Crowley, Reinhart, and Johnson 1993). Altogether, these results strongly suggested that co-translational translocation occurs through an aqueous channel made up of Sec61.

Definitive demonstration that the Sec61 protein complex was the long-sought channel would depend ultimately on additional studies, including high-resolution structural determinations (see next section). However, the quest for the channel was effectively over. Although Blobel had lost the battle to identify the channel, he had won the war because his original vision articulated in the 1975 signal hypothesis turned out to be true. The basic mechanism was as he imagined it, and, with the recognition that the SecY complex, the bacterial homologue of the Sec61 complex,[16] was also a channel, the universality of essential aspects of the signal hypothesis was proven.

A Biological Mechanism Explained at the Molecular Level

Shortly after the close of the twentieth century, the mechanisms of protein translocation in eukaryotes and in bacteria had been worked out in great molecular detail (Rapoport 2007; Rapoport, Li, and Park 2017) (fig. 32). Co-translational translocation in eukaryotes is initiated by the emergence of a signal sequence from a ribosome translating an mRNA for a secretory protein or a transmembrane protein destined for the secretory pathway. SRP binds to the signal sequence and large ribosomal subunit, temporarily arresting synthesis of the nascent polypeptide chain. When the nascent chain complex encounters the cytoplasmic surface of the ER, SRP docks with the SRP receptor, releasing the signal sequence from SRP and SRP from the ribosome. SRP is then free to bind another signal sequence and direct that nascent chain complex to the ER. Once released by SRP, the signal sequence binds to the Sec61 protein complex in the ER membrane, which also provides a binding site for the ribosome. The association of the signal sequence with Sec61 opens the aqueous channel through the membrane and, as protein synthesis proceeds, the growing polypeptide is pushed by the ribosome through the membrane. In most cases, the signal sequence is cleaved by signal peptidase located on the lumenal side of the ER membrane. In addition to signal peptidase, other enzymes, such as the one that adds chains

---

16. The protein originally designated Sec61 (or sometimes Sec61p) was renamed Sec61α when it was recognized that Sec61 was associated with two other subunits, designated Sec61β and Sec61γ (Görlich and Rapoport 1993). The bacterial protein SecY is homologous to Sec61α. SecY also associates with two other proteins called SecE and SecG (Matlack, Mothes, and Rapoport 1998).

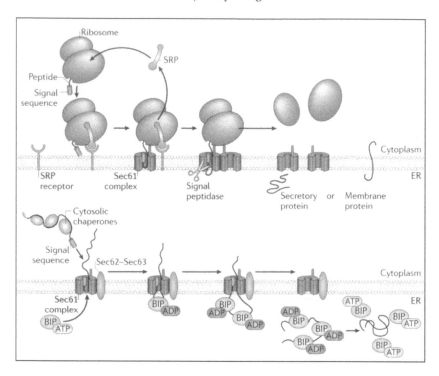

32. An overview of co-translational (upper panel) and post-translational (lower panel) translocation in eukaryotic cells. In the lower panel, BIP is a protein chaperone that helps pull the nascent polypeptide chain through the membrane through hydrolysis of ATP to ADP. See text for more details. Originally published as figure 1 in Matlin, K. S. 2011. *Nature Reviews Molecular Cell Biology* 12 (5): 333–40. © Karl S. Matlin. The illustrations were modified after figures in a review by Tom Rapoport (Rapoport 2007).

of sugars or oligosaccharides to the nascent polypeptide, are located at the lumenal exit site of Sec61.

Although the investigators studying prokaryotic secretion eventually believed that there was no co-translational mechanism in bacteria, the discovery of bacterial SRP by Dobberstein and Walter altered that view (Bernstein et al. 1989; Romisch et al. 1989; Poritz et al. 1990; Bassford et al. 1991). Translocation occurs through the SecY complex that is homologous to the Sec61 complex.

In post-translational translocation in eukaryotes, the new polypeptide is completely synthesized in the cytoplasm and released from the ribosome before translocation but remains associated with proteins known as chaperones that keep it from folding into a configuration incompatible with translocation (Rapoport 2007). Targeting of the polypeptide to the ER membrane occurs without the involvement of SRP and is dependent on binding of the signal sequence to Sec61. Translocation through the membrane proceeds through a Sec61 channel associated with two other membrane proteins, Sec62 and Sec63. Without a ribosome to push the polypeptide through the membrane, movement occurs by a process called ratcheting (Simon, Peskin, and Oster 1992; Nicchitta and Blobel 1993; Matlack et al. 1999). As the polypeptide moves in and out of the channel by Brownian motion, chaperones within the ER lumen bind to the protein and, through cycles of grabbing and releasing, pull the polypeptide through the membrane (see fig. 32).

In bacteria, post-translational translocation occurs through the SecY complex. A major difference with eukaryotes, however, is that the polypeptide is pushed through the bacterial plasma membrane from the cytoplasm by SecA and ATP hydrolysis (Rapoport 2007).

By the late 1990s, high-resolution molecular structures of the Sec61/SecY channels began to appear, culminating by 2009 in an X ray structure of the SecY complex at 3.2Å resolution (sufficient to see the folding of individual polypeptide chains), and a subnanometer resolution electron microscopic structure of the yeast and mammalian channels in association with a translating ribosome (Menetret et al. 2000; Beckmann et al. 2001; Van den Berg et al. 2004; Becker et al. 2009). From these, the functioning of the channel became clear. In the absence of a translocating polypeptide chain, the channel is closed on its distal end by a protein plug, keeping ions from leaking across the membrane. Even during translocation, the polypeptide chain is closely associated with Sec61 amino acids in the channel such that the membrane barrier is not compromised. The Sec61 structure also provides a final answer to the mystery of transmembrane protein translocation. Unlike secretory proteins, membrane proteins stop translocation when the hydrophobic transmembrane segment of the protein has been inserted into the channel. Based on cross-linking and other data, investigators had hypothesized that the channel somehow opens sideways in the plane of the membrane to allow the transmembrane segment to move out of the channel and into the lipid bilayer (Martoglio et al. 1995). From its structure, it appears that this is correct; the channel can indeed open laterally like a clamshell, allowing the transmembrane segment to diffuse into the membrane.

The descriptions of the mechanisms of secretory and membrane protein targeting and translocation at the molecular level are the culmination of the epistemic strategy devised initially by Albert Claude, developed further by George Palade and Philip Siekevitz, and then brought to the molecular level by Blobel. Their laboratories were not the only ones that contributed to these achievements, but their work provided the pathway that findings from the larger scientific community followed to reach the final molecular destination. Claude and in particular Palade relied on an integrated strategy that coupled electron microscopy with cell fractionation and biochemical analysis. By relating morphological parts of the cell to biochemical activities, they were able to determine the overall role of the ER in the first stages of the secretory process. Blobel then set the stage for discovery of the molecules that carry this out by devising an in vitro reconstituted system coupling protein synthesis to targeting to and translocation through the microsomal membrane. The key element in this system is the inclusion of microsomal vesicles because these provided the cellular context that enabled molecular events to be linked to the ER in the cell. In this way, the biochemical activities of SRP and its receptor, Sec61, and signal peptidase were directly connected to a biological function: secretion.

Blobel in Reflection

In 1999, Günter Blobel was awarded the Nobel Prize in Physiology or Medicine for the "discovery that 'proteins have intrinsic signals that govern their transport and localization in the cell' " ("The Nobel Prize in Physiology or Medicine 1999" 1999). As the sole recipient, the Nobel Assembly recognized that, despite the substantial contributions of Beckwith, Dobberstein, Gilmore, Meyer, Rapoport, Sabatini, Schekman, Walter, and many others, Blobel's vision, insights, and, particularly, his creation of the reconstituted system drove the discoveries from beginning to end. An editorial in *Science* announcing the prize was titled "Colleagues Say 'Amen' to This Year's Choices," and quotes from Schekman and Rothman suggested that it was long expected (Hagmann 1999).

There is no doubt that Blobel, who died in 2018, was a charismatic and complex person. At a 2019 memorial symposium in his honor, former colleagues often mentioned his passion and infectiously positive outlook. His focus was always on his work, and the limited separation between his personal and professional worlds sometimes got him into trouble with other scientists. He was not, however, one-sided. There was precision and rigor to his thinking and the execution of his experiments. But there was also an emotional side that was the source of his imagination and fantasies such as the 1971 predecessor of the signal hypothesis. He was a well-known romantic, with a love for classical music and Goethe's poetry acquired as a youth that never diminished during his lifetime. He was German but lived

33. The Frauenkirche in Dresden, ca. 2005. The original church was built in
the eighteenth century and destroyed during the firebombing of Dresden in 1945.
It was rebuilt in part with the proceeds of Günter Blobel's Nobel Prize and
contributions from around the world. Photograph by the author.

part-time in Italy. He used his Nobel Prize money to help rebuild the Frauenkirche in Dresden that was destroyed soon after he first saw it at the end of World War II, not only as a memorial to the tragedy that befell the city and a sister who died in the war, but also to restore a beautiful object and source of beautiful music from Bach's era (Ezzell 2000) (fig. 33). Blobel's two sides, precision and passion, were combined seamlessly. They were counterpoint, as described by one of his best friends, two independent melodies united into a harmonic whole.

The discovery of the channel and description of the molecular mechanisms of targeting and translocation to the secretory pathway are not the end of the story. There are implications of the work of Blobel and others on the signal hypothesis that extend to the present. Blobel's Nobel Prize was given not just for the signal hypothesis but also for the overall idea of "intrinsic signals" that determine where proteins, almost all of which are made in the cytoplasm, end up in their ultimate locations and final asymmetric dispositions in the cell, something Blobel referred to as protein topogenesis. Blobel's in vitro system containing a fragment of the ER spawned other systems to investigate a diverse array of cellular processes at the molecular level. And, finally, as I will claim, the epistemic strategy developed and exploited by Blobel and his predecessors that preserves the cellular context and hence is able to investigate biological phenomena at the molecular level continues to this day even while incorporating modern technical approaches. These subjects form the final part of the book.

PART THREE

Form Redux

NINE

Topogenesis and Spatial Information

The Future

The discoveries that led to a detailed understanding of the molecular mechanisms of targeting and translocation as described in previous chapters have continued, yielding findings that go beyond Blobel's original ideas. We now know, for example, that there are other ways of inserting proteins synthesized on free ribosomes into membranes that do not depend on Sec61, and even that ER bound ribosomes synthesize many cytoplasmic proteins in addition to those destined for the secretory pathway (Hegde and Keenan 2011; Reid and Nicchitta 2015; Borgese et al. 2019, 105). Beyond the ER, cell biologists understand much more about how proteins move to the Golgi complex and other parts of the secretory pathway via vesicular carriers and reenter the cell from the plasma membrane through a related vesicle-mediated process called endocytosis (Rothman and Wieland 1996; Schmid, Sorkin, and Zerial 2014). Cell biologists have now also learned much more about how proteins are imported into mitochondria and chloroplasts and how they reach subcompartments within the organelles (Fulgosi, Soll, and Inaba-Sulpice 2004; Mokranjac and Neupert 2009). These continued advances are part of the normal flow of science. Aside from any particular sets of discoveries, however, there are longer-term conceptual, technical, and strategic implications of the work of Blobel, his predecessors, and his contemporaries.

The core idea of the signal hypothesis is that a segment of the amino acid sequence of a protein, the signal, directs or *targets* the protein to the ER and that this sequence most often has little to do with the function of the protein once it reaches its destination (Blobel and Dobberstein 1975a). Blobel and his collaborators extended this basic concept in a slightly different form to mitochondria and chloroplasts. Although such sequences are encoded in the genome like any other protein sequence, their expression is elaborated spatially within the three-dimensional form of the cell. That is, without the cell, these sequences have no meaning. Indeed, the only way that the functions of these sequences were determined was through the use of in vitro assays that duplicated spatial events in the cell, the transfer of a protein from one cellular compartment, the cytoplasm, to others, the ER lumen or the interiors of mitochondria and chloroplasts (Blobel and Dobberstein 1975b; Chua and Schmidt 1978; Highfield and Ellis 1978; Maccecchini et al. 1979).

The use of in vitro or cell-free assays to investigate protein targeting was Blobel's extension to the molecular realm of the epistemic strategy devised originally by Claude and developed further by Palade. In all cases the result if not the goal of this strategy was to preserve the cellular context while asking questions that required the disruption of the cell to provide mechanistic answers at molecular resolution. This approach was a response to the dilemma that prevented cytologists in the early part of the twentieth century from breaking open cells to study their biochemistry, as well as a retort to biochemists who often ignored cell organization as they sought understanding of complex metabolic reactions (see chap. 1).

In the final three chapters of this book I explore the implications of Blobel's work in more detail. First, in this chapter, I describe Blobel's prediction in 1980 that signals like the ones that target proteins to the ER, mitochondria, and chloroplasts would be found for virtually every cellular destination (Blobel 1980). This idea, through the work of many other laboratories, turned out to be true. I then show how these signals map out spatial relationships in the cell with functional implications at both cellular and organismic dimensions. In chapter 10, I look at other Blobel-inspired cell-free assays that different laboratories used to probe mechanistic details of vesicular transport in the secretory pathway beyond the ER, endocytosis, and protein transport from the cytoplasm to the interior of the nucleus. I also show how these assays relate to Blobel's in vitro reconstituted system and describe why they conform to essential aspects of Claude and Palade's epistemic strategy. Finally, in chapter 11, I examine the epistemic strategy itself more closely. I argue that its strength is not only its integration of morphology and biochemical analysis but also its cyclic and recursive characteristics that permit functions to be examined at the molecular level without sacrificing the structural context of the whole cell. I also show how the strategy continues to be a powerful means to seek molecular explanations of biological phenomena, even after substitution of contemporary techniques for traditional approaches.

Topogenic Sequences

In 1977, the Blobel lab was in the midst of a variety of projects to extend and generalize the signal hypothesis. Work on the VSV G protein and secretory proteins from endocrine tissues was under way to determine if the signal hypothesis applied broadly to proteins synthesized by ER bound ribosomes (Katz et al. 1977; Shields and Blobel 1977) (see chap. 7). In an unexpected development, Blobel's collaborators at Rockefeller had discovered that a chloroplast protein was synthesized as a precursor with an amino-terminal extension that appeared to target the protein to the chloroplast after translation, and a new collaboration with the Schatz laboratory looking at mitochondrial proteins synthesized in the cytoplasm had also been initiated (Dobberstein, Blobel, and Chua 1977; Maccecchini et al. 1979). Blobel was also curious about the biosynthesis of proteins destined for two other organelles, the lysosome and peroxisome, and assigned new postdocs, Ann Erikson and Barbara Goldman, to the projects (Goldman and Blobel 1978; Erickson and Blobel 1979). Lysosomes are where cellular waste products, such as damaged proteins and other substances, are degraded by a complex mixture of proteases and other enzymes capable of breaking down almost any large molecule found in the cell or brought into the cell from the outside. The peroxisome is a center of oxygen metabolism responsible for neutralizing dangerous oxygen radicals, among other things (Alberts et al. 2015). At the time it was known that lysosomal enzymes, generically called hydrolases, are synthesized by ER bound ribosomes. Much less was known about the peroxisome, but the prevailing theory was that peroxisomes originated by budding from the ER, and presumably their enzymes were produced there as well.

In the context of all these experiments, it occurred to Blobel that if the diverse set of proteins synthesized on ER bound ribosomes had different *final* destinations in the cell, then there must be some way to target them there. In a paper published for a 1977 symposium, Blobel summarizes the problem: "If there is a common signal peptide for secretory, lysosomal, and peroxisomal proteins then these three groups of proteins would occur mixed within the cisternae of the RER and subsequent *sorting* would be required" (Blobel 1978, 104; emphasis added). In the same paper, he generalizes this sorting concept further, including the precursor segment on chloroplast proteins as a kind of sorting signal to the chloroplast and predicting that mitochondrial proteins might be similar. He also suggests that nuclear proteins, all of which are synthesized in the cytoplasm, might also have sequences that lead to their import into the nucleus. Transport into the nucleus does not require traversal of membranes but instead occurs through huge pores that penetrate the double membrane surrounding the nucleus, providing a pathway for communication with the cytoplasm. Blobel suggests that the pores might act in a "sorting capacity" to selectively pick out nuclear proteins with the correct signal from the cytoplasm (Blobel 1978, 105).

The following year, Blobel developed these ideas further for another symposium, now focusing on membranes and omitting discussion of the nucleus (Blobel et al. 1979). In a section titled "Sorting" near the end of the paper published in the symposium volume, Blobel is more explicit than before.

> Following translocation across membranes by signal sequences and orientation within membranes by both signal sequences and stop-transfer sequences, many proteins have not yet completed their intracellular pathway. For example, both secretory and lysosomal proteins . . . need to be separated within the lumen of the ER: many bitopic [transmembrane] integral membrane proteins that are biosynthetically inserted into the RER . . . by the co-translational formula need to be routed from there to various other cellular membranes. . . .
> We propose that the information for sorting resides in specific 'sorting' sequences. By analogy to signal and stop-transfer sequences it is likely that there are only a few group-specific sorting sequences that are common to a large number of proteins. (Blobel et al. 1979, 30)

He collectively refers to these sequences as "topological" because they determine where proteins are directed in the cell and, in the case of transmembrane proteins, how those proteins are oriented in the plane of the membrane.

In 1980, Blobel consolidated these ideas in a brief theoretical paper titled "Intracellular Protein Topogenesis" (Blobel 1980). As before, he extends the key concept of the signal hypothesis, the signal sequence, to account for the specific spatial distribution of proteins throughout the cell. At the same time, he also highlights the topographic relationships among membrane compartments in eukaryotic cells and proposes a scheme for their evolution.

The spatial organization of eukaryotic cells is complex (Alberts et al. 2015) (fig. 34). The cytoplasm is enclosed by the plasma membrane and filled with membrane-bounded organelles. The secretory pathway, as defined by Palade, consists of the ER, the Golgi complex, and vesicles for storing secretory proteins. Proteins translocated across the ER membrane proceed sequentially through these compartments. Almost all eukaryotic

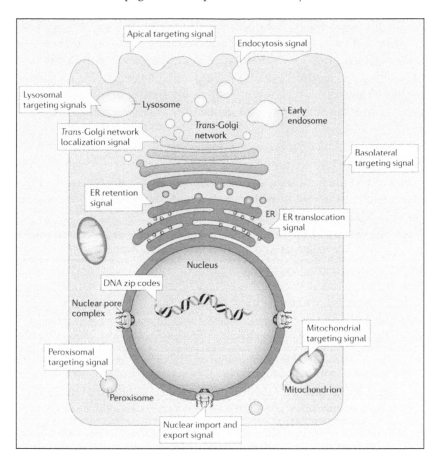

34. A simplified topogenic map of the cell indicating signals for different
intracellular destinations. Originally published as box 1 in Matlin, K. S. 2011.
Nature Reviews Molecular Cell Biology 12 (5): 333–40. © Karl S. Matlin.

cells also contain the vesicular organelles, lysosomes, and peroxisomes, as
well as mitochondria. Plant cells also contain chloroplasts. The organelles
of the secretory pathway, as well as lysosomes and peroxisomes, are demar-
cated by a single membrane, and, in 1980, it was generally believed that all
of these organelles communicated through transport vesicles that budded
from one organelle and then fused with the target organelle. Mitochondria
and chloroplasts are both surrounded by double membranes and do not
communicate through vesicles with other cellular organelles. As described
previously, the nucleus has a double membrane penetrated by nuclear pores
that is an outgrowth and specialization of the ER membrane.

The membrane-bounded organelles of the secretory pathway and the lysosome define a topography in which the inside of the organelles is spatially equivalent to the outside of the cell. Another way to express this is to say that once a protein crosses from the cytoplasm through the membrane of the ER by translocation, it can theoretically reach the outside of the cell without crossing another membrane because the transport vesicle membranes fuse with the plasma membrane and their contents are released to the outside as the two membranes become continuous (Palade 1975; Alberts et al. 2015). In contrast, the interior of the nucleus is topographically equivalent to the cytoplasm since substances can move between the cytoplasm and the interior of the nucleus through the pores without crossing the lipid bilayer of a membrane.

In his paper, Blobel calls the signals that designate protein locations in the cell *topogenic sequences* and refers to the process by which proteins are sorted from each other as *protein topogenesis*.

It is hypothesized here that the information for these processes, termed "protein topogenesis," is encoded in discrete "topogenic" sequences that constitute a permanent or transient part of the polypeptide chain. The repertoire of distinct topogenic sequences is predicted to be relatively small because many different proteins would be topologically equivalent—i.e., targeted to the same intracellular address. The information content of topogenic sequences would be decoded and processed by distinct effectors. Four types of topogenic sequences could be distinguished: signal sequences, stop-transfer sequences, sorting sequences, and insertion sequences. (Blobel 1980, 1496)

Blobel designates *signal sequences* as any amino acid sequence that results in the translocation of a protein bearing that sequence across a membrane, lumping together both ER signal sequences and the precursor sequences found on mitochondrial and chloroplast proteins. *Stop-transfer sequences* refers to the hydrophobic parts of transmembrane proteins that end up intercalated in the interior of the lipid bilayer (Blobel 1980). In Blobel's formulation, as these sequences are synthesized and enter the membrane, they cause translocation across the membrane (transfer) to stop and the putative (at the time) translocation channel to release the proteins into the lipid bilayer. Blobel labels the membranes of the ER, mitochondria, chloroplasts, and peroxisomes "translocation-competent" because they include machinery such as the postulated ER translocation channel that facilitates the movement of proteins across their membranes (Blobel 1980). Blobel has now included peroxisomes in this category because his postdoc Barbara Goldman had recently demonstrated that peroxisomal enzymes are synthesized on free ribosomes and thought that they likely entered peroxisomes directly through the membrane after translation (Goldman and Blobel 1978). The membranes of other organelles along the secretory pathway such as the Golgi complex, as well as the lysosome, are then translocation-incompetent because they lack translocation machinery and must receive their membrane and content proteins by vesicular transport from the ER. According to Blobel, these proteins must have *sorting sequences* that serve as addresses to their final destinations. The last category, *insertion sequences*, are reserved for membrane proteins that spontaneously insert in membranes from the cytoplasm, without the requirement for a translocation apparatus (Blobel 1980; Hegde and Keenan 2011; Borgese et al. 2019).[1]

As implied by experiments that had already been carried out in Blobel's lab, the various informational sequences can be combined in a single protein. Thus, the viral membrane protein VSV G has a signal sequence to initiate its translocation across the ER, a stop-transfer sequence to keep it from completely crossing the membrane, and, presumably, some kind of sorting sequence to target it ultimately to the plasma membrane where it can participate in virus assembly. Similarly, lysosomal hydrolases have signal sequences to get them into the ER lumen and, according to Blobel's proposal, must also have sorting sequences designating the lysosome as their final destination.

The ability of topogenic sequences to appropriately target proteins to cellular destinations implies, as Blobel says, that there are "distinct effectors" that can "decode and process" the information found in the sequences (Blobel 1980). By this, he means that something in the cell must recognize, for example, a sorting signal on a plasma membrane protein, facilitate its transport there, and make sure that it remains at the plasma membrane and does not then move somewhere else. Although Blobel does not explicitly say so, his idea presupposes the existence of a spatially organized cell with sorting machinery already installed in the correct locations (fig. 34). This machinery must have the capacity to not only read the information contained in each topogenic sequence but also transport the protein bearing this information

---

1. These are known as "tail-anchored" membrane proteins because the hydrophobic sequence that inserts into the lipid bilayer is typically found at the carboxy terminus of the polypeptide. One protein of this type that was known at the time that Blobel was formulating his ideas was cytochrome b5. It is now known that this class of proteins is widespread and uses the so-called GET pathway for insertion (Borgese et al. 2019).

to the correct destination. Of course, this means that the proteins making up the machinery must also have topogenic sequences that enabled them to get to the right locations in the first place. To prevent this conundrum from regressing into an insoluble chicken and egg problem, Blobel reminds the reader of Virchow's dictum that all cells are created by the division of existing cells and extends it to membranes, stating, "omnis membrana e membrana," all membranes come from membranes (Blobel 1980, 1498). This implies that when cells divide, the preexisting spatial organization of their membranes survives in the next generation, along with the machinery to read topogenic sequences and sort proteins throughout the cell.

In combination with the proposal of stop-transfer sequences in membrane proteins, Blobel's emphasis on the immortality of cellular spatial organization highlights another aspect of that organization with far-reaching consequences. As described previously (see chap. 2), biological membranes are asymmetric (Singer and Nicolson 1972; Rothman and Lenard 1977). The parts of transmembrane proteins that are exposed on the cytoplasmic side of the membrane are different from those exposed on the opposite, extracytoplasmic side. Proteins that are hormone receptors on the plasma membranes of cells, for example, bind hormones with the part of the receptor exposed on the outside of the cell and then signal this binding into the cell using the part of the receptor exposed to the cytoplasm. As Blobel reminds the reader, once a protein is inserted asymmetrically into the membrane through the use of signal and stop-transfer sequences, this orientation is fixed for the life of the protein (Blobel 1980).[2] Hence, the spatial organization of the cell is not limited to the existence of distinct organelles in different locations within its three-dimensional structure but also extends to a two-dimensional, inside-outside asymmetry in the membranes found in these locations. As I discuss further, spatial organization and asymmetry are essential characteristics of living organisms that not only affect their day-to-day functioning, but also, in the case of multicellular organisms, their development.

Signals Everywhere

Although Blobel's ideas about topogenic sequences were novel at the time, their direct impact on the research of others is difficult to pin down. By

---

2. As described in chapter 2, the concept of membrane asymmetry was well established by the early 1970s in conjunction with biochemical studies on specific proteins and lipids and development of the fluid mosaic membrane model. Blobel's integration of this concept with other aspects of protein topogenesis was novel (see also von Heijne 2006).

2020, the 1980 paper had been cited 1,246 times, a significant but certainly not unprecedented number.[3] If the simple phrase "protein sorting" is plugged into the search function of the PubMed database (again in 2020), then more than 11,700 results are returned.[4] Strikingly, almost all of these are after 1980. It is not as though Blobel invented the concept of protein sorting. From the growth of research on the topic immediately after 1980, however, it is possible that Blobel's articulation of the concept in specific and concrete terms created a vocabulary for laboratories working in diverse areas, enabling them to focus their work and test specific hypotheses, even if they did not cite Blobel.

The way this might have happened is illustrated by an example from the time of Blobel's publications. In 1978, Enrique Rodriguez-Boulan, working in Sabatini's lab at NYU, reported that when epithelial cells are infected with two different viruses, influenza virus and VSV, new virus particles emerge from either the top or the bottom of the cell, depending on the virus (Rodriguez-Boulan and Sabatini 1978). That virus budding occurs in such a polarized fashion in epithelial cells suggested that the virus somehow harnessed the cell's polarization machinery.

Epithelial cells, which will be important in a subsequent discussion, are a general cell type that lines the surfaces of all tissues in organisms in sheets of closely packed cells adherent to their neighbors. Such sheets are called epithelia (Fawcett 1986). Since the nineteenth century they have been known to be polarized with the top of the cell, called the apical surface, morphologically different from the bottom basal surface (Wilson 1925, 106–11). The apical surface is free of adhesive interactions and in tissues such as the intestine faces the open lumen. The basal surface, on the other hand, adheres to tissues and secreted proteins underlying the epithelium. Adhesive interactions with neighboring cells in the epithelium defines a part of the plasma membrane on the side of the cell known as the lateral surface and are responsible for creating the epithelial sheet. The locations of specific ion transporters and receptors uniquely in either the apical or basal plasma membrane domains reflect biochemically the morphological polarity, and this polar distribution of proteins in the plasma membrane is critical for most epithelial cell functions (Berridge and Oschman 1972; Fawcett 1986).

Rodriguez-Boulan wanted to investigate the origins of epithelial cell polarity and realized that for influenza virus to bud from the apical surface, the cell had to transport the two influenza membrane proteins specifically

3. Web of Science, https://login.webofknowledge.com; accessed July 2020.

4. PubMed, www.ncbi.nlm.nih.gov/pmc/.

to the apical plasma membrane. Conversely, for VSV to bud from the basal surface, the solitary VSV G membrane protein had to be transported to the basal plasma membrane. In his next publication in 1980, submitted before Blobel's topogenesis paper had appeared, Rodriguez-Boulan demonstrated that the influenza and VSV membrane proteins were indeed present exclusively on, respectively, the apical and basal plasma membranes (Rodriguez-Boulan and Pendergast 1980). In the discussion section of the paper he refers to postulated "sorting out signals." By the time a follow-up paper appeared in 1981, Rodriguez-Boulan made this idea more explicit and general, stating that the "information necessary for correct addressing of viral and intrinsic cellular plasma membrane proteins" will most likely be found in the polypeptide chains themselves (Green, Meiss, and Rodriguez-Boulan 1981, 238). His use of the term "information" to refer to sorting signals may reflect the influence of Blobel's 1980 paper, which he cited.

A few years after the publication of Blobel's paper, concrete examples of sorting sequences (or sorting signals, as they were often referred to) began to appear. I describe a few examples. The first is a most interesting and complicated story concerning the sorting of lysosomal hydrolases to the lysosome. By the early 1970s, several so-called lysosomal storage diseases had been described that were characterized by the accumulation of different metabolic breakdown products in lysosomes (Neufeld, Lim, and Shapiro 1975). One of these, called I-cell disease, results in the accumulation of undigested complex sugars and lipids in lysosomes, waste products normally degraded by the hydrolases that fill the organelle. When cells from diseased individuals were grown in culture in the laboratory, the results were puzzling. Many of the hydrolytic enzymes that should have been found in the lysosome were instead secreted and found in the culture medium surrounding the cells. All of the affected hydrolases are glycoproteins, meaning that enzymes covalently attach treelike chains of sugars, oligosaccharides, as they undergo translocation across the ER membrane during their synthesis (Kornfeld 1990, 2018) (see chaps. 7 and 10 and fig. 36). Investigators discovered that, normally, when the hydrolases are transported from the ER to the Golgi complex, their oligosaccharides are modified by phosphorylation of an existing mannose sugar, yielding 6-phosphomannose. Furthermore, sorting and transport of the hydrolases from the Golgi to the lysosome required a receptor protein that bound the 6-phosphomannose "recognition marker" on the hydrolases and then packed them into vesicles bound for the lysosome. In I-cell disease, cells lacked one of the two enzymes required for phosphorylating mannose, and the hydrolases were swept out of the

Golgi in vesicles with other secretory proteins and released from the cell instead of being sorted to the lysosome (Kornfeld 1990).

Superficially, it appears that the sorting signal for lysosomal hydrolases is the sugar 6-phosphomannose instead of an amino acid sequence, as predicted by Blobel. However, a closer look reveals that this is not the case. As the hydrolases move in vesicles from the ER to the Golgi, they must be recognized as lysosomal proteins by one of the enzymes that produces 6-phosphomannose. The enzyme's binding site on each of the hydrolases is part of the hydrolase polypeptide chain (Kornfeld 1990). This polypeptide sequence, which is folded into a kind of "sorting patch," as it was later called, is the actual sorting signal, consistent with Blobel's proposal (Pfeffer and Rothman 1987; Kornfeld 1990).

In 1984, another example of a sorting signal was reported, this time for localization of proteins to the nucleus. Daniel Kalderon and his colleagues at the Mill Hill laboratory of the National Institute for Medical Research in London studied the transport of a protein called the large T antigen into the nucleus (Kalderon et al. 1984). Large T, which is produced when the virus SV40 infects cells, concentrates in the nucleus after infection. Earlier studies from other laboratories had shown that deletion of amino acids near lysine 128 in large T using recombinant DNA techniques had partially impaired nuclear transport. To explore this in more detail, the Mill Hill group produced a series of mutants of large T, including some with deletion of lysine 128. When they expressed the mutants in cells, they found that mutant proteins lacking lysine 128 were excluded from the nucleus and remained completely in the cytoplasm. Eventually they narrowed down the sorting sequence to seven contiguous amino acids containing lysine 128. They also showed that this sequence could be moved to the amino terminus of large T antigen, away from its normal location, and still work. Most convincingly, they demonstrated that if they fused the putative signal to the bacterial protein β-galactosidase or to a cytoplasmic enzyme, pyruvate kinase, then both proteins were now transported into the nucleus. Although Blobel had mentioned localization of nuclear proteins in his 1978 symposium paper but not in his 1980 topogenesis paper, the results from Mill Hill were still a confirmation of the concept of protein topogenesis determined by a polypeptide sequence (Blobel 1978; Kalderon et al. 1984).

The next example concerns a kind of sorting signal referred to as a retention signal that Blobel did not explicitly anticipate. In his 1980 paper and its predecessors, Blobel focused on the transport of proteins from the ER and cytoplasm to distal locations in the cell such as lysosomes, mitochondria, and the nucleus. What he did not discuss was how proteins that are important for the functioning of organelles on the secretory pathway are retained in those organelles. For example, there are enzymes in the ER lumen that are involved in the modification and folding of newly synthesized proteins prior to their export to the Golgi complex and beyond. Similarly, as discussed previously, there are enzymes in the Golgi that modify lysosomal hydrolases. At the time of Blobel's paper, it may have been assumed that when such organellar proteins are translocated into the ER lumen or targeted to the Golgi complex, their final destinations, they simply remained there. This was easy to imagine if the proteins in question were membrane proteins because these were in some manner anchored in the organelle. But other organellar proteins are soluble; what is to keep them from being swept along the secretory pathway and released? After all, this is what happened to lysosomal hydrolases lacking their phosphomannose recognition marker in I-cell disease.

In 1987, Sean Munro and Hugh Pelham in Cambridge reported a study of a group of soluble proteins involved in protein folding that are found in the lumen of the ER (Munro and Pelham 1987). They discovered that all of the proteins have an identical sequence of four amino acids at their carboxy terminus. When these amino acids were removed, all of the proteins now left the ER and were secreted. When the short peptide was added to the carboxy terminus of a protein that is normally secreted, it was now retained in the ER. The peptide was a topogenic sequence as Blobel outlined it, but it kept proteins in the ER because the ER was their final destination.

A last example of Blobel's concept of topogenic sequences concerns sorting to the plasma membrane in polarized epithelial cells, the same problem investigated by Rodriguez-Boulan. In 1991, Jim Casanova from Keith Mostov's lab at the University of California, San Francisco, published a paper on the biosynthesis and transport of an integral membrane protein called the polymeric immunoglobulin receptor (Casanova, Apodaca, and Mostov 1991). Normally this protein is expressed by intestinal epithelial cells and is responsible for binding certain antibody molecules on the basal surface of the epithelium and carrying them in vesicles through the cell to be released on the apical surface, a process called transcytosis. These antibodies are part of the mucosal immune system that provide a first line of defense against pathogens that are found in the intestine (Apodaca et al. 1991). Mostov had earlier expressed the receptor protein in polarized epithelial cells in culture and found that after synthesis in the ER and transport to the Golgi, it is sorted directly to the basal plasma membrane where it remains until it binds antibodies. When he modified the receptor to remove the entire cytoplasmic part of the protein, it was delivered instead to the apical surface (Mostov, de Bruyn Kops, and Deitcher 1986). On this basis, Casanova reasoned that if there was a sorting signal on the receptor, then it might be located in the part of the protein facing the cytoplasm, where it could be recognized and segregated into vesicles headed for the basal surface. When he expressed a mutant lacking 14 of the 103 amino acids exposed on the cytoplasmic side of the membrane, the protein was targeted to the apical surface rather than the basal surface. To prove that the sequence of 14 amino acids was a basal sorting signal, he attached it to the cytoplasmic part of a known apical protein, alkaline phosphatase; the modified protein was then targeted to the basal surface. Thus, the receptor's sorting signal was both autonomous, because by itself it was able to target a protein to the basal surface, and portable, because it worked even when it was moved between different proteins (Casanova, Apodaca, and Mostov 1991).

Many other examples of sorting sequences or signals for nearly every conceivable cellular location were reported in the years following Blobel's 1980 paper (see, e.g., Mellman and Nelson 2008; Guo, Sirkis, and Schekman 2014). Often these did not specify fixed destinations but rather particular dynamic pathways along which proteins moved in vesicles within the cell, and their characterization helped map out complex intracellular traffic patterns. Looking back, it is now difficult to imagine a time when the concept of protein sorting driven by amino acid sequences did not exist.[5] Some researchers are now considering how to predict the location of a protein in the cell or its particular orientation in a membrane from its sequence alone, without observing it in the cell (Nakai 2000; von Heijne 2006; Nakai and Horton 2007). These efforts highlight two things. First, as expressed by Blobel, topogenic sequences are forms of information that help determine the spatial location of proteins in the cell. Second, because proteins need to be in the right place in the cell to work properly, cellular function cannot be considered apart from the cell's spatial organization. Cell structure and function are not independent entities.

---

5. A contemporary discussion of some of the problems addressed by Blobel's protein topogenesis idea but without use of the information concept is given in a long review article on "membrane flow" from 1979 (Morré, Kartenbeck, and Franke 1979). Although the article may not be completely representative of prevailing views at the time, it does illustrate the relative vagueness of the authors' idea of "membrane differentiation," the process by which different membranous organelles acquire their specific identities, in the absence of a specific, concrete proposal like Blobel's.

The Inheritance of Spatial Organization

In 1861, Ernst von Brücke postulated that the cell must be internally organized to account for the complex properties of protoplasm, and at the beginning of the twentieth century, Franz Hofmeister and Frederick Gowland Hopkins extended this with speculations on biochemical organization (Brücke 1898; Hofmeister 1901; Hopkins 1913) (see chap. 1). The development of cell fractionation and electron microscopy in the first half of the twentieth century demonstrated that cells are divided internally into membranous compartments, organelles, with distinct complements of enzymes but said nothing about how that organization comes about. The signal hypothesis provided the first clue by showing how particular proteins are targeted to one organelle, the ER. Then, Blobel's ideas about topogenesis and the discovery of topogenic sequences specifying all the cell's membranous compartments made it possible to conceive of the origins and perpetuation of cell organization.

The *internal* organization of eukaryotic cells is obvious if one considers particular biochemical reactions. Protein synthesis occurs almost exclusively in the cytoplasm, proteins are translocated and covalently linked to sugars in the ER, materials no longer needed by the cell are degraded in lysosomes, oxidative phosphorylation to produce ATP occurs in mitochondria, mRNA is transcribed in the nucleus, and so on. The individual functions of these organelles depend on specific enzymes. These are transported to the organelles mainly from the cytoplasm using processes dependent on topogenic sequences.

None of the cell's internal spatial organization is produced de novo. Instead, topogenic sequences on transported proteins are deciphered by prepositioned cellular machinery to target proteins to their correct destinations. When components of this sorting machinery are synthesized, they depend on preexisting copies of themselves to make sure that they end up in their correct locations. The translocation channel protein Sec61α, for example, uses SRP to get to the ER and a preexisting Sec61 channel to insert in the ER membrane (Knight and High 1998).

Cells are also spatially organized in relationship to their *external* environment. As in the example described previously, epithelial cells are polarized such that their apical plasma membrane is different both morphologically and biochemically from their basal plasma membrane. This asymmetric organization extends as well to the cytoplasm. The positions of organelles such as the nucleus and the Golgi complex are oriented along an imaginary axis running from the apical to basal surface. Although some cells may be able to polarize spontaneously by so-called symmetry-breaking events, polarization in general is dependent on *spatial cues* originating outside of the cell (Slack 1983; Nelson 2009; Goryachev and Leda 2017). Epithelial cells adhere to other tissues through specific receptor proteins targeted to their basal surface, and to adjacent cells through other plasma membrane proteins. The adhesive interactions at the basal and lateral surfaces spatially organize signaling molecules and the cytoskeleton in the cytoplasm, leading to polarization of biochemical reactions and positioning of organelles along the apical-basal axis. Mechanisms similar to those that polarize epithelial cells are also used to polarize most if not all other eukaryotic cells, although in these cases the polarity may be more transient (Mellman and Nelson 2008; Nelson 2009).

It is important to recognize that spatial organization of cells in response to external cues is still dependent on topogenic sequences. Sensing of the extracellular environment through, for example, adhesive interactions, depends on the correct asymmetric insertion of adhesion receptors into the membrane during biosynthesis in the ER such that the part of the receptor protein that recognizes another cell or tissue is ultimately oriented to the outside of the cell. This requires not only a signal sequence but also a stop-transfer sequence to position the receptor protein in the membrane. It also requires a specific sorting signal to target the protein to the plasma membrane. Once the cell is polarized, other sorting signals, such as the one described for Mostov's polymeric immunoglobulin receptor, are needed to make sure that the protein is transported to the basal and not the apical surface (Casanova, Apodaca, and Mostov 1991).

Consideration of epithelial cells provides an opportunity to emphasize that the organization of cells is three-dimensional, despite the common practice of representing cells as two-dimensional objects. Histologists categorize epithelial cells according to their three-dimensional shapes. Some types are cuboidal, others are columnar, and a third type is flat or squamous (Fawcett 1986). Even white blood cells, which appear spherical in the light microscope, do not have a random or radially symmetrical internal organization. Individual types of such leukocytes can be distinguished by the shape and eccentric location of their nuclei inside the sphere and in the electron microscope, by the particular distribution of granules within their cytoplasm.

When somatic cells divide, cellular organization is preserved by the division of the plasma membrane during cytokinesis and the roughly equal distribution of organelles between the two daughter cells (Warren 1993). Some membranes, such as the nuclear envelope, disperse into small vesicles and tubes and then reassemble on formation of daughter cells. Similarly, the Golgi breaks down into tubulovesicular components that distribute between daughter cells and then "coalesce" to re-form the organelle during telophase (Shorter and Warren 2002). Others, such as mitochondria, are present in multiple copies within the cell and divide into two populations during cell division (Alberts et al. 2015, 1001). They then regenerate to normal levels through biosynthetic activities once cytokinesis has occurred, exploiting preexisting sorting mechanisms passed down to daughter cells. In essence, the original cell and its membranes serve as templates for the production of new cells that retain the same organization. Thomas Cavalier-Smith has designated the membranes of the cell the *membranome* to emphasize the inheritance of membrane organization and properties, an inheritance that is independent of the genome of cells (Cavalier-Smith 2004). He has used this concept to trace a proposed pathway of cell evolution and has even designated "topologically and compositionally distinct membranes . . . that always grow from membranes of the same kind," such as the ER, plasma membrane, and mitochondria, as "genetic membranes" to emphasize that that they are inherited epigenetically, that is, independent of DNA (Cavalier-Smith 2004, 342). As he points out, if a type of genetic membrane is lost but the genes for its protein and lipid-synthesizing enzymes remain, then that membrane can still not be regenerated (see also Moss 1992; Cavalier-Smith 2004). In his book *What Genes Can't Do*, Lenny Moss says essentially the same thing: "If cellular membrane organization is ever lost, neither 'all the king's horses and all the king's men' *nor* any amount of DNA could put it back together again" (Moss 2003, 96). Another, rather dramatic way of expressing the same thing is to point out that the membranes of even our own cells are descended in a direct line from the first membrane-bounded cell that originated billions of years ago.

When polarized cells like epithelial cells divide, the resulting daughter cells may retain their polarized plasma membrane organization, in addition to their internal biochemical organization, if the plane of cytokinesis is parallel to the apical-basal axis along which the cell is polarized. If, on the other hand, the plane of division is orthogonal to the apical-basal axis, then the resulting daughter cells may consist of one polarized cell and one cell without an apical surface because the latter loses the asymmetric spatial cues that originally led to the polarized organization (Venkei and Yamashita 2018; Osswald and Morais-de-Sá 2019). Regardless of the orientation of cytokinesis, the division of polarized cells demonstrates that not only the internal organization of cells into biochemically distinct organelles but also the spatial organization of cells relative to their external environment are inherited.

In summary, it is evident that topogenic sequences are critical to the preservation of cellular spatial organization as somatic cells divide. Virchow's dictum that all new cells arise from other cells is thus manifested in a more concrete and specific sense. But what happens during development of multicellular organisms? New organisms arise from a single cell, the fertilized egg. Is spatial organization lost between generations? There is good evidence that it is not.

A Diversion into Development

A detailed examination of the subject of organismal development is far beyond the scope of this book. However, in my effort to establish the significance of the concept of protein topogenesis, it is worthwhile to see if the spatial information expressed through topogenic sequences has any bearing on the organization of multicellular organisms. On fertilization of the egg, the zygote undergoes a series of cell divisions, cell movements, and expansion that ultimately establishes stable axes that determine the organization of the adult animal. This patterning of the zygote is, in most if not all organisms, dependent on spatial information that originates in the mother and is then transmitted to the egg. In other words, this spatial information is inherited between generations. As stated by John Tyler Bonner:

> The commonest way in which a pattern appears in early development is by direct inheritance from the previous generation. For example, most eggs are asymmetrical and when they are shed from the ovary they already show a polarity. . . . The obvious cases of asymmetrical eggs are those in which there is a considerable amount of yolk which distinguishes the vegetal from the more protoplasmic animal pole. This ancient point is beautifully seen in the eggs of coelenterates, for when they cleave the nuclear, animal pole furrows first and slowly creases down the vegetal pole; the yolk distribution determines the direction of the first cleavage plane, and therefore the subsequent ones. . . . The classical and obvious explanation is that in the formation of the egg in the ovary, the deposition of yolk was one-sided; the axis of symmetry is directly acquired, or inherited from the mother. Here is one more example of the inheritance of information, other than direct DNA information, that is passed from one life cycle to the next. (Bonner 1974, 226)

One of the best understood examples of the inheritance of spatial information is formation of the anterior-posterior axis in the fruit fly *Drosophila melanogaster*. The *Drosophila* oocyte originates from a stem cell located in a

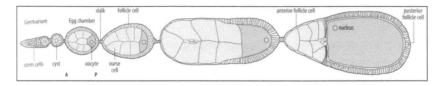

35. The sequential and polarized development of a *Drosophila* oocyte showing nurse and follicle cells. See text for more details. From *Principles of Development*, by Lewis Wolpert, Cheryll Tickle, and Alphonso Martinez Arias, 5th ed., 2015. © Oxford University Press. Used with permission.

part of the ovary called the germarium (fig. 35). Stem cells reside in various tissues of organisms and serve as unlimited sources of new cells that then go on to differentiate into gametes and specialized somatic cells. Stem cells accomplish this by dividing asymmetrically to yield another stem cell and a precursor of a differentiated cell (Slack 1983). In this way, new differentiated cells are produced without depleting the pool of stem cells. In the germarium of the ovary, the new cell produced by stem cell division continues to divide incompletely to form a germline cyst of sixteen cells connected by cytoplasmic bridges. One of these cells will become the oocyte under the influence of the other "nurse" cells and ovarian cells that encapsulate the new egg chamber and differentiate into polarized follicle cells. As successive oocytes begin to develop in the ovary, the individual egg chambers remain connected like ever enlarging beads on a string, with the most distal bead containing the most mature oocyte (Wolpert, Tickle, and Martinez-Arias 2015, 56–61) (see fig. 35).

During its development, the oocyte, which is located at one end of the egg chamber, remains in direct contact with several nurse cells and part of the follicle cell epithelium. These follicle cells are also connected to the previous egg chamber through a cellular stalk. The inherent polarity of the string of egg chambers in the ovary leads to signaling from the connected follicle cells to the nascent oocyte, resulting in the creation of an anterior-posterior axis. At the same time, the nurse cells surrounding the oocyte transport "maternal" mRNAs into the oocyte that will be used in the future development of the fly. These are distributed asymmetrically along the anterior-posterior axis by cellular processes in the forming oocyte. Thus, the overall architecture of the ovary and asymmetric processes such as the division of the stem cell and orientation of successive egg chambers transmits aspects of the mother fly's spatial organization to the oocyte (Wolpert, Tickle, and Martinez-Arias 2015; Wieschaus 2016).

When the egg begins to develop, translation of the asymmetrically distributed maternal mRNAs begins to set up the anterior-posterior axis of the embryo. Development of *Drosophila* at this early stage is unusual because replication of genetic material is not accompanied by cytokinesis, resulting in an embryo called a syncytial blastoderm (meaning that there are no cell borders) containing six thousand nuclei arranged under the outer surface. At this time the maternal mRNAs are translated in their polarized locations and the proteins produced at these sites diffuse within the embryo, creating a concentration gradient from one end of the embryo to the other. Many of these mRNAs code for transcription factors. Because the nuclei are all within a single continuous cytoplasm, nuclei closest to the translation site of a transcription factor will encounter higher concentrations of the protein than nuclei farther away, resulting in differential gene expression (Wieschaus 2016).

One example of this phenomenon is the transcription factor Bicoid. The mRNA for Bicoid is deposited by nurse cells at the anterior end of the oocyte during its maturation. After fertilization, Bicoid mRNA is translated and the Bicoid transcription factor diffuses from the anterior end of the embryo. As it encounters the nuclei in the syncytium, it turns on up to one thousand different genes, with the exact number and specificity dependent on the transcription factor concentration. Once the mitotic activity of the nuclei in the embryo has paused, the nuclei are divided into individual cells by invagination of the blastodermal plasma membrane, and further exposure to diffusing transcription factors is prevented. By this time, however, Bicoid and other diffusing transcription factors from the posterior pole of the embryo have affected the differentiation of individual cells as a function of their location along the embryonic anterior-posterior axis. These cells then continue to divide and differentiate, eventually resulting in the anterior-posterior axis and mature structures found in the adult fly (Wieschaus 2016).

Although the details are vastly different, a variety of other organisms, including experimental models of development such as *Xenopus* frogs and zebrafish, also develop their body plans in relationship to asymmetries built into the egg by the mother. Even in mammals, where the egg has been assumed to be radially symmetrical, there is some suspicion that axis development may depend on the spatial organization of the egg (Gardner 2001; Wolpert, Tickle, and Martinez-Arias 2015).

While these macro-developmental events might seem far away from any consideration of topogenic sequences, closer examination suggests that there is an important relationship. Certainly, when oocytes are produced, as in the case of *Drosophila*, typical cellular organelles with their characteristic enzymes and asymmetrically disposed membrane proteins are included as in any new cell produced by division. In addition, and importantly, the cellular asymmetries that play key roles in producing a polarized oocyte, have their origins in sorting sequences that place particular adhesive, signaling, and cytoskeletal proteins in different parts of the cell, whether that cell is the asymmetrically dividing stem cell, the polarized follicle epithelial cell, or the oocyte itself. These asymmetries do not arise randomly but are based on spatial cues originally emanating from the mother. As summarized by the developmental biologist Eric Wieschaus:

> In biology, spatial patterns build on previously existing patterns. This is a reflection of the fundamental principle of all life, which is not carried in specific molecules but in the self-propagating relationship between those molecules. At any given moment, those relationships are defined by and in their spatial distributions. . . .
>
> In a bacteria or any single-celled organism, the pattern that is transferred between generations is essentially the life-giving internal organization of the cell. Before each cell division, cellular components are duplicated and then partitioned by biological machines to daughter cells that are often essentially identical to the mother cell. In a multicellular organism, the internal cell organization of individual cells is also passed from one cell cycle to the next, but more interestingly, complex patterns of cells and organs are passed between generations. (Wieschaus 2016, 567–68)[6]

On the Nature of Spatial Information

Throughout this chapter, the term *information* has been used in discussions of topogenic sequences and the spatial expression of proteins throughout

6. In a similar statement: "If the living organism self-develops spatially, it requires certain mechanisms of *positional regulation* to control the differentiation, shape, size, and distribution of its parts, and in relation to the external environment. Spatial differentiation takes place by means of the creation of diverse asymmetries during the course of development, as in the development of different body axes. Cell polarization is a basic and general mechanism that underlies the processes of cell differentiation and spatial organization. The existence of poles in a cell involves an asymmetry in the distribution of components, and therefore its potential division into two different cells. A particular example, relevant for our purposes, is zygote polarization: in mammals, body axes are prepatterned early in development, as a consequence of an initial cell polarization that takes place during fertilization (López-Moratalla and Cerezo 2011, 190; emphasis original).

the cell. Blobel employed the word in his very first description of what would become the signal hypothesis in a grant application submitted in January 1971 (see chap. 5) and generalized its use as an element of topogenesis in the papers described earlier in this chapter. Questions about how information applies to biology have a long history and have been the subject of debates among philosophers of biology for some time (Godfrey-Smith and Sterelny 2007). What most agree on is that the concept of information used in biology originated in the work of Claude Shannon in 1948 on communication technology (Shannon 1948). He developed mathematical models to estimate the degree to which a signal emanating from a source can communicate something about that source to a receiver of the signal. By the late 1950s, scientists realized that DNA is the genetic material passed down through generations and that it encodes sequences of nucleotides corresponding to protein sequences. Because the latter process, in particular, resembled Shannon-like communication, with DNA sequence signals received and decoded in the cell by transcription and translation, it was natural to invoke informational concepts in the discussion of genetics and the central dogma (Crick 1958, 1970). Blobel's adoption of the term in the context of topogenesis seems an apparent extrapolation of its usage in genetics, particularly since topogenic protein sequences, like other protein sequences, originate in DNA. But is there more to it than an extension of information concepts from genetics to spatial determinants in topogenesis?

As the earlier discussion of *Drosophila* oogenesis illustrates, one way in which gene expression affects the phenotype of organisms is through the exertion of what Lewis Wolpert described as *positional information* (Wolpert 2016). By virtue of their location and concentration in the embryo, transcription factors effect the differentiation of cells that ultimately yields the adult animal. The source of positional information is not, however, the DNA sequence of the transcription factor but rather the spatial organization of the egg inherited from the mother in a process separate from DNA replication. Furthermore, as previously discussed, this spatial organization is ultimately vested in asymmetric processes in individual cells and when cells divide this asymmetry is passed to the daughter cells, again without any relationship to genetic information or its transmission. Hence, it is reasonable to look beyond genetic concepts to understand the nature of spatial information.

Because the inheritance of the spatial organization of cells and organisms is independent of DNA replication, it can be classed as a form of epigenetic inheritance (Jablonka and Lamb 2014; Sapp 2018). Specifically, it is a kind of structural inheritance or templating as defined by Eva Jablonka and

Marion J. Lamb, who include as examples not only Cavalier-Smith's membranome that was mentioned earlier and the inheritance of ciliary patterns in *Paramecium* as studied by Tracy Sonnenborn (Jablonka and Lamb 2014).[7] In an earlier publication, Jablonka includes the templating mode of epigenetic inheritance within a general definition of biological information that I believe helps to clarify the role of topogenic sequences in spatial organization. Jablonka's definition emphasizes the interpretative role of the receiver of information, stating, "It is . . . the receiver that makes the source . . . into an informational input" (Jablonka 2002, 583). This definition fits topogenic sequences. In general, topogenic sequences have no inherent function—no enzymatic activities, for example. While they are a potential source of information, they do not have any inherent informational content outside the context of the organized cell. As stated earlier, without the cell, topogenic sequences are meaningless. They only become an "informational input" when the cell interprets them. The cell is the receiver and interprets topogenic sequences to yield spatial information. Again, according to Jablonka: "Only living systems make a source an informational input. The more complex the organism, the more information it constructs" (Jablonka 2002, 588).

Thus, once again, the importance of the cellular context to understanding cellular functions at the molecular level is evident. Signal sequences carry no information directing proteins to the ER in the absence of an ER. Sorting signals carry no information targeting proteins to the basal surface of epithelial cells without a polarized epithelial cell. And transcription factors have no positional information in the absence of an egg with a polarized axis inherited from its mother. The biological phenomena of secretion, polarization, and cellular differentiation cannot be explained without the cell.

---

7. Sonnenborn was famous for an experiment in which he turned an ordered array of cilia in the *Paramecium* cortex 180° by microsurgery. When the altered cell divided, the new arrangement was inherited, indicating that the cilia pattern in daughter cells was duplicated from the mother cell template (Sapp 2018, 194–97).

TEN

In Vitro Veritas?

Epistemic Things

Blobel's insights into the mechanisms of protein targeting and translocation are clear accomplishments, but some of his contemporaries believed that his development of the cell-free or in vitro reconstituted system used in his experiments is an even greater contribution because it enabled certain cellular phenomena to be studied at the molecular level. The system he unveiled in the 1975 papers was exceedingly complex, with elements isolated from multiple species ranging from rabbits to dogs (see chap. 5). As studies in his lab progressed, his postdocs and students simplified the system until it consisted of just a few essential components. These included ribosomes and other factors needed for in vitro protein synthesis, an mRNA coding for the protein of interest, and microsomal vesicles. Of these, the microsomes were by far the most important element.

As described previously, Philip Siekevitz first accomplished protein synthesis in a cell-free system in 1952 while working with Paul Zamecnik at MGH (Siekevitz 1952; Rheinberger 1997) (see chap. 4). With further investigation, scientists discovered that the process required ribosomes, transfer RNAs, and initiation and elongation factors. Work on the topic of protein synthesis in the 1950s and 1960s sought to answer the biological question, how does a cell synthesize a protein? As such, the cell-free systems concocted to study protein synthesis at the time were *epistemic things*, defined by Hans-Jörg Rheinberger as "material entities or processes . . . that constitute the objects of inquiry" (Rheinberger 1997, 28). That is, the original focus of early investigators was on the mechanism of protein synthesis. As the basic steps of protein synthesis were established, the cell-free system for protein

The title of this chapter is borrowed from a paper by Ira Mellman and Kai Simons (Mellman and Simons 1992).

synthesis became a *technical* rather than epistemic object (Rheinberger 1997). It remained so when Blobel embarked on his investigation of the first steps in secretion; now, the mechanisms of targeting and translocation to the ER and the system used to study them were the epistemic thing, the object of inquiry, and the protein synthesis machinery just another reagent.

Blobel's inclusion of microsomes in his experiments opened up the possibility of investigating at the molecular level complex biological problems that are related to the form of the cell, as vested in its topographic organization. Microsomal vesicles embody fundamental aspects of cellular compartmentalization. The vesicles are closed, membrane-bounded structures with their insides topographically equivalent to the inside of the ER and, ultimately, the outside of the cell (see chap. 9). The outsides of the vesicles are topographically equivalent to the cytoplasmic side of the ER; in experiments, the protein synthesis machinery is located outside of the vesicle just as it is in the cell. Microsomal constituents are also organized in the same way as ER constituents. Proteins normally found in the ER lumen are present inside microsomal vesicles. In addition, the parts of ER transmembrane proteins exposed to the ER lumen or the cytoplasm are exposed in microsomal vesicles to, respectively, the inside and outside of the vesicles.

When experiments are conducted with Blobel's system, a series of biochemical events occur. These include in part the enzymatic reactions needed to synthesize proteins, the protein-protein interactions necessary for recognition of signal sequences leading to protein targeting and translocation across the microsomal membrane, and the proteolytic cleavage of signal sequences once the proteins have crossed the microsomal membrane. The microsomes act as a kind of cell surrogate in the assay. If microsomes are absent from the experiment, none of the biochemical events have any biological meaning beyond the simple synthesis and modification of the protein. The presence of microsomes provides the essential cellular context that enables the biochemical events in the experiment to be interpreted as a partial explanation of a biological phenomenon, the first step in secretion.

In this chapter, I describe three other cell-free experimental systems that were inspired by Blobel's to demonstrate the impact and power of his innovation. The first is one developed by James Rothman to study protein transport in the secretory pathway beyond the ER. The second is a system created by Kathryn Howell that used a novel method of cell fractionation to study the fusion of membrane-bounded compartments formed by endocytosis, the vesicular process that brings proteins into the cell from the plasma membrane. And the third is a system created by Larry Gerace that preserves much of the spatial organization of the cell but still permits molecular investigation of protein transport from the cytoplasm into the nucleus. Rothman's work is described in the greatest detail because it exemplifies most clearly the importance of morphological analysis in the pursuit of molecular explanations of cellular phenomena. At the end of the chapter I compare all these systems to Blobel's and highlight the essential features that all share.[1]

The Next Step

When Rothman went to the Lodish lab at MIT in the mid-1970s, he fell into Lodish's collaboration with Blobel to determine if G protein, the spike glycoprotein from the virus VSV, is inserted into the ER membrane during synthesis via a signal sequence–mediated mechanism (see chap. 7). Although the project had been initiated by the graduate students Flora Katz from the Lodish lab and Vishu Lingappa from the Blobel lab, Rothman jumped into the middle of things immediately (Katz et al. 1977) (see chap. 7). Based on his earlier work on membrane asymmetry, he had decided that the G protein experiment was among the most significant unanswered questions in membrane research and was disappointed that the project had gotten under way without him. As a biochemist, Rothman believed that the use of an in vitro system such as Blobel's was the best way to get at molecular mechanisms of complex cellular processes.

In 1978, Rothman joined the faculty of the biochemistry department at Stanford University and immediately began developing a cell-free approach to study protein transport from the ER to the Golgi complex, an organelle consisting of a complicated stack of flattened membrane-bounded vesicles. Based on Palade's analysis of the secretory pathway, Rothman and others believed that transport is mediated by small vesicles formed by the ER that then fuse with elements of the Golgi. While with Lodish, Rothman had examined the addition of oligosaccharides to VSV G during its translocation across the microsomal membrane, and that work gave him an idea for a biochemical assay for the transport of G protein from the ER to the Golgi (Rothman and Lodish 1977; Rothman, Katz, and Lodish 1978).

Many proteins that are translocated across the ER membrane are simultaneously glycosylated by the addition of what are called core oligosaccharides to asparagine amino acid residues in the protein sequence (Kornfeld and Kornfeld 1985) (fig. 36). These include not only viral membrane

---

1. Another good example that I have discussed elsewhere is the discovery of the motor protein kinesin using cell-free preparations of the squid giant axon and advanced video microscopy (Matlin 2020).

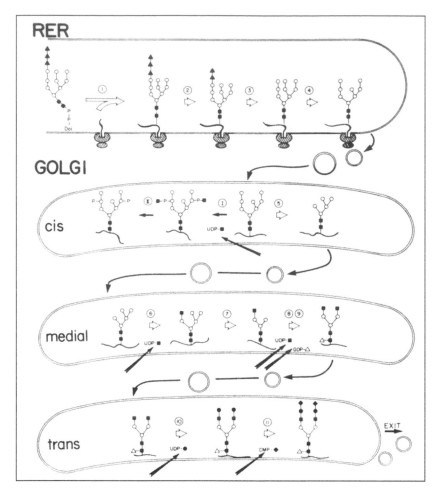

36. The addition of treelike oligosaccharides to proteins in the ER and their modification during transport to the Golgi complex, a series of flattened membrane-bounded vesicles consisting of *cis*, *medial*, and *trans* elements. Small filled squares represent N-acetylglucosamine, filled triangles represent glucose, and open circles represent mannose residues. The latter are trimmed both in the ER and in the proximal (*cis*) Golgi. Later stages in the *medial* and *trans* Golgi lead to addition of additional complex sugar residues (Kornfeld and Kornfeld 1985, fig. 3). © Annual Reviews. Used with permission.

proteins, but normal cellular plasma membrane proteins and lysosomal hydrolases, as described previously (see chap. 9). Core oligosaccharides consist of a treelike chain of the sugars N-acetylglucosamine (GlcNAc), mannose, and glucose. Two GlcNAcs form the base of the tree, followed by branching chains of mannose terminated by several glucose residues. This structure is preassembled in the ER and is transferred to proteins intact, with the first GlcNAc sugar at the base covalently linked to the protein. Immediately after glycosylation occurs, enzymes in the ER begin to remove or *trim* sugars, with the glucose residues cleaved almost immediately. At least one mannose residue is then removed prior to transfer of the glycoprotein to the Golgi complex. In the Golgi, several more mannose residues are trimmed, and then the oligosaccharide tree is built back up by the addition of more GlcNAc residues and several "complex" sugars including galactose and sialic acid to the distal ends of the remaining mannose branches.[2]

At the time that Rothman began to work on his cell-free system, the sequence of biochemical steps in assembly and processing of asparagine-linked oligosaccharides was fairly well known, but the distribution of modifying enzymes among organelles on the secretory pathway was not completely clear. One thing that appeared to be the case was that transfer of more GlcNAc to the mannose branches after trimming occurred in the Golgi complex (Goldberg and Kornfeld 1983). In glycoproteins, addition of this sugar can be easily distinguished biochemically through use of the enzyme Endo-β-N-acetylglucosaminidase H (endo H), a bacterial enzyme capable of cleaving off almost all of the oligosaccharide tree transferred to proteins in the ER when added after extraction of proteins from the cell (Hsieh, Rosner, and Robbins 1983). With SDS-gel electrophoresis, removal of the oligosaccharide by endo H can be easily detected as a shift in the mobility of the protein to a lower molecular weight. Once the glycoprotein is transported to the Golgi and GlcNAc added, endo H is incapable of removing the oligosaccharide. Thus, use of endo H provides a way to determine the degree of oligosaccharide processing in the cell, modifications that roughly correlate with protein transport from the ER to the Golgi complex. In practice, the glycoproteins whose oligosaccharides can be cleaved by endo H are referred to as endo H-sensitive and the ones with uncleavable oligosaccharides as endo H-resistant.

Rothman assigned the development of the new cell-free system to his postdoc, Eric Fries. Fries and Rothman decided to use extracts of VSV-infected Chinese hamster ovary (CHO) cells and endo H to determine if transport from the ER to the Golgi occurs in vitro. Their overall strategy to follow transport

---

2. The functional justification for this rather baroque series of subtractions and further additions of sugars to asparagine-linked oligosaccharides is not completely clear. Glucose removal (and readdition) in the ER is part of a quality control system that makes sure that the protein is properly folded (Helenius and Aebi 2001). As described previously, mannose residues are involved in hydrolase targeting to lysosomes.

of the G protein was related to so-called pulse-chase experiments normally conducted entirely in whole cells.[3] A cohort of protein is first labeled with a short "pulse" of a radioactive amino acid during its synthesis as it enters the ER. When the labeled cells are then incubated without the addition of additional radioactivity (the "chase"), the labeled protein is transported from the ER to the Golgi and then, in most cases, to the plasma membrane. With the right biochemical tools, such as endo H, the kinetics of protein transport through the cell can be determined. This kind of experiment had been conducted to measure the rate of VSV G transport in cells by David Knipe in the Lodish lab at the same time that Rothman was there. While he did not use endo H, Knipe found that G protein was converted to a form likely bearing complex sugars about fifteen to twenty minutes after synthesis (Knipe, Baltimore, and Lodish 1977a). Now, Rothman and Fries wanted to mimic this approach but with cell extracts instead of whole cells.

To construct the assay, Fries used a mutant CHO cell line called 15B that was incapable of adding GlcNAc to the trimmed oligosaccharides of proteins in the Golgi. He infected the 15B cells with VSV and, after the virus had taken over protein synthesis, labeled them for five minutes with radioactive methionine, an amino acid, reasoning that most of the G protein synthesized in this brief time would still reside in the ER. When VSV infects cells, the only membrane protein synthesized and inserted in the ER is the VSV G glycoprotein because infection shuts down the synthesis of all host cell proteins. This means that a large fraction of the radioactive methionine is incorporated into G protein during the five-minute labeling period. For the actual assay, Fries prepared cell extracts containing mixtures of ER and Golgi membranes from the infected and labeled 15B cells and from normal CHO cells that had neither been infected with VSV nor labeled. He then incubated these two extracts together. The idea was that G protein would never become endo H-resistant if it remained in the 15B cell extracts because those were unable to add GlcNAc. However, if transfer of the G protein contained in the 15B extracts (referred to as donor membranes) to the normal CHO cell extracts (the acceptor membranes) occurred, then G protein would become endo H-resistant because GlcNAc was added.

Fries's initial experiments were unsuccessful: almost none of the labeled G protein became endo H-resistant when the 15B extracts were incubated with the normal CHO extracts. To try to get the assay to work, he began playing around with the experimental conditions. At about that time, Randy

---

3. Palade and his associates had pioneered this approach in their studies of protein secretion in the pancreas (Jamieson and Palade 1967a, 1967b).

Schekman at Berkeley, having apparently heard of the problems, suggested that Fries incubate the infected 15B cells with a mitochondrial inhibitor to deplete ATP levels prior to preparation of extracts. By analogy with the pancreatic secretory pathway, Schekman reasoned that starving the cells of energy might cause the labeled G protein to accumulate in what are known as transitional elements of the ER, a stage in transport just before the Golgi.[4] Fries tried this, treating labeled 15B cells with CCCP (carbonyl cyanide m-chlorophenylhydrazone), an inhibitor of mitochondrial ATP production, for twenty minutes prior to preparing the donor extract. When he then incubated the CCCP-treated 15B cell extract with normal CHO (acceptor) extracts, labeled G protein now became endo H-resistant. Based on these results, Fries and Rothman quickly prepared a paper reporting the cell-free transport of G protein and submitted it to the *PNAS*, in which it was published in July 1980 (Fries and Rothman 1980).

All of Fries's experiments for the first paper had been conducted with a single 15B cell extract prepared from CCCP-treated cells that he had frozen in small amounts for use in multiple experiments. When he prepared new extracts using the same procedure, they were not active. Despairing that the first experiments might have been incorrect, Fries frantically tried to modify the extract preparation conditions. He finally discovered that if he did a chase incubation of the infected 15B cells for ten minutes after the initial five-minute labeling period, then he could reproducibly prepare active extracts. The problem was that after incubation for a total of fifteen minutes, the G protein was almost certainly already in the Golgi complex (Bergmann, Tokuyasu, and Singer 1981; Fries and Rothman 1981). This meant that he and Rothman were looking at transport between different parts of the Golgi rather than from the ER to the Golgi.[5] Fries and Rothman's dependence on the acquisition of endo H resistance as an indication that transport from the ER to the Golgi had occurred was part of the problem. As it turns out, the G protein arriving in the Golgi remains endo H-sensitive for some time as additional mannose residues are trimmed prior to GlcNAc addition (Kornfeld and Kornfeld 1985) (see fig. 36). That is, acquisition of endo H resistance

4. Schekman's suggestion was likely based on a paper by James Jamieson and George Palade from 1968. They observed that the transport of secretory proteins in pancreatic exocrine cells was arrested in transitional elements in the presence of inhibitors of oxidative phosphorylation (Jamieson and Palade 1968).

5. Fries suspected that the reason CCCP treatment had yielded that one active extract is that the drug had not completely inhibited ATP production. During the twenty minutes of exposure to CCCP the radioactively labeled G protein had moved from the ER to the Golgi.

is not a good marker for ER-to-Golgi transport. Rothman was not pleased with this development but soon made the best of it.

Rothman refocused his efforts on the Golgi complex and, specifically, transport through the parts of the Golgi. In most cells, the Golgi complex appears in electron micrographs as a stack of flattened membranes surrounded by a swarm of small vesicles. The individual flattened membranes that make up the Golgi, while morphologically similar, were believed to be functionally distinct based on histochemical reactions that labeled the membrane stack differentially (Farquhar and Palade 1981). Because the placement of the Golgi in the cytoplasm is generally not random, Palade had named one side of the Golgi stack the *cis* side, the one often nearest to the ER, and the other the *trans* side, where it was believed that proteins that had passed through the Golgi emerged on their way to other cellular destinations. By 1983 cell fractionation had indicated that the Golgi enzymes responsible for the final mannose trimming steps and the addition of complex sugars to oligosaccharides were located in separate parts of the Golgi (Goldberg and Kornfeld 1983). On this basis, Rothman now believed that he was looking at transport between Golgi stacks in his cell-free assay, presumably by a vesicular carrier.

Fries and Rothman were susceptible in part to misinterpretation of their initial results as ER-to-Golgi transport because they did not couple their biochemical experiments with morphological analysis of their donor and acceptor extracts. Indeed, a clever morphological study of VSV G transport in whole cells by others that used antibodies to localize G protein during transport by electron microscopy confirmed what Fries had surmised from his inactive cell extracts: G protein had entered the Golgi fifteen minutes after its synthesis (Bergmann, Tokuyasu, and Singer 1981). In Fries's experiments, if he had purified the active VSV G protein donor compartment and examined the purified material by electron microscopy, he might have distinguished that it was the rough ER and not some other membranes by the ribosomes attached to the membrane surface. Similarly, if the acceptor compartment was the Golgi, he might have recognized this because of the characteristic Golgi morphology. Fractionation of tissue culture cells like CHO to purify the ER and Golgi was, however, notoriously difficult, and, in any case, the extracts used in the cell-free assay were mixtures of barely purified membranes (Fries and Rothman 1980). Rothman and Fries were both aware of this problem and repeated their first experiments with highly purified Golgi membranes isolated from rat liver cells as a substitute for the CHO acceptor extract, in this way proving that G protein had entered the Golgi (Rothman and Fries 1981). While Rothman was not opposed to

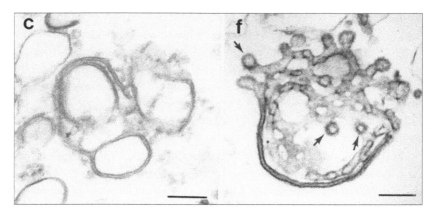

37. An "unprimed" (left) and "primed" donor (right) Golgi stack illustrating the arrested formation of transport vesicles in Rothman's cell-free assay for protein movement between parts of the Golgi (Balch et al. 1984, fig. 9). © Elsevier. Used with permission.

morphological analysis, his preference initially was for a purely biochemical approach. Such experiments moved much faster without parallel morphology and promised a way to get at the molecules both carrying out and regulating the transport process. As his work on intra-Golgi transport proceeded, however, Rothman could not do without morphology for long, and he soon incorporated it in his experiments.

Rothman revised his cell-free assay to enable not only a more direct way to measure changes in G protein glycosylation, but also parallel morphological analysis of transport. While he continued to use endo H, he also began adding radioactive GlcNAc and other sugars to his assay to directly measure glycosylation. At the same time, he developed more refined cell fractionation techniques that enabled him to purify morphologically recognizable Golgi membranes from CHO cells (Dunphy and Rothman 1983; Balch et al. 1984; Balch, Glick, and Rothman 1984; Dunphy, Brands, and Rothman 1985) (fig. 37).

Rothman was pushed along by the work of Graham Warren and Gareth Griffiths at the European Molecular Biology Laboratory in Heidelberg. They were refining views of the Golgi by pursuing Palade's integrated strategy using more sophisticated procedures to study transport along the secretory pathway. Warren and Griffiths investigated the transport of viral membrane proteins similar to G protein in cells infected with Semliki Forest Virus, combining pulse-chase experiments with cell fractionation and, significantly, immunoelectron microscopy, which enabled them to localize viral proteins in different parts of the Golgi complex during transport (Griffiths et al. 1983). Using this approach, they were able to define a third part of the Golgi membrane stack that they called the *medial* region because it was in the middle of the Golgi between the *cis* and *trans* faces (Griffiths, Quinn, and Warren 1983). Studies from others using the same immunoelectron microscopy technique were, at the same time, also beginning to directly localize the oligosaccharide-modifying enzymes within the Golgi stack (Slot and Geuze 1983).[6]

By 1984, Rothman and his colleagues, particularly William Balch, William Braell, and William Dunphy, had managed to develop the cell-free assay to the point that they could begin to dissect the steps of G protein transport from one part of the Golgi to another (Balch et al. 1984; Balch, Glick, and Rothman 1984; Braell et al. 1984). They determined that certain steps were dependent on soluble factors from the cytoplasm, one of which could be blocked by a specific protein-modifying agent N-ethylmaleimide (NEM). In experiments that were reminiscent of Palade's work in the 1960s, they used electron microscopic autoradiography to show that radioactive G protein was located in the Golgi stack of the acceptor membranes after in vitro transport (Caro and Palade 1964; Braell et al. 1984). Again with electron microscopy, they demonstrated that donor Golgi membranes incubated briefly in the absence of acceptor membranes developed round protrusions that appeared to be vesicles about to emerge from the Golgi stack (Balch, Glick, and Rothman 1984) (see fig. 37).

Morphology was now a permanent feature of Rothman's biochemical studies, and he soon formed a long-term partnership with Lelio Orci, a Swiss-Italian electron microscopist of great skill.[7] Eventually, Rothman's quest to understand molecular mechanisms of intracellular transport on the secretory pathway was very successful, and in 2013 he was awarded the

---

6. The immunoelectron microscopy technique is sophisticated and technically challenging. Tissue culture cells are mildly fixed and frozen to very low temperatures in dense sucrose solutions. This is in stark contrast to typical electron microscopy, which requires harsh fixatives and plastic embedding. The frozen sample is cut into very thin sections using a microtome held at low temperatures, and the sections are treated with specific antibodies marked by tiny gold particles visible in the electron microscope. Griffiths perfected the technique after learning "cryosectioning" (the cutting of ultrathin frozen sections) from Kiyoteru Tokuyasu in Singer's lab at the University of California, San Diego.

7. It is notable that at about the same time, Orci also began working with Randy Schekman at Berkeley to provide a morphological perspective on Schekman's studies of the secretory pathway in yeast (Orci et al. 1991). Whether the basic approach was genetic or biochemical, morphology was needed.

Nobel Prize together with Randy Schekman and Thomas Südhof. He began his Nobel Lecture by acknowledging his debt to Palade (Rothman 2014).[8]

A Diversity of Approaches

Under the influence of Blobel's in vitro system and Rothman's emerging success, many other laboratories began to tackle other cellular phenomena using cell-free approaches. The strategies were diverse but the goals remained the same: explain cellular events at the molecular level, with emphasis on the word *cellular*. While the investigators carrying out these studies wanted to know the biochemical reactions underpinning cellular events and felt that any explanations would be incomplete without this knowledge, these reactions were only significant when they intersected with aspects of cellular form that gave the reactions meaning. Below I briefly describe two examples, one focused on endocytosis and the other on transport of proteins into the nucleus, both of which incorporated a degree of morphological analysis.

Endocytosis

Endocytosis is the process by which material is brought into the cell from the outside by invagination of the plasma membrane to produce small membrane-bounded transport vesicles that then pinch off and travel through the cytoplasm (Silverstein, Steinman, and Cohn 1977; Schmid, Sorkin, and Zerial 2014). One type of endocytosis is called *receptor-mediated* because it uses plasma membrane receptors to bind extracellular substances and collect these substances in forming endocytic vesicles as they enter the cell. A well-known and early example of this is the uptake of low-density lipoprotein, or LDL, from the blood. LDL carries the lipid cholesterol through the bloodstream (Goldstein et al. 1985). When it binds to its receptor on the surface of liver and other cells, it is internalized by endocytosis and then delivered to acidic intracellular compartments where the cholesterol is released and metabolized. The LDL receptor is a transmembrane protein with an endocytosis sorting sequence in its cytoplasmic domain that enables it

---

8. Nobel Prize winners are asked to contribute something important to their development as a scientist or to their work to the Nobel Prize museum in Sweden. Rothman gave the glass tissue homogenizer used by Eric Fries to prepare the first active cell extracts, while Schekman donated the microscope that had influenced his interest in biology as a child. Thus, one more confluence of biochemistry, cell fractionation, and morphology.

to engage with the cell's endocytic machinery. Similar sequences are found in other plasma membrane receptors (Schmid, Sorkin, and Zerial 2014).

Scientists originally believed that the small endocytic vesicles went directly to lysosomes, which were for some time the only known acidic intracellular organelle. Some studies, however, recognized that there might be other "prelysosomal" compartments. In the 1980s, investigators discovered a series of membrane-bounded acidic compartments called *endosomes* that precede the lysosome on the endocytic pathway through the cell (Helenius et al. 1983). As with the secretory pathway directed to the plasma membrane, scientists believed that movement of material between different kinds of endosomes and the lysosome occurs by vesicle formation from one compartment and fusion with the succeeding compartment. However, the pathway of movement from the plasma membrane through endosomes is not linear but branched. In some cases, plasma membrane receptors enter endosomes and are ultimately degraded in the lysosome; in other cases, the receptors are recycled back to the plasma membrane for another round of endocytosis or sent to another endosome that routes them to other destinations in the cell. At the time that endosomes were first identified, little was known about the complexities of their traffic through the cytoplasm or their interactions with other membrane-bounded cellular compartments that were suspected to be part of the overall endocytic pathway.

When Kathryn Howell and her postdoctoral fellow Jean Gruenberg at the EMBL decided in the mid-1980s to study the movement of endocytosed proteins through endosomes, a major challenge was that endosomes were physically and to a certain extent morphologically indistinct (Helenius et al. 1983).[9] While the rough ER could be recognized by the ribosomes that studded its surface and the Golgi by its characteristic membrane stacks, endosomes appeared to be smooth-surfaced compartments of irregular size and shape. This created practical obstacles to their isolation and characterization and, consequently, investigation of their interactions with other compartments using a cell-free assay. Howell had previously worked on the Golgi complex with Palade at Rockefeller and then at Yale and was an expert in cell fractionation. To isolate endosomes she decided to employ a novel technique that she had helped develop, immunoisolation of membrane-bounded structures (Gruenberg et al. 1988; Howell et al. 1989) (fig. 38).

---

9. Howell and Gruenberg's work setting up a cell-free system to study endocytosis was based conceptually on the prior work by Graham Warren's lab, also at the EMBL (Davey, Hurtley, and Warren 1985). Warren used the endocytosis of two different viruses and an enzymatic assay to detect the fusion of endocytic vesicles.

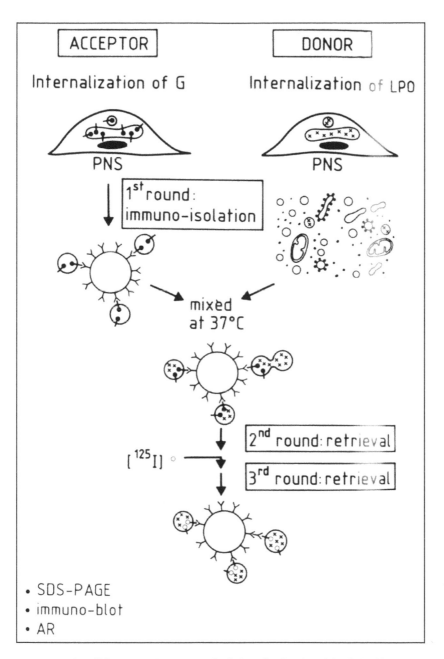

38. The cell-free reconstitution assay for fusion of endocytic vesicles devised by Kathryn Howell and Jean Gruenberg (Gruenberg and Howell 1986, fig. 7). See text for more details. © John Wiley and Sons. Used with permission.

The immunoisolation procedure requires a specific antibody that is directed to a protein exposed on the cytoplasmic surface of the compartment. Once the antibody binds to a membrane, the compartment is captured using plastic beads coated with a reagent that attaches to the bound antibody. While still tiny, the beads are large and dense enough to be easily collected by centrifugation and washed, in this way separating the captured compartments from other cellular material.

The problem with applying this technique to the isolation of endosomes is that it requires a transmembrane protein with a protruding cytoplasmic domain that can be shown to be definitively present in endosomes. The protein also needs to be at a sufficiently high concentration in the membrane to permit capture of the endosomes using an antibody against the exposed part of the protein. When Gruenberg and Howell began this work, no such naturally occurring endosomal protein was known. To solve this problem, they decided, like Rothman, to use VSV G protein as a probe. In this case, they took advantage of a property of G protein that allowed them to introduce it into the plasma membrane in the absence of any virus infection. When isolated and purified VSV virus particles are bound to the surface of a cell and the pH is dropped to a mildly acidic level, the G protein from the particles is introduced or "implanted" by membrane fusion into the plasma membrane in its characteristic transmembrane orientation (Matlin et al. 1983). If this procedure is conducted at ice cold temperatures, the G protein remains in the plasma membrane and does not enter the cell. However, when the cells with implanted G protein are warmed up, G protein enters the cell by endocytosis and begins appearing in endosomes and any subsequent endocytic compartments.[10] Thus, use of the G protein gave Gruenberg and Howell not only a way of isolating endosomes but also a dynamic marker of endocytosis.

Because so little was known about the steps in endocytosis and, particularly, the fate of endosomes, Gruenberg and Howell wanted to find out if the endosomes carrying the G protein communicated with other cellular endocytic compartments by vesicle fusion as they traveled through the cytoplasm.

---

10. The ability of G protein to cause fusion of the viral membrane with other membranes at acidic pH is related to the viral infection mechanism. When VSV (and many other viruses) infect cells, they bind to the cell surface and enter the cell intact in vesicles by a process similar to receptor-mediated endocytosis. When the viral particles enter acidic endocytic compartments, they fuse with the membranes of those compartments, delivering the viral genome to the cytoplasm (Matlin et al. 1982). Howell and Gruenberg exploited this property by binding the viral particles to the cell surface under conditions when endocytosis was blocked by low temperatures, and then inducing premature fusion of the viral membrane and plasma membrane by lowering the pH.

To do this, they incubated cells with implanted G protein at a warm temperature for five minutes to stimulate a wave of endocytosis, then immunoisolated G protein–containing endosomes from cell homogenates using an antibody against the cytoplasmic part of G protein now exposed on the outside of the endosome (Gruenberg and Howell 1986). They called this endosome preparation, which remained attached to the plastic beads used for immunoisolation, the *acceptor* fraction (see fig. 38). Electron microscopy of this material demonstrated that they had isolated irregularly shaped membrane-enclosed compartments resembling endosomes seen in whole cells. Next, they added high concentrations of the soluble enzyme lactoperoxidase to the culture medium surrounding a second set of cells and incubated them for thirty minutes. Because endocytosis was constantly taking place during this incubation, lactoperoxidase flooded the endocytic vesicles as they formed and gradually filled endosomes and most other compartments on the endocytic pathway with the enzyme. They designated a crude mixture of membranes isolated from these cells as the *donor* fraction for the cell-free assay. To determine if the donor membranes could fuse with the isolated acceptor endosomes, they incubated them together with ATP as an energy source and then reisolated the bead-bound, G protein–containing endosomes by centrifugation. They reasoned that if vesicle-mediated fusion between the donor membranes and the acceptor endosomes occurred, then lactoperoxidase from the donor would now be in the G protein–containing endosomes (see fig. 38). To detect this, they activated the enzyme so that it was capable of adding radioactive iodine (^{125}I) to any protein in its proximity and looked for iodination of G protein by SDS-gel electrophoresis and autoradiography. They observed radioactive G protein, indicating that donor membranes had indeed fused with the immunoisolated endosomes (Gruenberg and Howell 1986).

In subsequent experiments, they found that fusion could only occur soon after G protein had entered what they called "early" endosomes. If they permitted G protein to be endocytosed for more than fifteen minutes into "late" endosomes before immunoisolation, then fusion with the donor membranes did not occur (Gruenberg and Howell 1987). Gradually, Gruenberg and Howell used their cell-free system to map the entire pathway of endocytosed G protein through a set of endocytic compartments in the cell. From early endosomes, they found that some G protein recycled back to the cell surface, while the rest was routed through late endosomes to lysosomes for degradation (Gruenberg, Griffiths, and Howell 1989).[11] Eventually, their

11. The capacity of G protein to be endocytosed from the cell surface, recycled back to the plasma membrane, and otherwise progress through different endocytic compartments turned

work and that of others demonstrated that certain endosomes are specialized to recycle proteins to the cell surface, while others sort endocytosed
proteins to different destinations. Their cell-free approach also helped identify molecules carrying out the inter-endosome fusion reactions and regulating the process of endocytosis (Gruenberg and Howell 1989).

Nuclear Transport

Larry Gerace began studying the nucleus in Blobel's laboratory as a graduate student in the mid-1970s just as work on the signal hypothesis was
intensifying. Blobel developed a deep interest in the nucleus in graduate
school and had an ongoing project at Rockefeller focused on the nucleus
that paralleled his work on the signal hypothesis. Study of the nucleus, however, was temporarily pushed aside by the signal hypothesis, until Gerace
arrived in the lab.[12] With Blobel's attention largely elsewhere, Gerace was
left somewhat on his own. Nevertheless, he managed to make a number
of important discoveries, including identification and characterization of
the nuclear lamins, cytoskeletal proteins underlying the nuclear membrane,
and the mechanism of their mitotic disassembly (Gerace, Blum, and Blobel
1978; Gerace and Blobel 1980). His research on the nucleus continued in
his own independent laboratory after he left Rockefeller.

Although Gerace did not work on the signal hypothesis, he was impressed with the power of the in vitro assay devised to study it and felt
that cell-free systems were the best way to identify molecules involved in
complex cellular processes. He initially tried the approach to study dynamic
changes in the nucleus during mitosis. When cells begin to divide, chromosomes in the nucleus condense in preparation for duplication. At the same
time, the nuclear membrane, or *envelope*, breaks down into small cisternae
and tubules that are equally distributed to daughter cells during cytokinesis
(Warren 1993). As the new cells take shape, the nuclear remnants associate
with the chromosomes and then re-form into a complete membrane containing nuclear pores as the chromosomes decondense and normal cellular

---

out to be due to the existence of a topogenic sorting sequence in its cytoplasmic domain that
resembled similar signals used by the cell's own proteins, another example of the virus exploiting cellular mechanisms for its own benefit. As described in chapter nine, newly synthesized
VSV G is targeted to the basal plasma membrane in infected epithelial cells. This targeting is due
to the same sorting sequence that enabled G protein to be endocytosed (Hunziker et al. 1991).

12. After Gerace graduated, Blobel continued to study the nucleus for the rest of his career,
with the work on the nucleus eventually displacing other projects in the lab.

activities resume (Alberts et al. 2015, 656–57). Even though the breakdown and re-formation of the nuclear envelope are exceedingly complex processes, Gerace and his colleagues were able to reproduce these events in two separate cell-free assays and use the results to better understand the role of the nuclear lamins in disassembly and reassembly (Burke and Gerace 1986; Suprynowicz and Gerace 1986).

While Gerace continued to be interested in the lamins, he considered the problem of how proteins are transported into the nucleus of greater significance. As described previously (see chap. 9), nuclear proteins are synthesized in the cytoplasm and must pass through nuclear pores to reach the interior of the nucleus. While the pores are somewhat permeable to small molecules, most protein transport is regulated through a mechanism involving recognition of nuclear localization or sorting signals. While several laboratories were working on protein transport into the nucleus, there were technical difficulties that made their approaches problematic. As stated by Gerace, "All of the cell-free transport systems that have been described thus far have the limitation of utilizing nuclei that have been dissociated from other cellular structures. An intact nuclear envelope is essential to obtain transport-related nuclear accumulation of proteins, but it is difficult to isolate intact nuclei using mechanical homogenization of cells" (Adam, Marr, and Gerace 1990, 808).

Gerace decided to take a holistic approach. Instead of homogenizing cells, he treated cultured cells grown on glass coverslips with low concentrations of the detergent digitonin (Adam, Marr, and Gerace 1990). Digitonin is capable of permeabilizing membranes by binding to cholesterol. Because the plasma membrane is particularly rich in cholesterol, Gerace believed it likely that digitonin would mainly affect the plasma membrane but not other intracellular membranes like the nuclear envelope. By phase contrast light microscopy, he observed that the digitonin-treated cells neither detached from the substratum nor exhibited any gross changes in morphology. Nevertheless, biochemical analysis of the cells indicated that about 18% of the cellular protein was released from the cytoplasm by detergent treatment, suggesting that the plasma membrane was now permeable to proteins. To determine if only the plasma membrane and not the nuclear envelope was permeable, he added antibodies directed against proteins on the outer (cytoplasmic) surface of the nuclear envelope or against DNA on the inside of the nucleus to permeabilized cells, and localized the antibodies in the microscope by immunofluorescence. The former antibody highlighted the outer surface of the nuclear envelope, while the latter entered the cytoplasm but failed to stain DNA, indicating that no antibodies had entered the nucleus and that the nuclear envelope was, therefore, intact (Adam, Marr, and Gerace 1990).

To investigate transport into the nucleus using permeabilized cells, Gerace chemically attached the nuclear localization signal for the SV40 T antigen (see chap. 9) to a fluorescent protein that is normally unable to enter the nucleus. When he added this to permeabilized cells in the absence of any other proteins, it entered the cytoplasm but failed to enter the nucleus. However, when he added the fluorescent probe together with a mixture of cytoplasmic proteins, the probe concentrated inside of nuclei in the permeabilized cells within fifteen minutes. If the nuclear localization signal was not attached to the fluorescent protein, the protein did not enter the nucleus even in the presence of the added proteins. Apparently, a soluble cytoplasmic protein was absolutely required for transport of proteins carrying nuclear localization signals from the cytoplasm into the nucleus (Adam, Marr, and Gerace 1990). Over the next few years, Gerace used this assay to show that a regulatory protein called Ran controlled transport of proteins into the nucleus (Melchior et al. 1995). Later studies by Gerace and others demonstrated that Ran cycles through the nuclear pores to regulate not only protein transport into the nucleus but also transport from the nucleus to the cytoplasm (Alberts et al. 2015, 653–54).

The general strategy that Gerace used for his nuclear import assay was employed by other labs studying different problems, and cells treated with permeabilization agents were often referred to as *semi-intact* (Beckers, Keller, and Balch 1987; Grimes and Kelly 1992). In all cases what was required as a prerequisite for such studies was a good understanding of the cellular pathways involved, often garnered from more traditional combinations of cell fractionation and morphological analysis. While only nominally cell-free, one big advantage of the semi-intact cell approach is that many of the spatial relationships in cells are preserved, providing a robust cellular context for molecular analysis of cellular phenomena.

Why Do Cell-Free Assays Work?

Cell-free assays are used to investigate cellular functions. That is, they are designed to answer the question, How does the cell do *x*? Siekevitz and Zamecnik developed an in vitro assay for protein synthesis because they wanted to know how the cell synthesizes proteins. Blobel assembled his cell-free assay to determine how cells target proteins to the ER and translocate them across the ER membrane. Rothman mixed cell homogenates and purified organelles together to learn the way that cells transport proteins along the secretory pathway. Howell captured endosomes with antibodies because she wanted to find out about interconnections between endocytic compartments in cells. Gerace punched holes in cells with digitonin to determine how cells transport proteins from the cytoplasm through nuclear pores into the nucleus.

All of these cellular events involve at some level molecular interactions and reactions. While the goal is to identify and understand the roles of these biochemical processes, a constant concern is that the biochemistry discovered in the cell-free assays accurately represents what happens in intact cells. Rheinberger described these worries as they occurred in Zamecnik's lab: "The experimental alternative—cell homogenate and simpler model systems for peptide bond formation—were judged to be poor candidates for showing what went on within the intact cell. The first presented a 'biochemical bog' that would be difficult to drain; the second was prone to turn out to be artificial altogether. It might create conditions for a process not usually occurring in the cell. Alternatively, it might be emphasizing a reaction that was a metabolic sidepath not related to protein synthesis" (Rheinberger 1997, 56). Zamecnik's concerns were not unlike those expressed by cytologists in the late nineteenth and early twentieth century described earlier in this book as the "cytologist's dilemma" (see chap. 1). If one disrupts and kills the cell, how do you know that the biochemical reactions that you observe are the same as those that occur in the living cell?

The reason that the assays of Blobel and the others mentioned here were able to solve this problem is that the assays, while cell-free (though one can argue about Gerace's), still preserved a sufficient cellular context to *increase confidence* that results obtained with these systems replicated authentic cellular events. In the next chapter, I describe in general terms how this cellular context is preserved by scientists in situations where they use isolated parts of cells—microsomes, Golgi, endosomes—in the assays instead of whole cells. Suffice it to say here that organelles like this retain when isolated essential organizational features that are characteristic of their cells of origin to provide a certain *cellness* to the cell-free assays (Matlin 2016). Among these are the asymmetric disposition of their membranes and membrane proteins that are crucial to their functions.

Cell-free assays like the ones described here are designed to investigate molecular events that have no meaning in the absence of a cellular context. In describing the work of the Belgian biologist Jean Brachet, the philosopher Richard Burian refers to this phenomena when he says, "One particular matter is of key importance in understanding many of the molecular mechanisms of particular interest: *in many cases, it is only in the context of higher level structures, functioning properly, that molecular mechanisms operate 'correctly'"* (Burian 1996; emphasis original). Thus, Blobel's targeting and translocation assay is meaningless without closed microsomal vesicles that retain their asymmetric organization and lumenal contents, and Rothman's intra-Golgi vesicular traffic is meaningless in the absence of an organized Golgi complex. In essence, molecular events that acquire meaning only in a cellular context can be thought of as *emergent properties* of the cell-free assays.

The philosophical concept of ontological priority is relevant here. A simple definition of ontological priority is as a "dependence relation": "to say that x is ontologically prior to y is to say that (1) y is dependent on x for its existence, and (2) x is not dependent on y for its existence" (Gorman 1993, 460). How this applies to cells and their molecular components is expressed clearly by Wagner and Laubichler:

> Arguably the cell is the most fundamental biological unit. The cell is also the unit of ontological primacy in molecular biology. The subject matter of molecular biology is the investigation of the molecular mechanisms that underlie the fundamental processes of life, such as DNA replication, protein synthesis, regulation of gene expression, cross-membrane transport, metabolic pathways, and intracellular communication. All these processes take place within or between cells. Furthermore, they are both enabled and constrained by this cellular milieu. The cellular context not only guarantees the specific physico-chemical and spatial conditions that are required by these highly specific chemical interactions, it is also the level at which the *functional* roles of these processes can be assigned. The specific non-molecular, that is *functional* characterizations—*messenger* RNA, *metabolic* enzymes, *transcription factors, transport* molecules, etc.—of all objects in molecular biology represent exactly the kind of abstractions that are derived by means of a theory-guided conceptual decomposition of a higher level unit, the cell, that is ontologically prior to its component parts. Molecular biology therefore differs from biochemistry, in that it does not study interactions between these molecules *qua* molecules, but investigates them within the functional context of the cell. In this sense the level of the cell can be said to be ontologically prior to its component parts. (Wagner and Laubichler 2000, 23–24; emphasis original)

Cell-free assays are an evolved outcome of the epistemic strategy developed by Claude and Palade. Their evolutionary pathway and their possible supersession in the face of new technical developments to study cells are subjects of the next and final chapter.

ELEVEN

Form, Context, and the Epistemic Strategy of Cell Biology

Recapitulation

In the previous chapters I described the historical origins of molecular cell biology using as a case study the efforts of Albert Claude, Keith Porter, George Palade, Philip Siekevitz, David Sabatini, and Günter Blobel, among others, to decipher early steps in protein secretion. I concentrated mostly on Blobel, not only because his contributions have not been presented before in any detail, but also, more significantly, because of his use of a cell-free system to get at cellular mechanisms. Blobel's work studying protein targeting to the secretory pathway and translocation of proteins across a topographic cellular boundary, the membrane of the ER, helped bring the study of cellular phenomena to the molecular level. This historical narrative is fascinating, in part, because of its continuity; arguably the most significant findings occurred in a chain of interconnected discoveries spanning more than fifty years at the same institution, the Rockefeller Institute and, later, the Rockefeller University.

Aside from the history, however, there is another important aspect of the story. Why is it that this group of cell biologists was ultimately able to achieve molecular explanations of cell biological phenomena? What is it about the approach they developed that makes it so effective? Throughout much of the book I have referred to an epistemic strategy used by cell biologists to create new knowledge about the cell. My use of the term *epistemic* and its nonadjectival root, *epistemology*, follows the definition of Hans-Jörg Rheinberger in the sense that it is based on scientific practice rather than thinking of epistemology as a theory of knowledge: "The concept [of epistemology] is used here . . . for reflecting on the historical conditions *under* which, and the means *with* which, things are made into objects of knowledge. It focuses thus on the process of generating scientific knowledge and the ways in which it is generated and maintained" (Rheinberger 2010, 2–3;

emphasis original). The epistemic strategy I discuss is just that—a process for generating scientific knowledge.

It is worthwhile to remind the reader of how the cell biologists' epistemic strategy came about and evolved to a high level. As described early on, cytologists in the nineteenth century and in the first third of the twentieth century were highly reliant on the light microscope as their primary investigative tool. This served them well, particularly with regard to the functions of the nucleus, the most obvious and prominent of the visible microscopic structures in the cell. Morphological observations of mitosis, meiosis, and chromosomes, aided by parallel studies in genetics initiated after rediscovery of Mendel, provided significant correlations between structure and function. Discovery of the DNA double helix in 1953 then commenced efforts to explain the biological phenomena of inheritance, simultaneously spawning the field of molecular genetics (Kitcher 1984).

In other areas of cytology, reliance on light microscopy gradually became an impediment. As outlined in chapter 1, cellular functions other than those vested in the nucleus became associated by the 1920s with the amorphous protoplasm. Even though it was clear by that time that chemical reactions occurring in the protoplasm were responsible for cellular activities, cytologists balked at breaking open cells to examine these reactions out of fear that they would not accurately represent those of a living system. Attempts were made to map the chemistry of the cell in the microscope using histochemistry, but these efforts were only modestly successful at best (Schickore 2018). At the same time, scientists working in the new field of biochemistry believed that explanation of cellular functions could be achieved by a focus on enzyme specificity with little regard for cellular structure. This was successful to a point, so long as the reactions studied resided within a single cellular compartment.

This impasse began to be resolved by Claude who, as neither a cytologist nor a biochemist, may not have been inhibited by any preexisting notions held by either the cytology or biochemistry communities. He began breaking open cells and separating cell fragments by centrifugation and analyzing the resulting fractions chemically. To do this efficiently and quantitatively, he used the light microscope to guide his disruption of cells, permitting him to begin to link cell parts and their chemical composition to structures observed within whole cells (Claude 1948). The resolution and power of this approach was aided dramatically by the near-parallel development by Claude and Porter of biological electron microscopy. Soon they were able to demonstrate that the apparently amorphous protoplasm was an artifact of light microscopy as new cell parts were discovered and known cell parts were revealed in much greater detail. Claude's efforts were the first instantiation of an epistemic strategy that depended on the iterative comparison of parts isolated from disrupted cells to the localization of those parts in the whole cell. Ensuring that the morphology of isolated parts resembled their morphology in the intact cell helped generate confidence that the chemical activities identified in those parts reflected their activities within the cell.[1]

This epistemic strategy achieved a higher level of development through the efforts of Palade and his collaborator Siekevitz. With Porter's guidance, Palade became an excellent electron microscopist. However, Palade's first experiences in the Rockefeller laboratories were with Claude and cell fractionation, and he quickly realized the potential of combining morphology, cell fractionation, and biochemical analysis as a way of understanding how cells worked. Initially focusing on the ER, which Porter had discovered and Palade had characterized morphologically, he expanded his interests to the process of intracellular protein synthesis and transport. He added a dynamic aspect to the strategy through the use of radioactivity to track the transport of newly synthesized proteins through the cell both morphologically by using autoradiography to follow the movements of radioactive proteins in tissue sections and biochemically by analysis of labeled proteins in cell fractions. Palade called this particular combination of techniques his *integrated strategy* (Palade and Siekevitz 1956a, 1956b).

Palade was successful in localizing events during secretion to particular parts of the cell and establishing the general parameters of the secretory process. He, in particular, demonstrated the topography of secretion. Secretory proteins cross the ER membrane at the time of synthesis and remain within closed, membrane-bounded compartments until released from the cell by vesicle fusion with the plasma membrane. Palade, however, never achieved many mechanistic insights into the secretory process because his version of the epistemic strategy did not really extend to the molecular events underlying membrane translocation and transport vesicle formation and fusion.

Blobel and, initially, Sabatini, Palade's protégés, together with Colvin Redman, made the jump to the molecular level by using microsomes as a model for the ER. This enabled Blobel to dissect the steps initiating the secretory process, including protein targeting, translocation, and modification.

1. As mentioned in chapter 3, the use of sucrose in the isolation of mitochondria by Claude's associates, Palade, Hogeboom, and Schneider, in the 1940s was important because the isolated mitochondria both resembled mitochondria in intact cells and behaved like mitochondria in living cells when stained with a vital dye. This development then spurred investigation of the biochemistry of isolated mitochondria by others (Hogeboom, Schneider, and Pallade 1948; Matlin 2016).

Along with the contributions of other laboratories, he then identified the specific molecules responsible for these events. He could do this with the confidence that he was looking at biochemical events identical to those occurring in living cells because the direct linkage between microsomes and the ER had already been established by Claude, Porter, Palade, and Siekevitz. Palade and Siekevitz had first showed that the ER and rough microsomes resembled each other morphologically, including the attachment of ribosomes to their membranes and the presence of electron-dense content inside their lumenal spaces. They then demonstrated that secretory proteins synthesized in vivo appeared first in microsomes isolated from radioactively labeled cells, in this way functionally linking ribosomes and protein synthesis in intact cells with ribosomes and protein synthesis in microsomes.

The Blobel-era manifestation of the epistemic strategy became a powerful way to investigate molecular events in cells, an approach later exploited to study other biological phenomena, including, as described in the previous chapter, further steps in secretion, endocytosis, and nuclear transport, among others. In each case, the preservation in cell-free assays of relevant spatial relationships found in living cells, the cellular form, helped ensure that the questions asked and the answers obtained concerned cellular functions, even though results were expressed in molecular terms. This kind of cell biology is a real molecular biology in the sense that both the terms "molecular" and "biology" carry equal explanatory weight.

In this chapter, I first describe the epistemic strategy of cell biology schematically as a way of identifying and highlighting distinct steps. I then discuss the steps using concepts articulated by art historians and philosophers of biology that are relevant to mechanistic investigation of complex biological systems. I argue that the epistemic strategy is a reductionist approach that employs the heuristics of *cellular form* to help formulate mechanistic hypotheses about cellular phenomena and *decomposition and localization* to, ultimately, explain cellular phenomena at the molecular level. Finally, I ask whether the epistemic strategy continues to be relevant in the face of technical advances in biological research, including the reemergence of light microscopy, development of recombinant DNA, and application of computational modeling to biological systems.

Function Based on Form

Once a phenomenon to be investigated is chosen, the epistemic strategy unfolds with an examination of cell morphology using microscopy (Matlin 2018) (fig. 39). In parallel, cells are disrupted by homogenization and individual parts isolated by cell fractionation (*decomposition*). This is a guided disruption—visual representations of whole cells are compared at each stage from the initial "crude homogenate" to the purified fractions with visual representations of cell parts (see fig. 39). Typically cell disruption is followed by analysis of both the chemical and the enzymatic composition of fractions. From this point on, experiments are conducted to test specific hypotheses about the phenomenon under investigation. These hypotheses originate partly from the morphological representations of the cell and its parts, a critical aspect of the epistemic strategy that I describe in more detail later, and partly from prior ancillary studies by others on related or otherwise relevant problems.

The results of experiments are either consistent with the initial hypothesis or not. The latter case may require modification of the hypothesis or even alteration of the specific experimental setup (see fig. 39). The former case leads to a refinement and extension of the hypothesis to include more detailed predictions and the perturbation of the cell, or, in later stages, cell parts, to test the refined hypothesis. Such perturbations are designed to alter cell functions through, for example, the use of inhibitors, or to alter structures through chemical manipulations. Then the cycle of morphological representation, guided disruption, characterization of cell parts, and experimental testing of the hypothesis is repeated under the perturbed conditions. Multiple cycles with different perturbations to test more and more refined hypotheses eventually reach the level of molecular mechanisms. Crucially, these molecular explanations of the phenomenon under investigation are grounded in the cellular context, having grown from the original visual representation of the cell. To make these steps in the epistemic strategy more tangible, I review in more detail studies of ER function from Palade to Blobel.

Palade began his work combining morphology, cell fractionation, and biochemistry after some early experiences that may have convinced him of the limitations of morphological analysis alone. Two of his first papers with Claude in the late 1940s led him to the false conclusion that the Golgi apparatus as seen in the light microscope was a fixation artifact (Palade and Claude 1949a, 1949b).[2] Other experiments to localize the enzyme

---

2. Ironically, Palade later became a leading Golgi apparatus investigator. For obvious reasons he never referred to his early papers. He did, however, include an inside joke in a historical review of Golgi studies with Marilyn Farquhar, which they titled, "The Golgi Apparatus (Complex)—(1954–1981)—from Artifact to Center Stage." His two 1949 papers were not cited (Farquhar and Palade 1981).

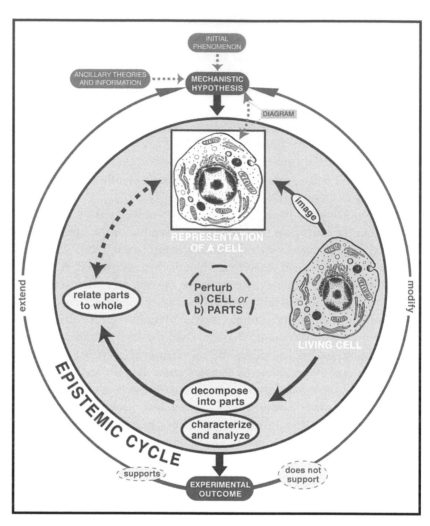

39. The epistemic strategy of cell biology. Once a phenomenon of interest is chosen, a mechanistic hypothesis is developed by reference to a representation of a cell (usually a microscopic image) where the phenomenon takes place (double-headed arrow). In some cases, the hypothesis is schematized in a diagram. Ancillary theories and information, such as previous work and characterizations of molecules hypothesized to be important, also contribute to the hypothesis. The hypothesis is then tested by cycles of cell fractionation into parts or other forms of decomposition, as well as characterization and analysis of the parts. Throughout, the relationship of those parts to the whole cell are compared, often by microscopic examination of the parts. Successive experiments proceed with perturbations of the cell or its parts. If experimental outcomes support the mechanistic hypothesis, then the hypothesis is extended, ultimately to a molecular level of analysis. If the outcomes do not support the hypothesis, then the hypothesis is modified. See text for more details.
Drawing © 2020 by Anna C. Guerrero. Used with permission.

acid phosphatase using histochemistry of whole cells and cell fractionation caused him to conclude that "the histochemical test is not reliable for
intracellular localization studies" and convinced him that this was not, in
general, the correct approach (Palade 1951, 535). After recruiting Siekevitz
to the lab because of his biochemical expertise, Palade soon began combining morphology and biochemistry. Their integrated studies of microsomes
ensued immediately, using Palade's morphological work on the ER as a
foundation (Palade and Siekevitz 1956a, 1956b).

The *phenomenon* that Palade decided to focus on was the synthesis, transport, and secretion of proteins in the cell (Palade 1975). Although some of
his first work with Siekevitz was on the liver, they quickly switched to the
guinea pig exocrine pancreas because its major function was the secretion
of large amounts of digestive enzymes. At the outset, Palade's *hypothesis* that
secretory proteins are transported through the cell in vesicles was based on
not only his own morphological work but also the findings of Rudolph
Heidenhain in the nineteenth century (Heidenhain 1875). Using histological and physiological approaches, Heidenhain noted that granules in exocrine pancreas cells had disappeared when he examined tissue from animals
that had just been fed but had apparently regenerated when he examined
tissue from animals sampled several hours later. Palade incorporated Heidenhain's *ancillary study* into his own ideas about the secretory process, describing in 1959 his work with Siekevitz as "a collaboration over almost a
century between Rudolph Heidenhain, Philip Siekevitz, and myself" (Palade 1959). The first few papers of what was to become a multipart sequence
of studies extending into the 1960s presented results in a rigidly defined
order. Detailed *morphology of whole cells and cell parts* is presented first in the
papers, followed in order by *chemical analysis of fractions* and the results of
several different *perturbations*. In the initial paper these included the "aging"
of microsomes at different temperatures to test their stability and treatments
with the enzyme RNase and the detergent DOC to dissect their chemical
composition (Palade and Siekevitz 1956b) (see chap. 4).

After concluding from the first experiments that pancreatic microsomes
are indeed fragments of the ER, as in liver, and that they contain RNA-rich
particles and a protein-rich content, Palade and Siekevitz proceeded in a
second round to analysis of the specific enzymatic content of microsomes
and zymogen granules, demonstrating that the zymogen granules in particular contain concentrated secretory proteins, consistent with the initial
hypothesis (Siekevitz and Palade 1958b). They then returned their focus
to microsomes in the third cycle after refining their hypothesis to propose
that microsomes are the source of secretory proteins found in the zymogen

granules.[3] To test this, they introduced a new perturbation modeled on Heidenhain's work. They fed animals that had been starved to stimulate the production of digestive enzymes and then sacrificed animals at several times after feeding to examine their microsomes (Siekevitz and Palade 1958c). They found that feeding leads to the appearance of "dense, cohesive content" in the microsomes that they identified as digestive enzymes. From here they proceeded to other refinements. In the fourth round, they injected animals with the radioactive amino acid leucine, allowing them to tentatively localize protein synthesis to the microsomes (Siekevitz and Palade 1958a) and then, in the fifth round, isolated a specific secretory enzyme, chymotrypsinogen, from radioactively labeled microsomes (Siekevitz and Palade 1960). They concluded that chymotrypsinogen is synthesized on ribosomes attached to the ER and then transported to subsequent membrane-bounded compartments in the cell.

At this point they had obtained sufficient evidence consistent with their refined hypothesis to be confident that the ER is where secretory proteins are synthesized. Now, with microsomes the established surrogate for the ER in intact cells, they began to work with microsomes directly to look at the mechanisms that initiate secretion. Morphological analysis by electron microscopy continued but was now focused on microsomes alone before and after perturbations. With Colvin Redman, Palade and Siekevitz showed that proteins can be synthesized and segregated inside microsomes in vitro (Redman, Siekevitz, and Palade 1966). With Sabatini, they then determined the orientation of ribosomes on the membrane morphologically and "stripped" ribosomes chemically from the ER membrane to find out how they are attached (Sabatini, Tashiro, and Palade 1966; Adelman, Sabatini, and Blobel 1973). Blobel, who had now joined the team, created his reconstituted cell-free system and began to pick apart mechanisms through a series of perturbations such as the high salt and protease treatments that led to the discovery of SRP (Blobel and Dobberstein 1975a; Walter and Blobel 1980). With ancillary findings from genetics, Tom Rapoport, Bernhard Dobberstein, and Blobel used biophysical, cross-linking, and reconstitution experiments to identify the translocation channel Sec61 (Simon and Blobel 1991; Görlich, Prehn, et al. 1992; High et al. 1993; Beckmann et al. 1997). Ultimately, the cycles of hypotheses and perturbations that started at a cellular level and then proceeded to the level of first cell fractions and then individual molecules uncovered the mechanisms responsible for the first step in secretion.

---

3. Beginning in the 1960s, Palade also began to study events in secretion after the ER with his colleagues Lucien Caro and James Jamieson (Matlin and Caplan 2017).

It is important to emphasize that, in general, there is no reason that the cycles of the epistemic strategy leading eventually to molecular analysis need to occur in the same laboratory or at the same institution. The case study from Rockefeller that I have employed here as a way of describing the epistemic strategy is, in fact, unusual in this regard. The investigations of other cell biological phenomena are typically spread over several different laboratories and institutions, with each successive cycle or two carried out independently from the cycles that initiated the study of the phenomena. I have, for example, separately described Peter Mitchell's work on oxidative phosphorylation in mitochondria as depending in part on Palade's morphological studies of mitochondria in intact cells; Mitchell used Palade's model of mitochondrial structure to help formulate his initial hypothesis (Matlin 2016). As with the ER and microsomes, the focus then shifted to isolated mitochondria for subsequent, more biochemical and biophysical experiments. The distribution of cycles of the epistemic strategy among diverse laboratories working at different but sequential times may *falsely* make the molecular work done near the end *appear to be independent* of morphological analysis. In fact, all cycles depend on the initial description of cell form as well as the retention of or reference to this form in biochemical assays to biologically validate the experimental results.

Observing, Describing, Imagining

As discussed in the previous section, the epistemic strategy as employed by Palade began with examination of the morphology of the pancreatic exocrine cell. Palade used these observations and the background information provided by the early studies of Heidenhain to formulate the hypothesis that drove all future work. Hence, encounters of investigators with microscopic representations of cells and cell parts are crucial aspects of the strategy. It is impossible to know precisely how Palade went from looking at thin sections of cells in the electron microscope to his hypothesis that secretion is accomplished through vesicular transport. Nevertheless, we may gain some insights by reflecting on the technique of electron microscopy as practiced by Palade and examining how he described electron micrographs in his papers and commented on microscopy in his correspondence.

The development of thin sectioning of plastic embedded samples in the early 1950s enabled biologists to investigate the morphology of tissues like the pancreas in the electron microscope. It is important to recognize that both thin sectioning and the high magnification of the electron microscope create an enormous sampling problem. If we make the reasonable assumption that the approximate size of a cell is a cube with fifteen micron sides and realize that the ultramicrotome developed by Porter and Joseph Blum at Rockefeller in the 1950s was capable of cutting sections that are 25 to 50 nanometers thick, then each cell can be sliced into hundreds of sections (Porter and Blum 1953). Palade and his colleagues at Rockefeller were careful to orient pieces of tissue during the embedding process so that they knew what macroscopic part of the tissue was being sectioned, and to collect sections from different areas of the embedded tissue block. Even so, the challenge of making general morphological conclusions based on electron microscopy is enormous. To view intracellular details, sections have to be magnified in the electron microscope to such an extent that only a part of each section can be viewed at a time. Adding to the difficulty, the sections appear as poorly contrasted patterns on a dim fluorescent screen. And viewing the sections in the electron microscope is not the end; selected areas of the specimen need to be photographed and the photographs printed so that they can be examined more closely.[4] This combination of challenges means that microscopists like Palade, who was known to take his own electron micrographs, have to view and photograph hundreds of sections to get a sense of the overall morphology of cells and their organelles. As stated in a well-known 1964 textbook on electron microscopic technique:

> It is obvious that only a small fraction of what the observer sees on the fluorescent screen of the microscope can be recorded on photographic film. Indeed, the time it takes to focus and otherwise adjust the microscope for picture taking, and to replace the exposed plates, is a major production bottleneck. Inevitably then, every electron micrograph of a tissue section represents a highly selected field. So much judgement is involved in the selection that it ordinarily cannot be delegated to technical help. . . . [T]he histologist who deals with ever changing patterns usually must expect to spend many hours looking at the fluorescent screen of the microscope, and take dozens of micrographs before he can feel confident that he has understanding and adequate records of any particular specimen. This makes any comparative study very slow; for days and even weeks may be devoted to a single sample. (Pease 1964, 299)[5]

4. Of course, the photographic process in even electron microscopes is now completely digital.

5. At the time that Pease's text was published, electron microscopes used glass photographic plates instead of film sheets or rolls to record images. Images were photographed by raising the fluorescent screen to expose the camera underneath. When Palade left Rockefeller in 1974 after more than twenty-five years, he left behind closets filled with exposed glass plates.

In his masterful history of biological electron microscopy, Nicolas Rasmussen talks about the standardization of vision required for consistent interpretation of results when electron microscopy gave unprecedented views of the cell (Rasmussen 1997, 103). At Rockefeller, this new visual culture required a transition from looking at cell parts in their entirety to looking at cell parts that had been sliced at what were often random angles by the ultramicrotome. Porter discovered the ER by peering through a thin part of the cytoplasm of an entire cultured cell (Porter, Claude, and Fullam 1945). Under these circumstances, the disposition of the ER in the image was evident: a reticulum collapsed into a single plane of focus due to the electron microscope's depth of field. Looking at thin sections of tissues was another matter. The parts of the tubular reticulum coursing through the cell might be cross-sectioned, yielding circles in electron micrographs, cut at an angle, producing ovals, cut longitudinally to appear as railroad track–like lines of membranes, or even grazed by the microtome knife to create what appears to be a solid object with indistinct borders floating in the cytoplasm. Palade was able to make this transition aided perhaps by his training in anatomy where translating histological sections into macroscopic organs was a necessary skill. This issue was highlighted in his competition with the Swedish electron microscopist Fritiof Sjöstrand, whom Rasmussen describes as the European version of Keith Porter. Sjöstrand and Palade both worked on the morphology of mitochondria in the 1950s. Palade favored a cigar-shaped model of mitochondria that included infoldings from the surface membrane that partially traversed the inside of the organelle, structures he referred to as *cristae*. Sjöstrand believed that these internal membranes were not infoldings from the surface but rather separate, coin-shaped membrane stacks that filled the interior. Part of Palade's argument for his model, which was eventually accepted by most biologists, was that Sjöstrand failed to take different orientations of mitochondria in thin sections into account. To illustrate his point of view, he even had a three-dimensional mock-up of a mitochondrion built and incorporated a photograph of it into a paper illustrating the effects of different sectioning angles on the appearance of mitochondria in images (Palade 1953; Rasmussen 1997).

But Sjöstrand and Palade had more fundamental disagreements. Sjöstrand believed that the high resolution of the electron microscope enabled him to make predictions about the organization of molecules that make up structures in the cell. For this reason he often sought out in his electron micrographs ideal orientations of organelles like the mitochondrion that enabled him to make precise measurements. While Palade also wanted high-resolution micrographs, he felt that inherent limitations in the technique of electron microscopy prevented definite conclusions about molecular structure. In discussing the issue of fixation at an international meeting of electron microscopists attended by Sjöstrand, Palade remarked that "Sjöstrand is of the opinion that fixation tends, in general, to disorganize the cytoplasm so, in comparison with the situation in vivo, no structure is added, but structure may be subtracted. Accordingly the best fixative is considered to be that which leaves the specimen with a maximum of organization. Sjöstrand's view . . . supposes that there is always more order in a living than in a fixed specimen, which may not be necessarily true" (cited in Rasmussen 1997, 139).

It was not that Sjöstrand relied exclusively on one approach to make his conclusions. It was that all of the technical variations that he used to strengthen his molecular predictions ultimately depended on morphology (Sjöstrand 1962). Palade, in contrast, believed that ideas about cell function suggested by electron microscopy needed to be buttressed by distinct biochemical approaches. As Rasmussen suggests, "Palade . . . refrained from addressing all questions of molecular structure by microscopy, leaving these to fractionation biochemistry, and based his interpretations on topography on a larger number of partially imperfect images. . . . Moreover, Palade referred problematic questions of interpretation not to any basic epistemological principle or criterion of microscopic veracity fully under the microscopists control. Palade would instead choose among possible interpretations of micrographs according to evidence from cell fractionation, and according to plausibility in accepted biochemical theory" (Rasmussen 1997, 140).[6] In other words, Palade employed his integrated epistemic strategy to gain functional insights.

If Palade did not completely trust the order that he observed in his electron micrographs of cells, then how exactly did they contribute to his integrated strategy? In the sections of his papers devoted to morphological results, he describes the micrographs that he chose for publication in great detail. For example, in his 1956 paper with Siekevitz that initiated their work on the exocrine pancreas, his section on the morphology of the guinea pig pancreatic exocrine cell runs for four pages (of a total of twenty-nine), followed by another page and a half on the morphology of pancreatic homogenates and isolated microsomes. This is in addition to three pages discussing the morphological observations and the figure legends

---

6. Palade's acceptance of the imperfection of micrographs and his embrace of biochemical analysis as an adjunct freed him from nineteenth century cytologists' overreliance on morphology, a factor that had limited progress (see chap. 1).

accompanying seven published electron micrographs of intact exocrine cells at various magnifications and the legends to multiple micrographs illustrating isolated and perturbed microsomes (Palade and Siekevitz 1956b).

Palade's dry and lengthy descriptions in the morphological sections of his papers reflect his desire to examine micrographs objectively, without apparently preordaining any particular functional conclusions.[7] As Lorraine Daston and Peter Galison point out, however, objectivity is not the same as truth or certainty: "Objectivity preserves the artifact or variation that would have been erased in the name of truth; it scruples to filter out the noise that undermines certainty. To be objective is to aspire to knowledge that bears no trace of the knower—knowledge unmarked by prejudice or skill, fantasy or judgement, wishing or striving. Objectivity is blind sight, seeing without inference, interpretation, or intelligence" (Daston and Galison 2010, 17). But they also state that objectivity in science has a history and is therefore not one stable, immutable state. Palade's objectivity was not the "mechanical objectivity" that some hoped could be achieved when photography was introduced to science in the nineteenth century but rather closer to the "trained judgement" developed by twentieth-century scientists (Daston and Galison 2010). While Palade's language suggests an unprejudiced view, considerable judgment was applied in the background as he spent his hours and weeks sitting at the electron microscope. Not all the images he examined were published; that would be impossible. Instead, Palade chose the images for his papers because he believed them to be representative of what he collectively observed in the microscope as well as a reflection of his functional hypothesis. What one sees and represents is not only dependent on vision, but also the mind. In *Art and Illusion* Ernst Gombrich tackles what he calls the riddle of style in paintings and other works of artists over centuries. Why is it that two artists from different times who see similar scenes represent them in such different ways? While technical ability or available materials can explain some of the differences, it is the mind of the observer that ultimately decides what is seen (Gombrich [1969] 2000, 15). There is no truly objective vision.

7. For example, in his 1956 paper with Siekevitz, Palade writes, "In the sections of the basal half of the cell, almost the entire cytoplasmic space is occupied by numerous, tightly packed profiles of the endoplasmic reticulum which are frequently disposed with remarkable regularity (Figs. 1 and 2). Practically all these profiles belong to the rough surfaced variety, i.e., bear small, dense particles, 10–15 mμ in diameter, attached to the outer aspect of their limiting membranes. The profiles measure 40 to 100 mμ in diameter and vary in shape from circular to elongated, the latter form being predominant . . . [etc.]" (Palade and Siekevitz 1956b, 673). This particular description of just the basal part of the exocrine cell goes on for another page and a half.

A certain aesthetic element may also play a role in the selection and interpretation of images. While not surrendering any ground in his claim of objectivity, Palade gives us a hint of this in a letter he sent to the electron microscopist Lelio Orci in September 2001, near the end of his life. Orci had been asked by *Nature* to write a perspectives article on whether microscopy is an art form and had sent a draft to Palade for comment. In his article, Orci presents a number of strikingly beautiful micrographs but ultimately concludes that microscopy is not an art form because it requires "uncompromising objectivity" (Orci and Pepper 2002). Palade, in his letter, concurs: "I agree with you on the general premise that microscopy is not an art form. As you put it, micrographs require ruthless insistence on objectivity, thereby eliminating any significant input from the microscopists. The situation is quite different in the arts, where subjectivity and imagination count more than reality."[8] However, in the next paragraph of his letter he goes on to say:

> We have to recognize that many biological constructs have inherent beauty when their structure is perfectly adopted [*sic*] to their specific function. It was easy to see that at the [macroscopic[9]] level when contemplating the shape of a dolphin or a gazelle; but, in fact, it becomes progressively clear that the same may apply to all subcellular components which are perfectly functional specific cell organs. So, on this account, beauty and aesthetics are contributed by nature to the work of microscopists. In addition, selecting a field of interest, guided by the balanced distribution of densities or by the alignment of major structural elements can add other elements of aesthetics to the products of microscopy. In the selection of fields, microscopists are in a situation comparable to that of a master landscape painter. And, in fact, many of the microscopists, of our generations, including yourself, have added this kind of aesthetic element to their scientific documentation.[10]

Klaus Hentschel in his book *Visual Cultures in Science and Technology* refers to "aesthetic fascination as visual culture's binding glue" (Hentschel 2014, 362). As with Palade's measured response to Orci, Hentschel points out that in science "remarks about 'beauty' are tucked away as irrelevant sidelines,

---

8. Letter dated September 5, 2001, from George Palade to Lelio Orci (Courtesy of the National Library of Medicine).

9. Here Palade writes "microscopic," not "macroscopic." It is clear from the context, though, that he meant macroscopic. Palade may use the dolphin as an example of a beast with "inherent beauty" out of nostalgia for his anatomical research on dolphins while still in Romania.

10. Letter dated September 5, 2001, from George Palade to Lelio Orci (Courtesy of the National Library of Medicine).

as ancillary feelings, as slightly embarrassing subjective deviations from the 'purely objective' main line of research" (Hentschel 2014, 362). This suggests that aesthetic impulses may be significant contributors to the epistemic roles played by images in science and, in our case, to the work of cell biologists, although they are hesitant to admit it.

Ultimately, for Palade and perhaps for others who joined the visual culture of the Rockefeller group, images played two distinct roles, first as representations of the biological objects under study and then as stimuli to the imagination. Svetlana Alpers distinguishes between seventeenth-century Dutch art and the art of the Italian Renaissance (Alpers 1983). Dutch art in her view is the "art of describing" linked to the simultaneous work by Antoni van Leeuwenhoek with his simple microscopes. Painters such as Vermeer "present their pictures as describing the world seen rather than as imitations of significant human actions" (xxxv). Italian art, on the other hand, is a narrative art, a story, "a stage on which human figures performed significant actions based on the texts of the poets" (xix). Palade was a combination of both the Dutch and Italian impulses. His images allowed him to describe the cell and its parts as objectively as he believed possible. However, what he saw in those images was filtered through past works that he had studied, the "texts of the poets," the Heidenhains, and also through his aesthetic sensibilities. This allowed his imagination to inject itself into his interpretations, leading to hypotheses about the cellular phenomena he intended to examine. While there was the possibility of overinterpretation, his acceptance of the imperfection of micrographs and intention to test his ideas biochemically made this use of imagination acceptable.

When Palade's imagination was stimulated by his micrographs, leading him to propose his overarching hypothesis of vesicular transport in secretion, his examination of images of the pancreatic exocrine cell was a heuristic process, something that I referred to elsewhere as a *heuristic of form* when describing the role of morphology in the search for the mechanism of oxidative phosphorylation (Matlin 2016). The sense of "heuristic" that I use here is the one developed by William Wimsatt based on the work of Herbert Simon (Wimsatt 2007, 76). Such heuristics are shortcuts to achieving solutions to problems posed by complex systems. Wimsatt rejects the determinism of LaPlacean demons that posits that, given sufficient detailed information, the behavior of a system can be explained and predicted. Instead, he favors what he calls "piecewise approximations of reality" as a more appropriate approach to scientific investigation, particularly in biology. According to Wimsatt, heuristics are not "truth preserving algorithms" but rather educated guesses that make "no guarantees . . . that they will produce a solution or the correct solution to the problem" (Wimsatt 2007, 76). Nevertheless, they are a "cost-effective" strategy because, in all likelihood, they will get you to a correct solution more quickly.

In the case of Palade and the other cell biologists that initiated their studies using morphology, the examination of cellular form, their images of cells served as a constraint on the types of explanations that were reasonable for them to propose. When Palade looked at the exocrine cell, one that he knew secreted large amounts of protein, he saw that the cytoplasm was filled with layers of rough ER and vesicles of all sizes, some with electron-dense material in their lumens. This, he felt, was a reasonable basis on which to propose that the secretory process involved the rough ER and vesicles. He could not be sure of this at this point; secretion may have occurred differently. Indeed, in a rather surprising article in *Science* published just after Palade received the Nobel Prize recognizing him for his work on vesicular transport, a physiologist disputed his findings, claiming that secretory proteins moved through the cytoplasm and were transported out of the cell directly through the plasma membrane, an assertion that was for almost all proteins demonstrably false (Rothman 1975). Nevertheless, Palade's educated guess formed a basis for further studies using cell fractionation and biochemistry. Palade was comfortable with his micrographs serving as "approximations of reality." They were, as Herbert Simon suggests, *satisficing*, good enough for him to formulate his hypothesis (Simon 1996, 27). There was no need for him to demand from his electron micrographs the precision desired by Sjöstrand; he had other goals. Palade leaned on morphology in a way that resembled but at the same time was distinct from how cytologists of an earlier era used images of cells. For him, morphology yielded clues to the way cells might work but not the final answer. More significantly, his images of cells represented the required cellular context that validated studies at higher and higher resolution through application of his epistemic strategy.

The Path to Mechanistic Explanation

The steps of the epistemic strategy that I outline resemble in many respects heuristics that are designed to discover mechanisms in complex systems such as the cell beyond the one based on form (see fig. 39). Indeed, the ultimate goal of the epistemic strategy followed by Palade and Blobel and their colleagues was to explain mechanisms of secretion at the molecular level. William Bechtel provides a succinct definition of mechanism: "A mechanism is a structure performing a function in virtue of its components [sic] parts, component operations, and their organization. The orchestrated functioning of the mechanism is responsible for one or more phenomena" (Bechtel 2006, 26). Explanations of mechanisms may be reductionist in both the ontological and methodological senses,[11] but they differ from what Wimsatt has called "nothing but reductionism" because the latter ignores the biological context. Extreme or "vulgar" reductionism suggests that a (biological) system is nothing but the sum of its parts, a condition that Wimsatt refers to as aggregative (Wimsatt 1997). This extreme reductionism implies that the system can ultimately be understood by analysis of its parts separate from the whole. Despite claims made in the past that the study of molecules alone can explain biological phenomena, aggregative biological systems are, at best, rare. Instead, most biological systems exhibit various forms of non-aggregativity, with parts more or less dependent on other parts for their functions. More important, according to Wimsatt, aggregative systems are inconsistent with emergent properties that are often characteristic of living organisms. Emergent system properties are context-sensitive, dependent on the organization of the system's parts (Wimsatt 1997). As pointed out previously, targeting and translocation of proteins in the ER is an emergent property: it cannot be explained mechanistically by the activities of isolated molecules. It requires the organized context of the ER compartment to make sense. The way the cell is organized is critical to understanding how it functions.[12]

Cell biologists needed a way to study cellular functions at the molecular level that did not sacrifice the organized context of the cell. To do so would have made it difficult for them to be sure that their molecular findings are biologically relevant. The epistemic strategy that they developed, which integrates morphology, cell fractionation, and biochemical analysis, has been described as *decomposition and localization*, a way of studying complex systems originally articulated by Bechtel and Robert Richardson (Bechtel and

11. According to Ingo Brigandt and Alan Love, ontological reductionism "is the idea that each particular biological system (e.g., an organism) is constituted by nothing but molecules and their interactions," while methodological reductionism "is the idea that biological systems are most fruitfully investigated at the lowest possible level, and that experimental studies should be aimed at uncovering molecular and biochemical causes" (Brigandt and Love 2017, 3). It is not a stretch to say that both of these ideas are readily accepted by nearly all cell biologists in the current era.

12. Bechtel and Robert Richardson state this quite explicitly: "Discovering the organized context within the cell permitted researchers to overcome the long-standing debates over whether processes in the living cell are common chemical effects or of some different nature. The apparent differences between ordinary chemical reactions and those in living systems are the result of the organization found in the cellular environment. The basic reactions are chemical, and organization serves to modulate their operation" (Bechtel and Richardson 1993, 170).

Richardson 1993). Decomposition is taking a system such as the cell apart, whereas localization is mapping particular operations to the decomposed parts (Bechtel 2006, 31–32). Decomposition may be structural, as when a system is physically broken into its component parts, or functional, as when a system is decomposed into component operations. According to Bechtel and Richardson, "*Decomposition* allows the subdivision of the explanatory task so that the task becomes manageable and the system intelligible. Decomposition assumes that one activity of a whole system is the product of a set of subordinate functions performed in the system" (Bechtel and Richardson 1993, 23). Systems that are completely decomposable are aggregative as defined previously. To the contrary, the cell is closer to a *nearly decomposable* system (Wimsatt 1972, 72–73; Bechtel and Richardson 1993; Simon 1996, 25, 27). Individual parts of such a system can be investigated *as if* they are independent of the rest of the system, as long as the intrinsic functions under investigation are localized to those parts.

When cell biologists decompose the cell into parts and localize functions to those parts, they do so in the context of the whole cell because the strategy for decomposition is based on examination of the cell in micrographs. They pursue both structural and functional decomposition. Palade and Siekevitz isolated microsomes, structures within the cell, using centrifugation, while localizing the functions of protein synthesis and secretory protein sequestration to microsomes, the ER surrogate. Later, Blobel decomposed the functions of the ER by separating in microsomes SRP-mediated targeting of proteins from Sec61-mediated translocation. Blobel was able to focus his work on microsomes because the cell is a near-decomposable system, at least with regard to aspects of the secretory process. The microsomes can be isolated and studied as if their functions are independent of other cellular components, even though, in fact, that is only true up to a point.

Built into the procedures of decomposition and localization is consideration of different levels of organization, and analysis normally proceeds from higher to lower levels.[13] For cell biologists, the cell is the highest level, while isolated parts like microsomes and then the molecules that carry out intrinsic functions of the parts are lower levels of organization. The epistemic strategy pursued by cell biologists has a cyclic and progressive character (see fig. 39). In the work of Palade and Blobel, the early cycles isolated

13. A definition of levels of organization provided by Wimsatt is appropriate here: "By level of organization, I mean here compositional levels—hierarchical divisions of stuff (paradigmatically, but not necessarily material stuff) organized by part-whole relations, in which wholes at one level function as parts of the next (and at all higher levels) (Wimsatt 2007, 201).

and characterized microsomes as parts of the whole cell; later cycles investigated the functions of microsomes in the targeting and translocation of proteins into the microsomal interior and then identified and characterized the molecules carrying out these functions. As mentioned in the previous chapter, Rheinberger describes such cycles as the production of *epistemic things*, defined as "material entities or processes—physical structures, chemical reactions, biological functions—that constitute the objects of inquiry" (Rheinberger 1997, 28). Production of epistemic things generates *traces* or *inscriptions*, essentially experimental results. Due to the progressive nature of the investigative process, epistemic things are transformed into technical objects, defined as "the technical repertoire of the experimental arrangement" (Rheinberger 1997, 29). Microsomes that were epistemic things to Palade became technical objects to Blobel, while the functions of protein targeting and translocation along with the protein structures and chemical reactions carrying out these functions became epistemic things. The production of epistemic things in such a cyclic and progressive process is by definition recursive: the results of one cycle are incorporated into and form the basis for the next. As Rheinberger states, "Whether the traces that are produced in an experiment will prove 'significant,' depends on their capacity to become reinserted into the experimental context and to produce further traces. No experimental work can escape this recursive action, this iterative process of detaching an inscription from its transient referent and turning the referent itself into an inscription" (Rheinberger 1997, 107).

Ultimately, the processes that characterize the epistemic strategy of cell biologists yield molecular, mechanistic explanations of biological functions. The starting point, the original referent, is the whole cell. Based on its representation in electron micrographs, it is decomposed, and functions and constituents are localized to cell parts. Recursively, these parts then become the starting point for another round of decomposition and localization, followed by others that decompose functions into the molecules and structures carrying out the functions. The strategy is reductionistic in the sense that it seeks explanations at the lowest possible (molecular) level, but those explanations are embedded in the structure and form of the cell. The organization that makes the explanations biological is never lost, forgotten, or ignored.[14]

14. My description of the epistemic strategy of cell biology resembles in many respects strategies articulated by Lindley Darden that are discussed by Marcel Weber (Darden 1991, 67–72; Weber 2005). What is unusual about the cell biology strategy is that Claude and Palade stated it explicitly and it did not have to be reconstructed post facto by philosophers. Palade was very

One can go even further to ask whether molecular explanations of cell biological phenomena *require* the use of an epistemic strategy like the one described here. Can we show that discoveries considered historically to be the result of solely molecular biological approaches are actually the ultimate outcomes of a cell biological strategy? Perhaps. The discoveries of mechanisms of DNA replication and gene expression are claimed as pure achievements of gene-centric molecular biology with the 1953 DNA structure determination as a starting point. However, it was the study of nuclear and chromosomal dynamics by cytologists, the linkage of these events to Mendel's work, and the localization of DNA to the nucleus, all accomplished with microscopy, that provided the foundation for the later molecular studies by establishing the cellular context of these events (see Kitcher 1984; and epilogue, below). This work may have progressed more rapidly to the molecular level than the investigation of cytoplasmic mechanisms because the nucleus and chromosomes are parts of the cell that are easily visible in the light microscope, and, therefore, the work clearly started earlier than the studies of secretion detailed here.

The Status of the Epistemic Strategy in Contemporary Cell Biology

The studies of protein targeting and translocation into the ER described here were largely complete at the end of the twentieth century. On reflection, it is of interest to ask whether the epistemic strategy developed by cell biologists is still a useful approach to the study of biological phenomena. After all, the techniques that led to breakthroughs and were key parts of the epistemic strategy, electron microscopy and cell fractionation, are now practiced in different and often less rigorous ways. Through developments in sample preparation and improvements in microscope technology, electron microscopy is now more frequently used by experimental biologists to determine the structures of proteins and complex cellular components than as a guide to cellular decomposition or as a way to imagine how cells work. Breaking cells open is, of course, common, but it is now done with little concern for

---

clear that from almost the beginning he pursued an integrated approach combining microscopy, cell fractionation, and biochemical analysis. As illustrated in figure 39, the epistemic strategy of cell biology is also remarkably similar to what James Griesemer described as an "Extended Giere Model of Scientific Reasoning" with the "Perspective" and "Image" that initiates the cycle of modeling equivalent to the morphological representation of the cell (Giere 1997; Griesemer 2000; see also Callebaut 2012). Space limitations do not allow me to do more than mention these insights.

the introduction of nonbiological artifacts and, sadly, no accounting for all of the cell's parts in a balance sheet. Have these changes affected the value of the heuristics of form and decomposition and localization to the discovery of cellular mechanisms?

As in the 1930s and 1940s, technologies have emerged that have changed the investigative practices of experimental cell biology. Among these are the ready availability of cheap and powerful computers, the development of recombinant DNA techniques that allow the sequencing, manipulation, and expression of genes in cells, and the discovery of jellyfish proteins that fluoresce when illuminated with particular wavelengths of light. These innovations resulted in the recent reemergence of light microscopy as the primary means of examining both fixed and living cells. Computers revolutionized the collection of images and their manipulation and quantitative analysis. The creation of hybrid proteins consisting of part of a fluorescent protein fused to any protein of interest using recombinant techniques then permitted individual proteins to be observed directly in living cells using a fluorescence microscope.

An example of the application of these techniques to a biological problem can be found in the work of Ed Munro, a cell biologist at the University of Chicago. Munro and his laboratory study the origins of cell polarity in the early embryo of the worm *Caenorhabditis elegans*. Susan Strome and William Wood first observed the polarization of *C. elegans* embryos in 1983 (Strome and Wood 1983) (fig. 40). Soon after fertilization but before the first cell division, they observed, germline or P granules move to what is defined as the posterior pole in the embryo, indicating polarization. The granules can be seen in living embryos using differential interference contrast microscopy[15] or by immunofluorescent staining. Sometime later, others observed that polarization was accompanied by ruffling of the plasma membrane and "cortical flow," the movement of cytoplasmic material along the periphery of the embryo just under the plasma membrane. Based on these observations, Steven Hird and John White hypothesized that the cytoskeletal protein actin might play a role in polarization (Hird and White 1993).

Munro adopted this hypothesis and proposed a mechanistic model suggesting that certain "PAR" proteins, known through genetic studies to be involved in polarization, were moved by actin-mediated cortical flow to initiate and then regulate polarization (Munro, Nance, and Priess 2004).

15. Differential interference contrast microscopy, like phase contrast microscopy, is an optical technique for making cellular organelles visible in living or fixed cells without chemical staining. Both techniques were introduced in the early to mid-twentieth century (James 1976).

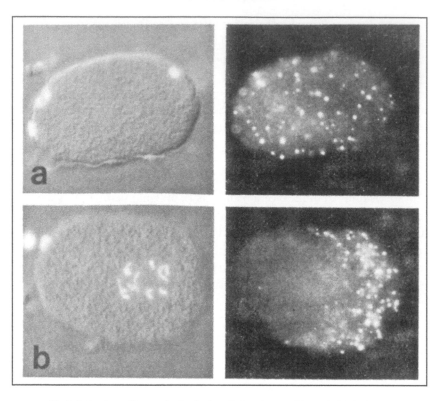

40. Polarization of P granules in single-celled embryos of *Caenorhabditis elegans*.
Left panels: Nomarski differential interference optical images with fluorescently
stained chromosomes. Right panels: immunofluorescence images of P granules. In **a**
the granules are randomly distributed; in **b** they have moved to one pole of the cell
(Strome and Wood 1983, fig. 2). © Elsevier. Used with permission.

To study this, he used recombinant forms of non-muscle myosin, a motor
protein that attaches to actin, and recombinant constructs of two PAR proteins, PAR-2 and PAR-6, fused to jellyfish green fluorescent protein (GFP)
to follow the dynamics of the proteins in living embryos using time-lapse
fluorescence microscopy. He also used a technique called RNA interference
to block the expression of normal PAR proteins in embryos. From his initial
observations, he produced a more detailed hypothesis for the interaction
and relocalization of PAR proteins by cortical actin flow that he illustrated
in diagrams (Munro, Nance, and Priess 2004). In subsequent studies, he and
his lab incorporated different regulatory proteins into his model and experiments, continuing to use fluorescent fusion proteins and live-cell imaging
to follow the dynamics of polarization. At this point, Munro's dynamic

models had become too complex to mentally rehearse the output of the models, and he resorted to computational models based on partial differential equations as an alternative (Bechtel and Abrahamsen 2010; Dawes and Munro 2011).[16] He then investigated his models by working back and forth between experimental perturbations and the predictions of his models to test his hypotheses. Ultimately, he concluded that cortical flow leads to the segregation of particular PAR proteins and their regulators to different poles of the embryo and that this polarity is stabilized by "dynamic opposition" between clusters of PAR proteins (Sailer et al. 2015).

While Munro's experiments are drastically different from the molecular studies of Blobel and his colleagues, there remain essential similarities. As with studies of secretion, the overall hypothesis of cortical flow and polarization was generated by microscopic observations of whole cells, *C. elegans* embryos. Decomposition occurred when the *function* of polarization was linked to actinomyosin and PAR proteins. This was a heuristic process—an educated guess—based on *ancillary studies* of the properties of actin and myosin and genetic experiments that had identified the PAR proteins and then linked them through mutational analysis to polarization. Observations carried out using fluorescent fusion proteins expressed in embryos were *by design* done directly in the context of the living cell, ensuring that the observations were biologically relevant. Perturbations included the blockage of protein expression in the embryos using RNA interference and the use of chemical inhibitors. Cycles of Munro's epistemic approach led to more refined (and complex) hypotheses that eventually could only be represented mathematically. Ultimately, the roles of particular proteins in a dynamic mechanistic model were established.

In the End

We have now traveled in this story from the first decades of the nineteenth century through the twentieth century to the present. Cells were established as the fundamental living units of organisms and their delicate structures described using improved microscopes. Nevertheless, the simplest living cells were judged to be similar to amoebae: amorphous, living blobs of protoplasm. Cellular functions, the basis of life, were known to depend on the organization of this protoplasm, but the chemical reactions that carried out these functions were inaccessible to cytologists because they were

16. An excellent description of the transition of models in cell biology from simple diagrams to computational models has been provided by Fridolin Gross (Gross 2018).

afraid to disrupt protoplasmic organization in their search for answers. Cytology reached such an impasse that E. B. Wilson declared in 1925 that while cytologists needed to consider both structure and function, which one was "the primary determining factor in vital phenomena" was an insoluble philosophical problem (Wilson 1925).

I have argued that the relationship between structure and function is no longer philosophically intractable. The epistemic strategy developed by cell biologists in the last half of the twentieth century provides a practical means to localize functions within the structure of the living cell and to extend this relationship down to the level of molecular parts. This does not mean that the relationship is either simple or completely knowable. There is no LaPlacean demon that will allow us to place molecular parts precisely within cellular structures to predict the future or recall the past. The best we can hope for are "piecewise approximations of reality." Despite this, as our experimental technologies improve, each turn of the epistemic cycle will bring us closer to our goal of understanding what is meant by a living system. Along the way, we will have to be content with satisficing solutions.

1975 and All That

Kitcher's Tale of Two Sciences

In 1984, the philosopher of science Philip Kitcher published the now-classic paper "1953 and All That" (Kitcher 1984). In it he discusses the relationship between classical genetics and molecular genetics. Kitcher considers classical genetics to be studies of heredity exemplified by the work of Thomas Hunt Morgan and others after the rediscovery of Gregor Mendel's experiments at the beginning of the twentieth century. He links molecular genetics, on the other hand, to the discovery of the structure of DNA by James Watson and Francis Crick in 1953. Classical genetics relies on the chromosomal theory of heredity to explain certain genetic concepts such as the transmission of genes and recombination. Classical genetics was effective not only in clarifying Mendel's so-called laws but also in showing why Mendel's findings did not always hold, as when two genes are located on the same chromosome. Molecular genetics attempts to account for these concepts through analysis of DNA segments identified as gene equivalents.

While Kitcher acknowledges the many significant contributions of molecular genetics to our understanding of mechanisms of DNA replication and gene expression, among others, he argues that molecular genetics is not better than classical genetics at explaining key genetic concepts. He asserts that cytological analysis of chromosomes provides the clearest and most succinct answers to fundamental questions, as in this simple explanation of an "amended" version of one of Mendel's laws.

> Why do genes on nonhomologous chromosomes assort independently? Cytology provides the answer. At meiosis, chromosomes line up with their homologues. It is then possible for homologous chromosomes to exchange some genetic material, producing pairs of recombinant chromosomes. In the meiotic division, one member of each recombinant pair goes to each gamete,

and the assignment of one member of one pair to a gamete is probabilistically independent of the assignment of a member of another pair to that gamete. Genes which occur close on the same chromosome are likely to be transmitted together (recombination is not likely to occur between them), but genes on nonhomologous chromosomes will assort independently. (Kitcher 1984, 347)

Kitcher does not dismiss the importance of molecular studies, stating that "to emphasize the adequacy of the explanation is not to deny that it could be extended in certain ways" (Kitcher 1984, 347). For example, processes such as spindle formation and chromosome condensation, both visible cytologically, can be more completely understood if molecular details of the phenomena are also uncovered.

In the terms that I used in previous chapters, the cytological analysis that Kitcher refers to provides a *cellular context* that ensures that subsequent molecular studies are biologically relevant and actually explain *biological* phenomena. This is, essentially, an application of the cell biologists' epistemic strategy to problems in genetics. The inheritance of genes is explained first at the cytological level by the behavior of chromosomes and then, in later steps, by the interactions of DNA with the enzymes and regulatory proteins that drive replication and chromosome condensation and with the microtubule spindle fibers and motor proteins that separate chromosomes prior to cytokinesis. The embrace of an epistemic strategy that preserves cellular context while ultimately enabling molecular explanations is reductionist, even if stages in the explanatory cycle are cytological and not molecular. Essentially, the strategy provides a blueprint for studies ranging over several levels of biological organization from the cell on down. As William Wimsatt states, "We must work back and forth between ontologies of different levels to check that features crucial to upper level phenomena are not simplified out of existence when modelling it at the lower level" (Wimsatt 1997, S374). On its own, investigation of DNA does not illuminate the biological processes of genetics if what we have learned from cytology is "simplified out of existence." Kitcher says something similar in his paper, acknowledging that the "explanatory structure within biology" encompasses "several explanatory levels" and that none should be dismissed out of hand. When applied to the study of cells, both Wimsatt's and Kitcher's statements are reminding us that we should not forget the cellular context in our search for molecular mechanisms. Indeed, I argue that the incorporation of upper levels into the process of securing lower-level explanations enhances the power of the latter. As Kitcher concludes, "Even if geneticists must become 'physiological chemists' they should not give up being embryologists, physiologists, and cytologists" (Kitcher 1984, 373).

Molecular Myths

As I mentioned in the prologue, in the heady days following the discovery of the structure of the DNA molecule, a kind of extreme or nothing-but reductionism became a founding principle of molecular biology (Wimsatt 2007, 308–9). Essentially, this principle stated that almost all biological phenomena could ultimately be explained by examination of molecules alone. Watson and Crick famously said in their 1953 paper, "It has not escaped our notice that the specific pairing we have postulated immediately suggests a possible copying mechanism for the genetic material" (Watson and Crick 1953b). In a second paper published shortly thereafter, they discussed some of these implications in more detail (Watson and Crick 1953a). What they were saying with regard to DNA was true to some extent. Hydrogen-bonded base pairing between the two antiparallel strands does suggest a way to make exact copies, although Watson's and Crick's imaginations may have also been stimulated by cytological images of chromosome pairs. DNA, however, is an exceptionally unusual molecule, and similar insights into mechanism are not often granted by examination of molecular structures alone.

The history of molecular biology after the discovery of the DNA structure can be divided into two phases. The first followed up on the initial implications of the 1953 papers, culminating in formulation of the central dogma stating that information encoded in DNA sequences flows unidirectionally from DNA to RNA and then to protein (Crick 1958, 1970). Even here, it is arguable that the extreme reductionist approach bore fruit. Much progress, including deciphering the genetic code, depended on understanding the mechanism of protein synthesis, and the study of protein synthesis substantially depended on the work of the biochemists in Paul Zamecnik's lab at MGH and the cell biologists at Rockefeller in the 1940s and 1950s (see chap. 4). Hans-Jörg Rheinberger, in his book detailing Zamecnik's studies of protein synthesis, reports that Watson made many forays from his lab at Harvard across the Charles River to MGH to learn what was going on (Rheinberger 1997). After their work on protein synthesis using liver preparations, Zamecnik's group developed one of the first successful in vitro systems based on *Escherichia coli*, the key model organism of the molecular biologists. This system was quickly adopted and optimized by Watson and his colleagues (Rheinberger 1997, 199). According to Nicolas Rasmussen, by the early 1960s, molecular biology had become a Kuhnian "normal science," entering a period of "puzzle solving" that built on the paradigmatic central dogma without challenging it (Rasmussen 2014, 28).[1,2,3] By 1968, Gunther Stent, one of the chief practitioners of molecular biology, signed off on the last phase of the "informational school," a strand in the development of molecular biology woven by physicists like Max Delbrück and central to Watson's formative education (Kendrew 1967; Watson 2007). Seeming to throw up his hands in frustration, Stent declares at the end of his article, "Perhaps *this* then is the paradox: there exist processes which, though they clearly obey the laws of physics, can *never* be explained" (Stent 1968, emphasis original). Even Crick, in referring to Marshall Nirenberg and Heinrich Matthaei's work on the genetic code, remarked in his 1962 Nobel Lecture, "We are coming to the end of an era in molecular biology. If the DNA structure was the end of the beginning the discovery of Nirenberg and Matthaei is the beginning of the end" (quoted in Kay 2000, 271).

The second phase in the history of molecular biology is a technical phase leading to the invention of recombinant DNA procedures. As described by Rasmussen, these developments began with the discovery of restriction enzymes (Rasmussen 2014). Restriction enzymes are part of the bacterial immune system that are capable of cleaving invading DNA at specific sequences. Some of these produce staggered cuts in the two strands of the DNA double helix, creating "sticky ends" that can anneal by base pairing with another DNA molecule cut using the same enzyme. This enabled the

---

1. Michel Morange makes a similar point in *A History of Molecular Biology* (Morange 1998, 167).

2. "Kuhnian" refers to Thomas Kuhn's 1962 *Structure of Scientific Revolutions*, which introduced the notion of scientific paradigms as well as normal science (Kuhn [1962] 1970).

3. To state that molecular biology had become a normal science in no way diminishes its contributions to our understanding of essential processes related to the central dogma. A particular focus has been on the regulation of gene expression initiated through the work of Jacob and Monod on the *E. coli lac* operon. In eukaryotes, the regulation of gene expression is incredibly complex, involving huge numbers of transcription factors and other regulators that bind to DNA, as well as the splicing of primary RNA transcripts into multiple different mRNAs. In many if not most cases these problems were treated as mainly one-dimensional, involving differential protein binding to the linear DNA polymer. More recently, in attempts to apply regulatory insights gained from these studies to biological problems, understanding the spatial constraints of cells were critical to explanations of phenomena. For example, to make sense of how gene regulatory networks control early sea urchin development, it was necessary for Eric Davidson and his colleagues to include the effects of cell-cell interactions and cellular position within the embryo in models of the developmental process (Davidson 1991; Peter, Faure, and Davidson 2012). At the level of the individual cell, biologists have discovered that the three-dimensional disposition of chromatin within the nucleus is an essential aspect of transcriptional regulation (Van Bortle and Corces 2012).

first genetic engineering when the circular DNA of the eukaryotic virus SV40 and a circular DNA plasmid from E. *coli* were separately cut with a restriction enzyme and then recombined into a single hybrid circle. When bacteria took up the SV40-plasmid hybrid, signals in the plasmid sequence induced the bacteria to replicate the molecule, permitting production of the SV40 genome in large quantities. Through the use of an enzyme called reverse transcriptase prepared from certain animal viruses, it was then possible to start with an mRNA and turn it into a complementary DNA molecule that could then be introduced into bacterial plasmids using restriction enzymes. All of these developments were essentially complete by the late 1970s, including the discovery of techniques to more easily sequence large stretches of DNA.

Recombinant DNA technology rapidly disseminated into all biological disciplines, including cell biology. Because it was invented by scientists who viewed themselves as molecular biologists, other researchers who began using these procedures identified as molecular biologists as well, even if the investigative strategies that they pursued in their laboratories were not based on the extreme reductionism that the founders of molecular biology imagined. Some cell biologists, for example, while still pursuing an epistemic strategy that preserved the cellular context, now considered themselves molecular biologists, conflating the use of recombinant techniques with adoption of the alternative investigative strategy of reductive molecular biology that was not, in fact, that effective in deciphering complex biological problems.

This confusion was also fed by another development that followed right on the heels of the invention of recombinant DNA techniques, the creation of the biotechnology industry in the 1970s and 1980s. Most of the early companies were founded by molecular biologists who pursued the production of synthetic hormones and, along with it, substantial financial rewards (Rasmussen 2014). Now the success of molecular biology, writ large, was incentivized. The more that the general public, and especially the investment community, was convinced that support of academic molecular biology and biotech would lead to therapies and cures for the most recalcitrant diseases, the more likely it was that venture capital dollars, buyouts, and improved stock prices would increase the net worth of the founder scientists. These were not, of course, completely hollow promises; many new drugs and therapies have been and continue to be created.[4] However, the monetization of

4. A dramatic current example of this is, of course, the Covid-19 vaccine. Sequencing of the viral genome led in less than a year to vaccine production, an astounding accomplishment. We should not forget, however, that cell biological studies on mechanisms of virus infection created a foundation of knowledge on which vaccine development rested.

molecular biology distorted how we think about the history of modern biology and obscured our understanding of the ways that biological investigation actually works.

As recombinant technologies improved in the late 1980s, molecular biologists began pushing for the sequencing of the entire human genome. From a purely academic perspective, the availability of such information for not only humans but also common laboratory model organisms is of obvious value to disciplines ranging from cell and developmental biology to evolutionary biology. After all, if one's research seeks the gene products carrying out biological mechanisms, the availability of a genome sequence that seems to provide structural information for all possible proteins expressed in organisms is seductive. But the main selling point of the Human Genome Project (HGP), as it was called, was its medical potential. Inherited diseases that pedigree analysis indicated were due to single gene defects were obvious targets. However, the claims that were made to justify the HGP went far beyond that, bordering on the preposterous. As the geneticist Jon Beckwith noted in a 1996 paper, scientists referred to the human genome sequence as a potential "Rosetta stone," something that "defines a human being" and might "transform medicine in the 21st century into a preventive mode, where genetic predispositions are identified and treated before the onset of illness," leading even to a cure for homelessness (quoted in Beckwith 1996, 177). What was ignored in the public relations campaign in favor of the project was that the tracing of a disease to a mutation in a single gene, such as sickle cell disease linked to a defect in hemoglobin, does not easily lead to a cure. Indeed, after the HGP was completed around the turn of the twenty-first century, academic physicians soon realized that most diseases could not be linked to single genes and that establishment of even strong correlations between multiple genes and a disease phenotype, let alone devising a therapy based on sequence information, was incredibly difficult and complex (Strohman 2002; Manolio et al. 2009).

From an epistemological perspective, there were also problems, despite claims to the contrary. Watson, who first led the US sequencing effort, stated in a book modestly titled *The Code of Codes*, "I have spent my career trying to get a chemical explanation for life, the explanation of why we are human beings and not monkeys. The reason, of course, is our DNA. If you can study life from the level of DNA, you have a real explanation for its processes" (Watson 1992, 164). However, DNA sequences do not explain anything: sequencing provides valuable ancillary information to an epistemic strategy but is not in itself an epistemic strategy.[5] Furthermore, the failure of the vast reservoir of structural molecular information provided by sequencing to directly solve biological puzzles casts doubt on the validity of the extreme reductionist approach. In fact, at the time of Watson's statement, molecular biology had reached a point of reckoning. According to Walter Gilbert, a Nobel laureate and biotech pioneer:[6]

> As a successful science, molecular biology has become a set of cookbook techniques. Its very success is producing an odd sort of reaction: all these wonderful techniques can simply be looked up in a handbook, and biologists seem to spend their time reading techniques and then cloning genes, or reading techniques and then sequencing DNA. Where is the biology? We are witnessing the last stage of the development of a technology and, as has happened a number of times before, many of the techniques of molecular biology will very soon leave the research laboratories entirely. We will purchase them externally as services; they will not be performed by research scientists. . . . The science will have moved on to the problem of what a sequence *means*, what the gene actually does. (Gilbert 1992, 93; emphasis original)

Some might argue that other scientists, cell biologists in particular, had already been studying what genes actually do when Gilbert made these remarks.

Nouveau Holism

In reaction to the overwhelming complexity of living systems uncovered by genomic sequencing projects as well as the difficulties of explaining those systems using sequencing data alone, some scientists embarked on a new approach called *systems biology*. There are various interpretations of what systems biology is by both scientists and philosophers. I will not engage in these discussions here because they are beyond the scope of what I hope to accomplish. Instead I will focus on one kind of systems biology created by two of its earliest practitioners, Trey Ideker and Lee Hood. Hood, with

---

5. This conclusion is shared by Ulrich Krohs and Werner Callebaut: 'The huge amounts of data produced by the genome projects were in fact collected almost free of any theoretical burden; as could have been expected, they turned out to explain almost nothing" (Krohs and Callebaut 2007, 184).

6. Gilbert shared the 1980 Nobel Prize for devising methods to sequence DNA.

Ideker as a student, founded the first institute focused on systems biology in 2000, and Ideker continues to run one of the major laboratories in the field (Ideker and Hood 2019).

Systems biology was built on not only genomics but also other -*omic* projects that followed and were dependent on the probes and reagents created using DNA sequences. All of these relied on so-called high-throughput procedures that automated technical steps and were thus able to process hundreds if not thousands of individual samples at a time. The most important of these was the comprehensive analysis of gene expression as defined by the production of mRNAs. Such *transcriptomes* were developed using microarrays of thousands of short DNA sequences attached to a solid support that could detect the binding of individual mRNAs that had been converted into complementary DNA using reverse transcriptase. Other procedures produced *proteomes* listing all the proteins synthesized by transcribed mRNAs from a tissue or group of cells and *interactomes* that cataloged all *possible* protein interactions. Notably, the latter did not typically measure real protein interactions in cells or tissues but rather interactions between proteins expressed artificially from sequences inserted into model organisms (usually yeast cells).[7] Later efforts created *metabolomes* that identified and measured the amounts of small molecule metabolites. Each of these -*omic* projects generated enormous amounts of data that required computers for its collection, storage, display, and interpretation. Big data biology was born.

At the outset Hood and Ideker considered systems biology an information science that would "investigate the behavior and relationships of *all* the elements in a particular biological system while it is functioning" (Ideker, Galitski, and Hood 2001, 343; emphasis added). In this sense, systems biology is *holistic* because it attempts to look at the whole biological system at once (Krohs and Callebaut 2007, 204). Their intent was to integrate all the datasets produced from a particular biological system under one set of conditions and model the system computationally. Then the system would be perturbed, new datasets collected, and experimental results compared to predicted results generated from the original model. Since the model represents a hypothesis about how the system should operate, discrepancies between the experimental results and the model required reconciliation

7. In a story that may be apocryphal, a high-throughput screen for protein-protein interactions identified two proteins whose association would have never been suspected on the basis of their known activities. Although originally claimed as a new and important discovery, when each protein was finally localized within cells by immunofluorescence microscopy, they were found to be in separate compartments that had no chance of intermixing. Apparently the cellular context had some role in determining function.

of the two and generation of new hypotheses and revised models.[8] Ideker and his associates considered systems biology to "practice discovery science" by "defining all elements of the system" (Ideker, Galitski, and Hood 2001, 344). As they state, "Discovery science lies in contrast to hypothesis-driven science, which creates hypotheses and attempts to distinguish them experimentally. The integration of these two approaches, discovery and hypothesis-driven science, is one of the mandates of systems biology" (Ideker, Galitski, and Hood 2001, 344).

As systems biology developed in multiple laboratories, the amount of data collected began to be a problem. There were initially few standards for the formats used to record the data, and meetings among systems biologists sometimes devolved into discussions about how to make one lab's dataset "talk" to another lab's dataset, with no reference to biology at all.[9] Eventually, standards were established and data began to be represented using graph theory, with relationships between components displayed as a series of "nodes," or vertices, and "edges," or lines, connecting the vertices.

Systems biologists such as Ideker are interested in using the vast amounts of data available to them to discover biological mechanisms. The problem, however, is that the interactions between genes and gene products represented in network diagrams are flat or nonhierarchical, while mechanisms are inherently hierarchical in their organization. The solution to this, as described by William Bechtel in his analysis of this issue, is to utilize gene ontologies as a way of seeding mechanistic insights into the networks generated by systems biology (Bechtel 2017, 2020). Gene ontologies, or GOs, are data bases that were established beginning in the late 1990s as a way of systematically linking and clustering newly sequenced genes and gene products from a variety of organisms around particular cellular functions based on the published literature (Ashburner et al. 2000; Leonelli 2016; Bechtel 2017). As stated in the paper announcing the formation of the Gene Ontology Consortium, "The goal of the Consortium is to produce a structured, precisely defined, common, controlled vocabulary for describing the roles of genes and gene products in any organism" (Ashburner et al. 2000, 26). The "biological annotations" of gene products in the GO data bases are organized according to three categories: biological process, molecular

8. There appear to be similarities between this approach and the strategy used by Ed Munro that was described in chapter 11. One major difference, however, is that Munro's models are initially built with a minimum number of components while the Ideker-Hood models attempt to use every component (genes, transcripts, proteins, etc.) that can be detected and catalogued.

9. I personally witnessed this at a meeting organized by Hood in about 2005.

function, and cellular location. For example, biological processes might be DNA replication or DNA repair, organized under the overall category of DNA metabolism. Molecular functions, in reference to these same processes, include DNA binding or particular enzymatic activities. Cellular location, as might be expected in this case, is mainly in the nucleus. According to Bechtel:

> The knowledge represented in GO is mechanistic knowledge. The Cell Component ontology identifies different components of the cell that are associated with specific cell activities and so the locus of one or several mechanisms. The part_of relation between different components at different levels of the hierarchy corresponds to the fact that parts . . . are structural constituents of particular mechanisms. . . . Identifying a gene as active in a part of the cell that can be situated among other parts, as engaging in a molecular reaction that is categorized among the various reactions in the cell, and as contributing to a specific biological process that falls under more general biological process, specifies its role in a mechanism. (Bechtel 2017, 636)

When coupled with the gene and gene product associations identified by the high-throughput technologies of systems biology and represented in flat networks, GO provides a way to link new genes to those that through the annotation process of GO are known parts of identified mechanisms. The new genes then become candidate component parts of the same mechanism (something called *guilt by association*) (Bechtel 2017, 632). Through iterative procedures such as Active Interaction Mapping developed by the Ideker lab, the new candidates are then annotated in GO data bases through literature searches and analysis that reinforce the functional hypotheses generated by the initial network interactions (Bechtel 2017; Farré et al. 2017; Kramer et al. 2017; Bechtel 2020).

While my brief description of GO and its use in expanding mechanistic understanding of biological processes in conjunction with the large datasets generated by systems biology does not really do it justice, my final point is straightforward. Even with all the information generated by advanced molecular techniques, the key elements in mechanistic explanations informed by this information are conventional experimental studies carried out using epistemic strategies similar to those pioneered by cell biologists and described in preceding chapters. If one traces the GO annotations for particular gene products back into the experimental literature, then you often find studies integrating microscopic localization with biochemical analysis. Systems biology has not replaced conventional (cell) biology. It has just dramatically improved the pool of molecular information available and provided powerful technologies that make conventional procedures even more effective in explaining biological phenomena.

Cells and Molecules

In the prologue I asked whether cell biology was the most significant molecular biology of the twentieth century (and the twenty-first). The answer to this question is for the reader to decide. My goal has been to describe how integration of the disciplines of biochemistry, biophysics, and cytology led to a way to discover molecular mechanisms that explain biological phenomena. I have done this through a focus on the 1975 discoveries of Günter Blobel and their aftermath, discoveries that I believe deserve as much attention as those made in 1953. More briefly, I also tried to point out how the inherent productivity of this approach, this epistemic strategy, continues to provide mechanistic understanding of a multitude of cellular processes at the molecular level, even when there are no more molecules to be discovered.

My descriptions of many of the post-1970 events are based in part on conversations and correspondence with scientists mentioned (and not mentioned) in the text. A complete list of this group is included in the preface. I typically do not cite these sources directly, mainly because any statements that I make are not derived from a single source but are distilled from multiple sources.

I was fortunate that when I began research for this book most of the key individuals were not only alive but also still scientifically active, a fact that, inevitably, is gradually changing. The downside of having direct access to living participants is that to my knowledge almost no archival sources for this period exist. Blobel, whom I spoke with three times and corresponded with extensively, provided me with a few documents, including the annotated manuscripts for the seminal 1975 papers and associated correspondence. He claimed that he had thrown out most other material related to the early work on the signal hypothesis, but I always believed that much more was preserved than he either remembered or was willing to admit. After his death in 2018, his papers became inaccessible pending their transfer to the Rockefeller Archive Center. If and when this occurs, they will not be available for examination until they are cataloged, a process that will take several years. A small amount of correspondence was also shared with me by the late Donald Steiner of the University of Chicago. Specific references to materials from Blobel and Steiner are given in footnotes.

An uncurated archive of George Palade's papers exists in the National Library of Medicine. I am most grateful to the library staff for permitting me to review this material in its current state. I also made limited use of the Bernard Davis papers archived at Harvard Medical School. References to both the Palade and Davis papers are given in footnotes. Albert Claude's papers are held privately by his daughter Philippa Claude, with the goal of transferring them to the Rockefeller archive at some point. I did not examine these papers directly, but Carol Moberg's references to them in her 2012 book, *Seeing an Unseen World*, were very useful (Moberg 2012). Moberg's book, which contains a number of first-person accounts of discoveries by cell biologists at Rockefeller University, is, in general, a very valuable resource for further historical research.

REFERENCES

Adam, S. A., R. S. Marr, and L. Gerace. 1990. "Nuclear Protein Import in Permeabilized Mammalian Cells Requires Soluble Cytoplasmic Factors." *Journal of Cell Biology* 111 (3): 807–16.

Adelman, M. R., G. Blobel, and D. D. Sabatini. 1973. "An Improved Cell Fractionation Procedure for the Preparation of Rat Liver Membrane-Bound Ribosomes." *Journal of Cell Biology* 56 (1): 191–205.

Adelman, M. R., D. D. Sabatini, and G. Blobel. 1973. "Ribosome-Membrane Interaction: Nondestructive Disassembly of Rat Liver Rough Microsomes into Ribosomal and Membranous Components." *Journal of Cell Biology* 56 (1): 206–29.

Adesnik, M., M. Lande, T. Martin, and D. D. Sabatini. 1976. "Retention of mRNA on the Endoplasmic Reticulum Membranes after In Vivo Disassembly of Polysomes by an Inhibitor of Initiation." *Journal of Cell Biology* 71 (1): 307–13.

Akopian, D., K. Shen, X. Zhang, and S.-o. Shan. 2013. "Signal Recognition Particle: An Essential Protein-Targeting Machine." *Annual Review of Biochemistry* 82 (1): 693–721.

Alberts, B., A. Johnson, J. Lewis, D. Morgan, M. Raff, K. Roberts, and P. Walter. 2015. *Molecular Biology of the Cell.* 6th ed. New York: Garland Science, Taylor and Francis.

Allen, D. W., and P. C. Zamecnik. 1962. "The Effect of Puromycin on Rabbit Reticulocyte Ribosomes." *Biochimica et Biophysica Acta* 55: 865–74.

Allen, G. 1975. *Life Science in the Twentieth Century.* New York: John Wiley & Sons.

Allfrey, V., M. M. Daly, and A. E. Mirsky. 1953. "Synthesis of Protein in the Pancreas. II. The Role of Ribonucleoprotein in Protein Synthesis." *Journal of General Physiology* 37 (2): 157–75.

Allman, G. J. 1879. "Protoplasm and Life." *Popular Science Monthly:* 722–49.

Alpers, S. 1983. *The Art of Describing.* Chicago: University of Chicago Press.

Altman, L. K. 1999. "Nobel Prize Winner." *New York Times,* Oct. 12, 1999, A1.

Altmann, R. 1890. *Die Elementarorganismen und ihre Beziehungen zu den Zellen.* Leipzig: Verlag Von Veit und Comp.

Apodaca, G., M. Bomsel, J. Arden, P. P. Breitfeld, K. Tang, and K. E. Mostov. 1991. "The Polymeric Immunoglobulin Receptor: A Model Protein to Study Transcytosis." *Journal of Clinical Investigation* 87 (6): 1877–82.

Ashburner, M., C. A. Ball, J. A. Blake, D. Botstein, H. Butler, J. M. Cherry, A. P. Davis, K. Dolinski, S. S. Dwight, J. T. Eppig, M. A. Harris, D. P. Hill, L. Issel-Tarver, A. Kasarskis, S. Lewis, J. C. Matese, J. E. Richardson, M. Ringwald, G. M. Rubin, and G. Sherlock. 2000.

"Gene Ontology: Tool for the Unification of Biology. The Gene Ontology Consortium." *Nature Genetics* 25 (1): 25–29.

Aviv, H., and P. Leder. 1972. "Purification of Biologically Active Globin Messenger RNA by Chromatography on Oligothymidylic Acid-Cellulose." *Proceedings of the National Academy of Sciences* 69 (6): 1408–12.

Bacallao, R., K. Shiba, W. Wickner, and K. Ito. 1986. "The SecY Protein Can Act Post-Translationally to Promote Bacterial Protein Export*." *Journal of Biological Chemistry* 261 (27): 12907–10.

Baker, J. R. 1988. *The Cell Theory: A Restatement, History, and Critique.* Edited by John A. Moore. New York: Garland Publishing.

Balch, W., W. Dunphy, W. Braell, and J. E. Rothman. 1984. "Reconstitution of the Transport of Protein between Successive Compartments of the Golgi Measured by the Coupled Incorporation of N-Acetylglucosamine." *Cell* 39 (2 Pt. 1): 405–16.

Balch, W., B. Glick, and J. E. Rothman. 1984. "Sequential Intermediates in the Pathway of Intercompartmental Transport in a Cell-Free System." *Cell* 39 (3 Pt. 2): 525–36.

Bar, R. S., D. W. Deamer, and D. G. Cornwell. 1966. "Surface Area of Human Erythrocyte Lipids: Reinvestigation of Experiments on Plasma Membrane." *Science* 153 (3739): 1010–12.

Bassford, P., J. Beckwith, K. Ito, C. Kumamoto, S. Mizushima, D. Oliver, L. Randall, T. Silhavy, P. C. Tai, and B. Wickner. 1991. "The Primary Pathway of Protein Export in *E. coli.*" *Cell* 65 (3): 367–68.

Bayliss, W. M. 1918. *Principles of General Physiology.* Rev. 2nd ed. London: Longmans, Green, and Co.

Beale, L. 1869. "Proceedings of the Royal Microscopical Society: Protoplasm and Living Matter." *Quarterly Journal of Microscopical Science* s2–s9: 324–25.

Beams, H. W. 1943. "Ultracentrifugal Studies on Cytoplasmic Components and Inclusions." In *Biological Symposia: Frontiers in Cytochemistry,* edited by N. L. Hoerr, 71–90. Lancaster, PA: Jacques Cattell Press.

Bearn, A. G., and D. G. James. 1978. "Dr. William Harvey (1578–1657) and the Harvey Society." *Perspectives in Biology and Medicine* 21 (4): 524–35.

Bechtel, W. 2006. *Discovering Cell Mechanisms: The Creation of Modern Cell Biology.* Cambridge: Cambridge University Press.

———. 2011. "Mechanism and Biological Explanation." *Philosophy of Science* 78 (4): 533–57.

———. 2012. "Addressing the Vitalists' Challenge to Mechanistic Science: Dynamic Mechanistic Explanation." In *Vitalism and the Scientific Image in Post-Enlightenment Science, 1810–2010,* edited by S. Normandin and C. Wolfe, 1–25. Dordrecht: Springer.

———. 2017. "Using the Hierarchy of Biological Ontologies to Identify Mechanisms in Flat Networks." *Biology and Philosophy* 32 (5): 627–49.

———. 2020. "Hierarchy and Levels: Analysing Networks to Study Mechanisms in Molecular Biology." *Philosophical Transactions of the Royal Society of London B Biological Sciences* 375 (1796): 20190320–9. Accessed Feb. 24. https://doi.org/10.1098/rstb.2019.0320.

Bechtel, W., and A. Abrahamsen. 2010. "Dynamic Mechanistic Explanation: Computational Modeling of Circadian Rhythms as an Exemplar for Cognitive Science." *Studies in History and Philosophy of Science* 41 (3): 321–33.

Bechtel, W., and R. C. Richardson. 1993. *Discovering Complexity.* Princeton, NJ: Princeton University Press.

Becker, T., S. Bhushan, A. Jarasch, J.-P. Armache, S. Funes, F. Jossinet, J. Gumbart, T. Mielke, O. Berninghausen, K. Schulten, E. Westhof, R. Gilmore, E. C. Mandon, and

R. Beckmann. 2009. "Structure of Monomeric Yeast and Mammalian Sec61 Complexes Interacting with the Translating Ribosome." *Science* 326 (5958): 1369–73.

Beckers, C. J., D. S. Keller, and W. E. Balch. 1987. "Semi-Intact Cells Permeable to Macromolecules: Use in Reconstitution of Protein Transport from the Endoplasmic Reticulum to the Golgi Complex." *Cell* 50 (4): 523–34.

Beckmann, R., D. Bubeck, R. Grassucci, P. Penczek, A. Verschoor, G. Blobel, and J. Frank. 1997. "Alignment of Conduits for the Nascent Polypeptide Chain in the Ribosome-Sec61 Complex." *Science* 278 (5346): 2123–26.

Beckmann, R., C. M. Spahn, N. Eswar, J. Helmers, P. A. Penczek, A. Sali, J. Frank, and G. Blobel. 2001. "Architecture of the Protein-Conducting Channel Associated with the Translating 80S Ribosome." *Cell* 107 (3): 361–72.

Beckwith, J. 1996. "The Hegemony of the Gene: Reductionism in Molecular Biology." In *The Philosophy and History of Molecular Biology: New Perspectives*, edited by S. Sarkar, 171–83. Dordrecht: Kluwer Academic Publishers.

———. 2002. *Making Genes, Making Waves.* Cambridge, MA: Harvard University Press.

Bennett, V. 1978. "Purification of an Active Proteolytic Fragment of the Membrane Attachment Site for Human Erythrocyte Spectrin." *Journal of Biological Chemistry* 253 (7): 2292–99.

Bensley, R. R. 1933. "Studies on Cell Structure by the Freezing-Drying Method. IV. The Structure of the Interkinetic and Resting Nuclei." *Anatomical Record* 58 (1): 1–15.

Bensley, R. R., and I. Gersh. 1933a. "Studies on Cell Structure by the Freezing-Drying Method. I. Introduction." *Anatomical Record* 57 (3): 205–15.

———. 1933b. "Studies on Cell Structure by the Freezing-Drying Method. II. The Nature of the Mitochondria in the Hepatic Cell of Amblyostoma." *Anatomical Record* 57 (3): 217–33.

———. 1933c. "Studies on Cell Structure by the Freezing-Drying Method. III. The Distribution in Cells of the Basophil Substances, in Particular the Nissl Substance of the Nerve Cell." *Anatomical Record* 57 (4): 369–85.

Bensley, R. R., and N. L. Hoerr. 1934a. "Studies on Cell Structure by the Freezing-Drying Method V. The Chemical Basis of the Organization of the Cell." *Anatomical Record* 60 (3): 251–66.

———. 1934b. "Studies on Cell Structure by the Freezing-Drying Method VI. The Preparation and Properties of Mitochondria." *Anatomical Record* 60 (4): 449–55.

Bergmann, J. E., K. T. Tokuyasu, and S. J. Singer. 1981. "Passage of an Integral Membrane Protein, the Vesicular Stomatitis Virus Glycoprotein, through the Golgi Apparatus en Route to the Plasma Membrane." *Proceedings of the National Academy of Sciences* 78 (3): 1746–50.

Bernstein, H. D., M. A. Poritz, K. Strub, P. J. Hoben, S. Brenner, and P. Walter. 1989. "Model for Signal Sequence Recognition from Amino-Acid Sequence of 54k Subunit of Signal Recognition Particle." *Nature* 340 (6233): 482–86.

Berridge, M. J., and J. L. Oschman. 1972. *Transporting Epithelia.* New York: Academic Press.

Blackburn, P., G. Wilson, and S. Moore. 1977. "Ribonuclease Inhibitor from Placenta: Purification and Properties." *Journal of Biological Chemistry* 252: 5904–10.

Blobel, G. 1971a. "Isolation of a 5S RNA-Protein Complex from Mammalian Ribosomes." *Proceedings of the National Academy of Sciences* 68 (8): 1881–85.

———. 1971b. "Release, Identification, and Isolation of Messenger RNA from Mammalian Ribosomes." *Proceedings of the National Academy of Sciences* 68 (4): 832–35.

———. 1972. "Protein Tightly Bound to Globin mRNA." *Biochemical and Biophysical Research Communications* 47 (1): 88–95.

———. 1973. "A Protein of Molecular Weight 78,000 Bound to the Polyadenylate Region of Eukaryotic Messenger RNAs." *Proceedings of the National Academy of Sciences* 70 (3): 924–28.

———. 1974. "Ribosome-Membrane Interaction." *Acta biologica et medica germanica Zeitschrift für funktionelle Biowissenschaften* 33 (5–6): 711–13.

———. 1976. "Extraction from Free Ribosomes of a Factor Mediating Ribosome Detachment from Rough Microsomes." *Biochemical and Biophysical Research Communications* 68 (1): 1–7.

———. 1977. "Synthesis and Segregation of Secretory Proteins: The Signal Hypothesis." In *International Cell Biology, 1976–1977*, edited by K. R. Porter and B. R. Brinkley. New York: Rockefeller University Press.

———. 1978. "Mechanisms for the Intracellular Compartmentation of Newly Synthesized Proteins." edited by B. F. C. Clark, H. Klenow, and J. Zeuthen, 99–108. Oxford: Pergamon Press.

———. 1980. "Intracellular Protein Topogenesis." *Proceedings of the National Academy of Sciences* 77 (3): 1496–1500.

Blobel, G., and B. Dobberstein. 1975a. "Transfer of Proteins across Membranes. I. Presence of Proteolytically Processed and Unprocessed Nascent Immunoglobulin Light Chains on Membrane-Bound Ribosomes of Murine Myeloma." *Journal of Cell Biology* 67 (3): 835–51.

———. 1975b. "Transfer of Proteins across Membranes. II. Reconstitution of Functional Rough Microsomes from Heterologous Components." *Journal of Cell Biology* 67 (3): 852–62.

Blobel, G., and V. R. Potter. 1966. "Nuclei from Rat Liver: Isolation Method That Combines Purity with High Yield." *Science* 154 (757): 1662–65.

———. 1967a. "Ribosomes in Rat Liver: An Estimate of the Percentage of Free and Membrane-Bound Ribosomes Interacting with Messenger RNA In Vivo." *Journal of Molecular Biology* 28 (3): 539–42.

———. 1967b. "Studies on Free and Membrane-Bound Ribosomes in Rat Liver. II. Interaction of Ribosomes and Membranes." *Journal of Molecular Biology* 26 (2): 293–301.

———. 1967c. "Studies on Free and Membrane-Bound Ribosomes in Rat Liver. I. Distribution as Related to Total Cellular RNA." *Journal of Molecular Biology* 26 (2): 279–92.

Blobel, G., and D. D. Sabatini. 1970. "Controlled Proteolysis of Nascent Polypeptides in Rat Liver Cell Fractions. I. Location of the Polypeptides within Ribosomes." *Journal of Cell Biology* 45 (1): 130–45.

———. 1971a. "Dissociation of Mammalian Polyribosomes into Subunits by Puromycin." *Proceedings of the National Academy of Sciences* 68 (2): 390–94.

———. 1971b. "Ribosome-Membrane Interaction in Eukaryotic Cells." In *Biomembranes*, edited by L. A. Manson, 193–95. New York: Plenum.

Blobel, G., P. Walter, C. Chang, B. Goldman, A. Erickson, and V. Lingappa. 1979. "Translocation of Proteins across Membranes: The Signal Hypothesis and Beyond." *Symposium of the Society for Experimental Biology* 33: 9–36.

Bölter, B., and J. Soll. 2016. "Once Upon a Time—Chloroplast Protein Import Research from Infancy to Future Challenges." *Molecular Plant* 9 (6): 798–812.

Bonner, J. T. 1974. *On Development: The Biology of Form.* Cambridge, MA: Harvard University Press.

Borgese, D., G. Blobel, and D. D. Sabatini. 1973. "In Vitro Exchange of Ribosomal Subunits between Free and Membrane-Bound Ribosomes." *Journal of Molecular Biology* 74 (4): 415–38.

Borgese, N., J. Coy-Vergara, S. F. Colombo, and B. Schwappach. 2019. "The Ways of Tails: The Get Pathway and More." *Protein Journal* 38 (3): 289–305.

Borgese, N., W. Mok, G. Kreibich, and D. D. Sabatini. 1974. "Ribosomal-Membrane Interaction: In Vitro Binding of Ribosomes to Microsomal Membranes." *Journal of Molecular Biology* 88 (3): 559–80.

Bowen, R. H. 1926. "Studies on the Golgi Apparatus in Gland Cells. IV. A Critique of the Topography, Structure, and Function of the Golgi Apparatus in Glandular Tissue." *Quarterly Journal of Microscopical Science* 70: 419–49.

Bracegirdle, B. 1978. *A History of Microtechnique.* Ithaca, NY: Cornell University Press.

Brachet, J. 1960. *The Biological Role of Ribonucleic Acids.* Weizmann Memorial Lecture Series. Amsterdam: Elsevier.

Bradbury, S. 1967. *The Evolution of the Microscope.* Oxford: Pergamon Press.

Braell, W., W. Balch, D. Dobbertin, and J. E. Rothman. 1984. "The Glycoprotein That Is Transported between Successive Compartments of the Golgi in a Cell-Free System Resides in Stacks of Cisternae." *Cell* 39 (3 Pt. 2): 511–24.

Branton, D. 1966. "Fracture Faces of Frozen Membranes." *Proceedings of the National Academy of Sciences* 55 (5): 1048–56.

Branton, D., and R. B. Park, eds. 1968. *Papers on Biological Membrane Structure.* Boston: Little, Brown.

Bretscher, M. S. 1971a. "Major Human Erythrocyte Glycoprotein Spans the Cell Membrane." *Nature New Biology* 231 (25): 229–32.

———. 1971b. "A Major Protein Which Spans the Human Erythrocyte Membrane." *Journal of Molecular Biology* 59 (2): 351–57.

———. 1972. "Asymmetrical Lipid Bilayer Structure for Biological-Membranes." *Nature New Biology* 236 (61): 11–12.

Brigandt, I., and A. Love. 2017. "Reductionism in Biology." In *Stanford Encyclopedia of Philosophy (Spring 2017 Edition),* edited by Edward N. Zalta. https://plato.stanford.edu /archives/spr2017/entries/reduction-biology/.

Britten, R. J., and R. B. Roberts. 1960. "High-Resolution Density Gradient Sedimentation Analysis." *Science* 131 (3392): 32–33.

Brownlee, G., T. Harrison, M. Mathews, and C. Milstein. 1972. "Translation of Messenger RNA for Immunoglobulin Light Chains in a Cell-Free System from Krebs II Ascites Cells." *FEBS Letters* 23 (2): 244–48.

Brücke, E. v. 1898. "Die Elementarorganismen." In *Planzenphysiologische Abteilungen Von Ernst Von Brücke. 1844–1862,* edited by Alfred Fischer, 54–79. Leipzig: Wilhelm Engelmann.

Brundage, L., J. P. Hendrick, E. Schiebel, A. J. Driessen, and W. Wickner. 1990. "The Purified *E. coli* Integral Membrane Protein SecY/E Is Sufficient for Reconstitution of SecA-Dependent Precursor Protein Translocation." *Cell* 62 (4): 649–57.

Burian, R. M. 1996. "Underappreciated Pathways toward Molecular Genetics as Illustrated by Jean Brachet's Cytochemical Embryology." In *The Philosophy and History of Molecular Biology: New Perspectives,* edited by S. Sarkar, 67–85. Dordrecht: Kluwer Academic Publishers.

Burke, B., and L. Gerace. 1986. "A Cell Free System to Study Reassembly of the Nuclear-Envelope at the End of Mitosis." *Cell* 44 (4): 639–52.

Burnette, W. N. 1981. "'Western Blotting': Electrophoretic Transfer of Proteins from Sodium Dodecyl Sulfate-Polyacrylamide Gels to Unmodified Nitrocellulose and Radiographic Detection with Antibody and Radioiodinated Protein A." *Analytical Biochemistry* 112: 195–203.

Bütschli, O. 1894. *Investigations on Microscopic Foams and Protoplasm*. Translated by E. A. Minchin. London: Adam and Charles Black.

Callebaut, W. 2012. "Scientific Perspectivism: A Philosopher of Science's Response to the Challenge of Big Data Biology." *Studies in History and Philosophy of Biological and Biomedical Sciences* 43 (1): 69–80.

Caro, L. G., and G. E. Palade. 1964. "Protein Synthesis, Storage, and Discharge in the Pancreatic Exocrine Cell: An Autoradiographic Study." *Journal of Cell Biology* 20 (3): 473–95.

Casanova, J. E., G. Apodaca, and K. E. Mostov. 1991. "An Autonomous Signal for Basolateral Sorting in the Cytoplasmic Domain of the Polymeric Immunoglobulin Receptor." *Cell* 66 (1): 65–75.

Caspersson, T. 1947. "The Relations between Nucleic Acid and Protein Synthesis." *Symposia of the Society of Experimental Biology* 1: 127–51.

Cavalier-Smith, T. 2004. "The Membranome and Membrane Heredity in Development and Evolution." In *Organelles, Genomes, and Eukaryote Phylogeny*, edited by R. P. Hirt and D. S. Horner, 335–51. Boca Raton, FL: CRC Press.

Chambers, R. 1915. "Microdissection Studies on the Germ Cell." *Science* 41 (1051): 290–93.

———. 1918. "The Microvivisection Method." *Biological Bulletin* 34 (2): 121–36.

———. 1922. "A Micro Injection Study on the Permeability of the Starfish Egg." *Journal of General Physiology* 5 (2): 189–193.

———. 1924. "The Physical Structure of Protoplasm as Determined by Microdissection and Injection." In *General Cytology: A Textbook of Cellular Structure and Function for Students of Biology and Medicine*, edited by E. V. Cowdry, 235–310. Chicago: University of Chicago Press.

Chambers, R., and E. L. Chambers. 1961. *Explorations into the Nature of the Living Cell*. Cambridge, MA: Harvard University Press.

Chan, S. J., P. Keim, and D. F. Steiner. 1976. "Cell-Free Synthesis of Rat Preproinsulins: Characterization and Partial Amino Acid Sequence Determination." *Proceedings of the National Academy of Sciences* 73 (6): 1964–68.

Chang, C., P. Model, and G. Blobel. 1979. "Membrane Biogenesis: Cotranslational Integration of the Bacteriophage F1 Coat Protein into an *Escherichia coli* Membrane Fraction." *Proceedings of the National Academy of Sciences* 76 (3): 1251–55.

Chua, N.-H., G. Blobel, and P. Siekevitz. 1973. "Isolation of Cytoplasmic and Chloroplast Ribosomes and Their Dissociation into Active Subunits from *Chlamydomonas reinhardtii*." *Journal of Cell Biology* 57 (3): 798–814.

Chua, N.-H., G. Blobel, P. Siekevitz, and G. E. Palade. 1973. "Attachment of Chloroplast Polysomes to Thylakoid Membranes in *Chlamydomonas reinhardtii*." *Proceedings of the National Academy of Sciences* 70 (5): 1554–58.

———. 1976. "Periodic Variations in the Ratio of Free to Thylakoid-Bound Chloroplast Ribosomes during the Cell Cycle of *Chlamydomonas reinhardtii*." *Journal of Cell Biology* 71 (2): 497–514.

Chua, N.-H., and G. W. Schmidt. 1978. "Post-Translational Transport into Intact Chloroplasts of a Precursor to the Small Subunit of Ribulose-1,5-Bisphosphate Carboxylase." *Proceedings of the National Academy of Sciences* 75: 6110–14.

Claude, A. 1937. "Fractionation of Chicken Tumor Extracts by High Speed Centrifugation." *American Journal of Cancer* 30: 742–45.

———. 1938a. "Concentration and Purification of Chicken Tumor I Agent." *Science* 87 (2264): 467–68.

———. 1938b. "A Fraction from Normal Chick Embryo Similar to the Tumor-Producing Fraction of Chicken Tumor I." *Proceedings of the Society for Experimental Biology and Medicine* 39: 398–403.

———. 1939. "Chemical Composition of the Tumor-Producing Fraction of Chicken Tumor I." *Science* 90 (2331): 213–14.

———. 1940. "Particulate Components of Normal and Tumor Cells." *Science* 91 (2351): 77–78.

———. 1941. "Particulate Components of Cytoplasm." *Cold Spring Harbor Symposia on Quantitative Biology* 9: 263–71.

———. 1943a. "The Constitution of Protoplasm." *Science* 97 (2525): 451–56.

———. 1943b. "Distribution of Nucleic Acids in the Cell and the Morphological Constitution of the Cytoplasm." In *Frontiers in Cytochemistry*, edited by N. L. Hoerr, 111–29. Biological Symposia. Lancaster, PA: Jacques Cattell Press.

———. 1944. "The Constituion of Mitochondria and Microsomes, and the Distribution of Nucleic Acid in the Cytoplasm of a Leukemic Cell." *Journal of Experimental Medicine* 80 (1): 19–29.

———. 1946a. "Fractionation of Mammalian Liver Cells by Differential Centrifugation: I. Problems, Methods, and Preparation of Extract." *Journal of Experimental Medicine* 84 (1): 51–59.

———. 1946b. "Fractionation of Mammalian Liver Cells by Differential Centrifugation: II. Experimental Procedures and Results" *Journal of Experimental Medicine* 84 (1): 61–89.

———. 1948. "Studies on Cells: Morphology, Chemical Constitution and Distribution of Biochemical Functions." *Harvey Lectures* 43: 121–64.

Claude, A., and E. F. Fullam. 1945. "An Electron Microscope Study of Isolated Mitochondria: Method and Preliminary Results." *Journal of Experimental Medicine* 81 (1): 51–62.

———. 1946. "The Preparation of Sections of Guinea Pig Liver for Electron Microscopy." *Journal of Experimental Medicine* 83 (6): 499–503.

Claude, A., K. R. Porter, and E. G. Pickels. 1947. "Electron Microscope Study of Chicken Tumor Cells*." *Cancer Research* 7: 421–30.

Coleman, W. 1965. "Cell, Nucleus, and Inheritance: An Historical Study." *Proceedings of the American Philosophical Society* 109 (3): 124–58.

———. 1977. *Biology in the Nineteenth Century: Problems of Form, Function, and Transformation*. Cambridge: Cambridge University Press.

Cowdry, E. V., ed. 1924a. *General Cytology: A Textbook of Cellular Structure and Function for Students of Biology and Medicine*. Chicago: University of Chicago Press.

———. 1924b. "Mitochondria, Golgi Apparatus, and Chromidial Substance." In *General Cytology: A Textbook of Cellular Structure and Function for Students of Biology and Medicine*, edited by E. V. Cowdry, 311–82. Chicago: University of Chicago Press.

Creager, A. N. H. 2002. *The Life of a Virus*. Chicago: University of Chicago Press.

Crick, F. H. C. 1958. "On Protein Synthesis." *Symposium of the Society for Experimental Biology* 12: 138–63.

———. 1970. "Central Dogma of Molecular Biology." *Nature* 227 (5258): 561–63.

Crowley, K. S., G. D. Reinhart, and A. E. Johnson. 1993. "The Signal Sequence Moves through a Ribosomal Tunnel into a Noncytoplasmic Aqueous Environment at the ER Membrane Early in Translocation." *Cell* 73 (6): 1101–15.

Dallner, G., S. Orrenius, and A. Bergstrand. 1963. "Isolation and Properties of Rough and Smooth Vesicles from Rat Liver." *Journal of Cell Biology* 16 (2): 426–30.

Dallner, G., P. Siekevitz, and G. E. Palade. 1966a. "Biogenesis of Endoplasmic Reticulum Membranes. I. Structural and Chemical Differentiation in Developing Rat Hepatocyte." *Journal of Cell Biology* 30 (1): 73–96.

———. 1966b. "Biogenesis of Endoplasmic Reticulum Membranes. II. Synthesis of Constitutive Microsomal Enzymes in Developing Rat Hepatocyte." *Journal of Cell Biology* 30 (1): 97–117.

Danielli, J. F. 1953. *Cytochemistry: A Critical Approach*. New York: John Wiley & Sons.

———. 1958. "Surface Chemistry and Cell Membranes." In *Surface Phenomena in Chemistry and Biology*, edited by J. F. Danielli, K. G. A. Pankhurst and A. C. Riddiford, 246–65. New York: Pergamon Press.

———. 1975. "The Bilayer Hypothesis of Membrane Structure." In *Cell Membranes: Biochemistry, Cell Biology, and Pathology*, edited by G. Weissman and R. Claiborne, 3–12. New York: HP Publishing.

Danielli, J. F., and H. Davson. 1935. "A Contribution to the Theory of Permeability of Thin Films." *Journal of Cellular Physiology* 5 (4): 495–508.

Danielli, J. F., and E. N. Harvey. 1935. "The Tension at the Surface of Mackerel Egg Oil, with Remarks on the Nature of the Cell Surface." *Journal of Cellular Physiology* 5 (4): 483–94.

Darden, L. 1991. *Theory Change in Science*. Oxford: Oxford University Press.

Daston, L., and P. Galison. 2010. *Objectivity*. New York: Zone Books.

Date, T., J. M. Goodman, and W. T. Wickner. 1980. "Procoat, the Precursor of M13 Coat Protein, Requires an Electrochemical Potential for Membrane Insertion." *Proceedings of the National Academy of Sciences* 77 (8): 4669–73.

Davey, J., S. M. Hurtley, and G. Warren. 1985. "Reconstitution of an Endocytic Fusion Event in a Cell-Free System." *Cell* 42 (Pt. 2): 643–52.

Davidson, E. H. 1991. "Spatial Mechanisms of Gene Regulation in Metazoan Embryos." *Development* 113 (1): 1–26.

Davis, B. D., and P. C. Tai. 1980. "The Mechanism of Protein Secretion across Membranes." *Nature* 283 (5746): 433–38.

Davson, H., and J. F. Danielli. 1943. *The Permeability of Natural Membranes*. Cambridge: Cambridge University Press.

———. 1952. *The Permeability of Natural Membranes*. 2nd ed. Cambridge: Cambridge University Press.

Dawes, A. T., and E. M. Munro. 2011. "Par-3 Oligomerization May Provide an Actin-Independent Mechanism to Maintain Distinct Par Protein Domains in the Early Caenorhabditis Elegans Embryo." *Biophysical Journal* 101 (6): 1412–22.

De Robertis, E. D. P., W. W. Nowinski, and F. A. Saez. 1948. *General Cytology*. Philadelphia: W. B. Saunders.

Deshaies, R. J., and R. Schekman. 1987. "A Yeast Mutant Defective at an Early Stage in Import of Secretory Protein Precursors into the Endoplasmic Reticulum." *Journal of Cell Biology* 105 (2): 633–45.

Devillers-Thiery, A., T. Kindt, G. Scheele, and G. Blobel. 1975. "Homology in Amino-Terminal Sequence of Precursors to Pancreatic Secretory Proteins." *Proceedings of the National Academy of Sciences* 72 (12): 5016–20.

Dickman, S. R., and E. Bruenger. 1962. "Incorporation of Amino Acids into Protein by Intracellular Particles from Dog Pancreas." *Biochimica Biophysica Acta* 63 (3): 522–24.

Dobberstein, B., and G. Blobel. 1977. "Functional Interaction of Plant Ribosomes with Animal Microsomal Membranes." *Biochemical and Biophysical Research Communications* 74 (4): 1675–82.

Dobberstein, B., G. Blobel, and N.-H. Chua. 1977. "In Vitro Synthesis and Processing of a Putative Precursor for the Small Subunit of Ribulose-1,5-Bisphosphate Carboxylase of Chlamydomonas reinhardtii." *Proceedings of the National Academy of Sciences* 74 (3): 1082–85.

Driessen, A. J., and W. Wickner. 1990. "Solubilization and Functional Reconstitution of the Protein-Translocation Enzymes of *Escherichia coli.*" *Proceedings of the National Academy of Sciences* 87 (8): 3107–11.

Dröscher, A. 1999. "From the "Apparato Reticolare Interno" to "the Golgi": 100 Years of Golgi Apparatus Research." *Virchows Archiv* 434: 103–7.

Dunphy, W., R. Brands, and J. E. Rothman. 1985. "Attachment of Terminal N-Acetylglucosamine to Asparagine-Linked Oligosaccharides Occurs in Central Cisternae of the Golgi Stack." *Cell* 40 (2): 463–72.

Dunphy, W., and J. E. Rothman. 1983. "Compartmentation of Asparagine-Linked Oligosaccharide Processing in the Golgi Apparatus." *Journal of Cell Biology* 97 (1): 270–75.

Elzen, B. 1988. *Scientists and Rotors.* Enschede, Netherlands: Alfa.

Emr, S. D., S. Hanley-Way, and T. J. Silhavy. 1981. "Suppressor Mutations That Restore Export of a Protein with a Defective Signal Sequence." *Cell* 23 (1): 79–88.

Engelman, D. M. 1971. "Lipid Bilayer Structure in the Membrane of Mycoplasm laidlawii." *Journal of Molecular Biology* 58: 153–65.

Engelman, D. M., and J. E. Rothman. 1972. "The Planar Organization of Lecithin-Cholesterol Bilayers." *Journal of Biological Chemistry* 247 (11): 3694–97.

Engelman, D. M., and T. A. Steitz. 1981. "The Spontaneous Insertion of Proteins into and across Membranes: The Helical Hairpin Hypothesis." *Cell* 23 (2): 411–22.

Erickson, A. H., and G. Blobel. 1979. "Early Events in the Biosynthesis of the Lysosomal Enzyme Cathepsin D." *Journal of Biological Chemistry* 254 (23): 11771–74.

Erickson, A. H., G. E. Conner, and G. Blobel. 1981. "Biosynthesis of a Lysosomal Enzyme: Partial Structure of Two Transient and Functionally Distinct NH2-Terminal Sequences in Cathepsin D." *Journal of Biological Chemistry* 256 (21): 11224–31.

Ernster, L., P. Siekevitz, and G. E. Palade. 1962. "Enzyme-Structure Relationships in the Endoplasmic Reticulum: A Morphological and Biochemical Study." *Journal of Cell Biology* 15 (3): 541–62.

Evans, E. A., R. Gilmore, and G. Blobel. 1986. "Purification of Microsomal Signal Peptidase as a Complex." *Proceedings of the National Academy of Sciences* 83: 581–85.

Ezzell, C. 2000. "The Biologist and the Cathedral." *Scientific American* 282 (5): 38–40.

Farquhar, M. G. 2012. "A Man for All Seasons: Reflections on the Life and Legacy of George Palade." *Annual Review of Cell and Developmental Biology* 28: 1–28.

Farquhar, M. G., and G. E. Palade. 1981. "The Golgi Apparatus (Complex)—(1954–1981)—from Artifact to Center Stage." *Journal of Cell Biology* 91, no. 2 (1): 77s–103s.

Farré, J.-C., M. Kramer, T. Ideker, and S. Subramani. 2017. "Active Interaction Mapping as a Tool to Elucidate Hierarchical Functions of Biological Processes." *Autophagy* 13 (7): 1248–49.

Fawcett, D. W. 1986. *A Textbook of Histology.* 11th ed. Philadelphia: W. B. Saunders.

Flemming, W. 1882. *Zellsubstanz, Kern und Zelltheilung.* Leipzig: Verlag von F. C. W. Vogel.

Freienstein, C., and G. Blobel. 1974. "Use of Eukaryotic Native Small Ribosomal Subunits for the Translation of Globin Messenger RNA." *Proceedings of the National Academy of Sciences* 71 (9): 3435–39.

Fries, E., and J. E. Rothman. 1980. "Transport of Vesicular Stomatitis Virus Glycoprotein in a Cell-Free Extract." *Proceedings of the National Academy of Sciences* 77 (7): 3870–74.

———. 1981. "Transient Activity of Golgi-Like Membranes as Donors of Vesicular Stomatitis Viral Glycoprotein in Vitro." *Journal of Cell Biology* 90 (3): 697–704.

Fruton, J. S. 1999. *Proteins, Enzymes, and Genes: The Interplay of Chemistry and Biology*. New Haven, CT: Yale University Press.

Frye, L. D., and M. Edidin. 1970. "The Rapid Intermixing of Cell Surface Antigens after Formation of Mouse-Human Heterokaryons." *Journal of Cell Science* 7 (2): 319–35.

Fulgosi, H., J. Soll, and M. Inaba-Sulpice. 2004. "Protein Translocation Machinery in Chloroplasts and Mitochondria: Structure, Function, and Evolution." In *Organelles, Genomes, and Eukaryote Phylogeny*, edited by R. P. Hirt and D. S. Horner, 259–87. Boca Raton, FL: CRC Press.

Gardner, R. L. 2001. "The Initial Phase of Embryonic Patterning in Mammals." *International Review of Cytology* 203: 233–90.

Garnier, C. J. 1900. "Contribution a l'edude de la structure et du fonctionnement des cellules glandulaires: Du rôle de l'ergastoplasme dans la sécrétion." *Journal de l'anatomie et de la physiologie normales et pathologiques de l'homme et des animaux* 36: 22–98.

Garvey, J. S., N. E. Cremer, and D. H. Sussdorf. 1977. *Methods in Immunology*. 3rd ed. Reading, MA: Benjamin/Cummings.

Geison, G. L. 1969. "The Protoplasmic Theory of Life and the Vitalist-Mechanist Debate." *Isis* 60 (3): 272–92.

Gerace, L., and G. Blobel. 1980. "The Nuclear Envelope Lamina Is Reversibly Depolymerized during Mitosis." *Cell* 19 (1): 277–87.

Gerace, L., A. Blum, and G. Blobel. 1978. "Immunocytochemical Localization of the Major Polypeptides of the Nuclear Pore Complex-Lamina Fraction: Interphase and Mitotic Distribution." *Journal of Cell Biology* 79 (2 Pt. 1): 546–66.

Gersh, I. 1932. "The Altmann Technique for Fixation by Drying While Freezing." *Anatomical Record* 53 (3): 309–37.

Giere, R. N. 1997. *Understanding Scientific Reasoning*. 4th ed. Fort Worth, TX: Harcourt Brace College Publishers.

Gilbert, W. 1992. "A Vision of the Grail." In *The Code of Codes*, edited by D. Kevles and L. Hood. Cambridge, MA: Harvard University Press.

Gilmore, R., and G. Blobel. 1983. "Transient Involvement of Signal Recognition Particle and Its Receptor in the Microsomal Membrane Prior to Protein Translocation." *Cell* 35 (3 Pt. 2): 677–85.

———. 1985. "Translocation of Secretory Proteins across the Microsomal Membrane Occurs through an Environment Accessible to Aqueous Perturbants." *Cell* 42 (2): 497–505.

Gilmore, R., G. Blobel, and P. Walter. 1982. "Protein Translocation across the Endoplasmic Reticulum. I. Detection in the Microsomal Membrane of a Receptor for the Signal Recognition Particle." *Journal of Cell Biology* 95 (2 Pt. 1): 463–69.

Gilmore, R., P. Walter, and G. Blobel. 1982. "Protein Translocation across the Endoplasmic Reticulum. II. Isolation and Characterization of the Signal Recognition Particle Receptor." *Journal of Cell Biology* 95 (2 Pt. 1): 470–77.

Godfrey-Smith, P. 2014. *Philosophy of Biology*. Princeton, NJ: Princeton University Press.

Godfrey-Smith, P., and K. Sterelny. 2007. "Biological Information." In *The Stanford Encyclopedia of Philosophy* (Summer 2016 Edition). https://plato.stanford.edu/archives/sum2016/entries/information-biological/.

Goethe, J. W. v. 2016. "On Morphology (Excerpt)." In *The Essential Goethe*, edited by M. Bell, 977–983. Princeton, NJ: Princeton University Press.

Goldberg, D. E., and S. Kornfeld. 1983. "Evidence for Extensive Subcellular Organization of Asparagine-Linked Oligosaccharide Processing and Lysosomal Enzyme Phosphorylation." *Journal of Biological Chemistry* 258 (5): 3159–65.

Goldman, B., and G. Blobel. 1978. "Biogenesis of Peroxisomes: Intracellular Site of Synthesis of Catalase and Uricase." *Proceedings of the National Academy of Sciences* 75 (10): 5066–70.

———. 1981. "In Vitro Biosynthesis, Core Glycosylation, and Membrane Integration of Opsin." *Journal of Cell Biology* 90 (1): 236–42.

Goldstein, J. L., M. S. Brown, R. G. Anderson, D. W. Russell, and W. J. Schneider. 1985. "Receptor-Mediated Endocytosis: Concepts Emerging from the Ldl Receptor System." *Annual Review of Cell Biology* 1 (1): 1–39.

Gombrich, E. H. [1969] 2000. *Art and Illusion.* Millenium ed. Princeton, NJ: Princeton University Press.

Görlich, D., E. Hartmann, S. Prehn, and T. A. Rapoport. 1992. "A Protein of the Endoplasmic Reticulum Involved Early in Polypeptide Translocation." *Nature* 357 (6373): 47–52.

Görlich, D., S. Prehn, E. Hartmann, K. U. Kalies, and T. A. Rapoport. 1992. "A Mammalian Homolog of Sec61p and SecYp Is Associated with Ribosomes and Nascent Polypeptides during Translocation." *Cell* 71 (3): 489–503.

Görlich, D., and T. A. Rapoport. 1993. "Protein Translocation into Proteoliposomes Reconstituted from Purified Components of the Endoplasmic Reticulum Membrane." *Cell* 75 (4): 615–30.

Gorman, M. M. 1993. "Ontological Priority and John Duns Scotus." *Philosophical Quarterly* 43 (173): 460–71.

Gorter, E., and F. Grendel. 1925. "On Bimolecular Layers of Lipoids on the Chromocytes of the Blood." *Journal of Experimental Medicine* 41 (4): 439–43.

Goryachev, A. B., and M. Leda. 2017. "Many Roads to Symmetry Breaking: Molecular Mechanisms and Theoretical Models of Yeast Cell Polarity." *Molecular Biology of the Cell* 28 (3): 370–80.

Graham, T. 1861. "Liquid Diffusion Applied to Analysis." *Philosophical Transactions* 151: 183–224.

Green, D. E. 1937. "Reconstruction of the Chemical Events in the Living Cell." In *Perspectives in Biochemistry,* edited by J. Needham and D. E. Green, 175–86. Cambridge: Cambridge University Press.

Green, R. F., H. K. Meiss, and E. Rodriguez-Boulan. 1981. "Glycosylation Does Not Determine Segregation of Viral Envelope Proteins in the Plasma Membrane of Epithelial Cells." *Journal of Cell Biology* 89 (2): 230–39.

Green, R. H., T. E. Anderson, and J. E. Smadel. 1942. "Morphological Structure of the Virus of Vaccinia." *Journal of Experimental Medicine* 75: 651–56.

Griesemer, J. 2000. "Development, Culture, and the Units of Inheritance." *Philosophy of Science* 67 (Supplement): S348–S368.

Griffiths, G., P. Quinn, and G. Warren. 1983. "Dissection of the Golgi Complex. I. Monensin Inhibits the Transport of Viral Membrane Proteins from Medial to Trans Golgi Cisternae in Baby Hamster Kidney Cells Infected with Semliki Forest Virus." *Journal of Cell Biology* 96 (3): 835–50.

Griffiths, G., K. Simons, G. Warren, and K. T. Tokuyasu. 1983. "Immunoelectron Microscopy Using Thin, Frozen Sections: Application to Studies of the Intracellular Transport of Semliki Forest Virus Spike Glycoproteins." *Methods in Enzymology* 96: 466–85.

Grimes, M., and R. B. Kelly. 1992. "Intermediates in the Constitutive and Regulated Secretory Pathways Released In Vitro from Semi-Intact Cells." *Journal of Cell Biology* 117 (3): 539–49.

Gross, F. 2018. "Updating Cowdry's Theories: The Role of Models in Contemporary Experimental and Computational Biology." In *Visions of Cell Biology: Reflections Inspired* by Cowdry's *"General Cytology,"* edited by K. S. Matlin, J. Maienschein, and M. D. Laubichler, 326–50. Chicago: University of Chicago Press.

Grote, M. 2019. *Membranes to Molecular Machines.* Chicago: University of Chicago Press.

Grubman, M. J., E. Ehrenfeld, and D. F. Summers. 1974. "In Vitro Synthesis of Proteins by Membrane-Bound Polyribosomes from Vesicular Stomatitis Virus-Infected Hela Cells." *Journal of Virology* 14 (3): 560–71.

Grubman, M. J., S. A. Moyer, A. K. Banerjee, and E. Ehrenfeld. 1975. "Sub-Cellular Localization of Vesicular Stomatitis Virus Messenger RNAs." *Biochemical and Biophysical Research Communications* 62 (3): 531–38.

Gruenberg, J., G. Griffiths, and K. E. Howell. 1989. "Characterization of the Early Endosome and Putative Endocytic Carrier Vesicles In Vivo and with an Assay of Vesicle Fusion In Vitro." *Journal of Cell Biology* 108 (4): 1301–16.

Gruenberg, J., and K. E. Howell. 1986. "Reconstitution of Vesicle Fusions Occurring in Endocytosis with a Cell-Free System." *EMBO Journal* 5 (12): 3091–3101.

———. 1987. "An Internalized Transmembrane Protein Resides in a Fusion-Competent Endosome for Less than 5 Minutes." *Proceedings of the National Academy of Sciences* 84 (16): 5758–62.

———. 1989. "Membrane Traffic in Endocytosis: Insights from Cell-Free Assays." *Annual Review of Cell Biology* 5 (1): 453–81.

Gruenberg, J., K. E. Howell, A. Ito, and G. E. Palade. 1988. "Immuno-Isolation of Subcellular Components." *Progress in Clinical and Biological Research* 270: 77–90.

Guo, Y., D. W. Sirkis, and R. Schekman. 2014. "Protein Sorting at the Trans-Golgi Network." *Annual Review of Cell and Developmental Biology* 30 (1): 169–206.

Hagmann, M. 1999. "Colleagues Say 'Amen' to This Year's Choices." *Science* 286 (5440): 666.

Haguenau, F. 1958. "The Ergastoplasm: Its History, Ultrastructure, and Biochemistry." *International Review of Cytology* 7: 425–83.

Halic, M., T. Becker, M. R. Pool, C. M. T. Spahn, R. A. Grassucci, J. Frank, and R. Beckmann. 2004. "Structure of the Signal Recognition Particle Interacting with the Elongation-Arrested Ribosome." *Nature* 427 (6977): 808–14.

Hann, B. C., M. A. Poritz, and P. Walter. 1989. "*Saccharomyces cerevisiae* and *Schizosaccharomyces pombe* Contain a Homologue to the 54-Kd Subunit of the Signal Recognition Particle That in *S. cerevisiae* Is Essential for Growth." *Journal of Cell Biology* 109 (6 Pt. 2): 3223–30.

Hann, B. C., and P. Walter. 1991. "The Signal Recognition Particle in *S. cerevisiae*." *Cell* 67 (1): 131–44.

Hansen, W., P. D. Garcia, and P. Walter. 1986. "In Vitro Protein Translocation across the Yeast Endoplasmic Reticulum: ATP-Dependent Post-Translational Translocation of the Prepro-A-Factor." *Cell* 45: 397–406.

Hardy, W. B. 1899. "On the Structure of Cell Protoplasm." *Journal of Physiology* 24 (2): 158–210.

Harris, H. 1999. *The Birth of the Cell.* New Haven, CT: Yale University Press.

Harrison, T. M. 1972. "Messenger Ribonucleic Acid and Polysomes in a Mouse Plasmacytoma." PhD dissertation, Darwin College, Cambridge University.

Harrison, T. M., G. Brownlee, and C. Milstein. 1974. "Studies on Polysome-Membrane Interactions in Mouse Myeloma Cells." *European Journal of Biochemistry* 47 (3): 613–20.

Harvey, E. N., and A. L. Loomis. 1930. "A Microscope Centrifuge." *Science* 72 (1854): 42–44.

Harvey, E. N., and H. Shapiro. 1934. "The Interfacial Tension between Oil and Protoplasm within Living Cells." *Journal of Cellular and Comparative Physiology* 5: 255–67.

Hegde, R. S., and R. J. Keenan. 2011. "Tail-Anchored Membrane Protein Insertion into the Endoplasmic Reticulum." *Nature Reviews Molecular Cell Biology* 12 (12): 1–12.

Heidenhain, R. 1875. "Beiträge zur Kenntnis des Pankreas." *Archiv für die gesamte Physiologie des Menschen und der Tiere* 10: 557–632.

Helenius, A., and M. Aebi. 2001. "Intracellular Functions of N-Linked Glycans." *Science* 291 (5512): 2364–69.

Helenius, A., I. Mellman, D. Wall, and A. L. Hubbard. 1983. "Endosomes." *Trends in Biochemical Sciences* 1983 (July): 245–50.

Helenius, A., and K. Simons. 1975. "Solubilization of Membranes by Detergents." *Biochimica et Biophysica Acta* 415: 29–79.

Hentschel, K. 2014. *Visual Cultures in Science and Technology.* Oxford: Oxford University Press.

Hertwig, O. 1893. *Die Zelle und die Gewebe: Grundzüge der allgemeinen Anatomie und Physiologie.* Jena: Gustav Fischer.

———. 1895. *The Cell: Outlines of General Anatomy and Physiology.* London: Swan Sonnenschein & Co.

Hicks, S. J., J. W. Drysdale, and H. N. Munro. 1969. "Preferential Synthesis of Ferritin and Albumin by Different Populations of Liver Polysomes." *Science* 164 (3879): 584–85.

High, S., S. S. Andersen, D. Görlich, E. Hartmann, S. Prehn, T. A. Rapoport, and B. Dobberstein. 1993. "Sec61p Is Adjacent to Nascent Type I and Type II Signal-Anchor Proteins during Their Membrane Insertion." *Journal of Cell Biology* 121 (4): 743–50.

Highfield, P. E., and R. J. Ellis. 1978. "Synthesis and Transport of the Small Subunit of Chloroplast Ribulose Bisphosphate Carboxylase." *Nature* 271: 420–24.

Hille, B. 1992. *Ionic Channels of Excitable Membranes.* 2nd ed. Sunderland, MA: Sinauer Associates.

Hird, S. N., and J. G. White. 1993. "Cortical and Cytoplasmic Flow Polarity in Early Embryonic Cells of Caenorhabditis elegans." *Journal of Cell Biology* 121 (6): 1343–55.

Hirs, C. H. W., S. Moore, and W. H. Stein. 1953. "A Chromatographic Investigation of Pancreatic Ribonuclease." *Journal of Biological Chemistry* 200 (2): 493–506.

Höber, R. 1945. *Physical Chemistry of Cells and Tissues.* Philadelphia: Blakiston Co.

Hoerr, N. L., ed. 1943a. *Frontiers in Cytochemistry. Biological Symposia.* Lancaster, PA: Jacques Cattell Press.

———. 1943b. "Methods of Isolation of Morphological Constituents of the Liver Cell." In *Frontiers in Cytochemistry,* edited by N. L. Hoerr, 185–231. Biological Symposia. Lancaster, PA: Jacques Cattell Press.

———. 1957. "Robert Russell Bensley." *Anatomical Record* 128 (1): 1–18.

Hoffmann, D. 2017. "Gelehrtendynastien der DDR: Verfolgt, Geehrt und Umstritten." *Frankfurter Allgemeine Zeitung,* February 1. Accessed June 29, 2020. https://www.faz.net/-ibq-8rh11.

Hofmeister, F. 1901. *Die chemische Organisation der Zelle.* Braunschweig: Friedrich Verlag.

Hogeboom, G. H. 1949. "Cytochemical Studies of Mammalian Tissues; the Distribution of Diphosphopyridine Nucleotide-Cytochrome C Reductase in Rat Liver Fractions." *Journal of Biological Chemistry* 177 (2): 847–58.

———. 1951. "Separation and Properties of Cell Components." *Federation Proceedings* 10 (3): 640–45.

Hogeboom, G. H., A. Claude, and R. Hotchkiss. 1946. "The Distribution of Cytochrome Oxidase and Succinoxidase in the Cytoplasm of the Mammalian Liver Cell." *Journal of Biological Chemistry* 165 (2): 615–29.

Hogeboom, G. H., W. C. Schneider, and G. E. Pallade. 1947. "The Isolation of Morphologically Intact Mitochondria from Rat Liver." *Proceedings of the Society for Experimental Biology and Medicine* 65 (2): 320–21.

———. 1948. "Cytochemical Studies of Mammalian Tissues I. Isolation of Intact Mitochondria from Rat Liver; Some Biochemical Properties of Mitochondria and Submicroscopic Particulate Material." *Journal of Biological Chemistry* 172 (2): 619–35.

Hoober, J. K., and G. Blobel. 1969. "Characterization of the Chloroplastic and Cytoplasmic Ribosomes of *Chlamydomonas reinhardi.*" *Journal of Molecular Biology* 41 (1): 121–38.

Hoober, J. K., P. Siekevitz, and G. E. Palade. 1969. "Formation of Chloroplast Membranes in *Chlamydomonas reinhardi* Y-1: Effects of Inhibitors of Protein Synthesis." *Journal of Biological Chemistry* 244 (10): 2621–31.

Hopkins, F. G. 1913. "The Dynamic Side of Biochemistry." *Report of the British Association for the Advancement of Science* 83: 652–68.

Howell, K. E., R. Schmid, J. Ugelstad, and J. Gruenberg. 1989. "Immunoisolation Using Magnetic Solid Supports: Subcellular Fractionation for Cell-Free Functional Studies." *Methods in Cell Biology* 31: 265–92. New York: Academic Press.

Hsieh, P., M. R. Rosner, and P. W. Robbins. 1983. "Selective Cleavage by Endo-Beta-N-Acetylglucosaminidase H at Individual Glycosylation Sites of Sindbis Virion Envelope Glycoproteins." *Journal of Biological Chemistry* 258 (4): 2555–61.

Hunziker, W., C. Harter, K. Matter, and I. Mellman. 1991. "Basolateral Sorting in MDCK Cells Requires a Distinct Cytoplasmic Determinant." *Cell* 66: 907–20.

Huxley, T. H. 1868. "On Some Organisms Living at Great Depths in the North Atlantic Ocean." *Quarterly Journal of Microscopical Science* s2–s8: 203–12.

———. 1869. *On the Physical Basis of Life.* New Haven, CT: College Courant.

Ideker, T., T. Galitski, and L. Hood. 2001. "Systems Biology." *Annual Review of Genomics and Human Genetics* 2: 343–72.

Ideker, T., and L. Hood. 2019. "A Blueprint for Systems Biology." *Clinical Chemistry* 65 (2): 342–44.

Inouye, H., and J. Beckwith. 1977. "Synthesis and Processing of an *Escherichia coli* Alkaline Phosphatase Precursor *In Vitro.*" *Proceedings of the National Academy of Sciences* 74 (4): 1440–44.

Ito, K., G. Mandel, and W. Wickner. 1979. "Soluble Precursor of an Integral Membrane Protein: Synthesis of Procoat Protein in *Escherichia coli* Infected with Bacteriophage M13." *Proceedings of the National Academy of Sciences* 76 (3): 1199–1203.

Ito, K., M. Wittekind, M. Nomura, K. Shiba, T. Yura, A. Miura, and H. Nashimoto. 1983. "A Temperature-Sensitive Mutant of *E. coli* Exhibiting Slow Processing of Exported Proteins." *Cell* 32 (3): 789–97.

Jablonka, E. 2002. "Information: Its Interpretation, Its Inheritance, and Its Sharing." *Philosophy of Science* 69 (4): 578–605.

Jablonka, E., and M. J. Lamb. 2014. *Evolution in Four Dimensions.* Rev. ed. Cambridge, MA: MIT Press.

Jackson, R., and G. Blobel. 1977. "Post-Translational Cleavage of Presecretory Proteins with an Extract of Rough Microsomes from Dog Pancreas Containing Signal Peptidase Activity." *Proceedings of the National Academy of Sciences* 74 (12): 5598–5602.

———. 1980. "Post-Translational Processing of Full-Length Presecretory Proteins with Canine Pancreatic Signal Peptidase." *Annals of the New York Academy of Science* 343: 391–404.

Jacobs, M. H. 1924. "Permeability of the Cell to Diffusing Substances." In *General Cytology: A Textbook of Cellular Structure and Function for Students of Biology and Medicine,* edited by E. V. Cowdry, 97–164. Chicago: University of Chicago Press.

James, J. 1976. *Light Microscopic Techniques in Biology and Medicine*. Leiden: Martinus Nijhoff.

Jamieson, J. D., and G. E. Palade. 1967a. "Intracellular Transport of Secretory Proteins in the Pancreatic Exocrine Cell. I. Role of the Peripheral Elements of the Golgi Complex." *Journal of Cell Biology* 34 (2): 577–96.

———. 1967b. "Intracellular Transport of Secretory Proteins in the Pancreatic Exocrine Cell. II. Transport to Condensing Vacuoles and Zymogen Granules." *Journal of Cell Biology* 34 (2): 597–615.

———. 1968. "Intracellular Transport of Secretory Proteins in the Pancreatic Exocrine Cell. IV. Metabolic Requirements." *Journal of Cell Biology* 39 (3): 589–603.

Judson, H. F. 1979. *The Eighth Day of Creation*. New York: Simon and Schuster.

Kalderon, D., B. L. Roberts, W. D. Richardson, and A. E. Smith. 1984. "A Short Amino Acid Sequence Able to Specify Nuclear Location." *Cell* 39 (3 Pt. 2): 499–509.

Kant, I. 1987. *Critique of Judgement*. Translated by W. S. Pluhar. Indianapolis, IN: Hackett.

Katz, F. N., J. E. Rothman, V. R. Lingappa, G. Blobel, and H. F. Lodish. 1977. "Membrane Assembly In Vitro: Synthesis, Glycosylation, and Asymmetric Insertion of a Transmembrane Protein." *Proceedings of the National Academy of Sciences* 74 (8): 3278–82.

Kay, L. E. 2000. *Who Wrote the Book of Life*. Stanford, CA: Stanford University Press.

Ke, B. 1965. "Optical Rotary Dispersion of Chloroplast-Lamellae Fragments." *Archives of Biochemistry and Biophysics* 112: 554–61.

Keenan, R. J., D. M. Freymann, P. Walter, and R. M. Stroud. 1998. "Crystal Structure of the Signal Sequence Binding Subunit of the Signal Recognition Particle." *Cell* 94 (2): 181–91.

Keller, E. B. 1951. "Turnover of Proteins of Cell Fractions of Adult Liver In Vivo." *Federation Proceedings* 10: 206.

Keller, E. B., P. C. Zamecnik, and R. B. Loftfield. 1954. "The Role of Microsomes in the Incorporation of Amino Acids into Proteins." *Journal of Histochemistry and Cytochemistry* 2 (5): 378–86.

Kemper, B., J. F. Habener, R. C. Mulligan, J. T. Potts, and A. Rich. 1974. "Pre-Proparathyroid Hormone: A Direct Translation Product of Parathyroid Messenger RNA." *Proceedings of the National Academy of Sciences* 71 (9): 3731–35.

Kendrew, J. C. 1967. "How Molecular Biology Started." *Scientific American* 216 (3): 141–44.

———. 1980. "The European Molecular Biology Laboratory." *Endeavour* 4 (4): 166–70.

Kessler, S. W. 1975. "Rapid Isolation of Antigens from Cells with a Staphylococcal Protein A-Antibody Adsorbent: Parameters of the Interaction of Antibody-Antigen Complexes with Protein A." *Journal of Immunology* 115 (6): 1617–24.

Kincaid, H. 1990. "Molecular Biology and the Unity of Science." *Philosophy of Science* 57 (4): 575–93.

Kirsch, J. F., P. Siekevitz, and G. E. Palade. 1960. "Amino Acid Incorporation In Vitro by Ribonucleoprotein Particles Detached from Guinea Pig Liver Microsomes." *Journal of Biological Chemistry* 235: 1419–24.

Kitcher, P. 1984. "1953 and All That: A Tale of Two Sciences." *Philosophical Review* 93 (3): 335–73.

Kite, G. L. 1913. "Studies on the Physical Properties of Protoplasm. I. The Physical Properties of the Protoplasm of Certain Animal and Plant Cells." *American Journal of Physiology* 32: 146–64.

Kite, G. L., and R. Chambers. 1912. "Vital Staining of Chromosomes and the Function and Structure of the Nucleus." *Science* 36 (932): 639–41.

Kleinzeller, A. 1997. "Ernest Overton's Contribution to the Cell Membrane Concept: A Centennial Appreciation." *News in Physiological Science* 12 (1): 49–53.

Knight, B. C., and S. High. 1998. "Membrane Integration of Sec61α: A Core Component of the Endoplasmic Reticulum Translocation Complex." *Biochemical Journal* 331: 161–67.

Knipe, D. M., D. Baltimore, and H. F. Lodish. 1977a. "Maturation of Viral Proteins in Cells Infected with Temperature-Sensitive Mutants of Vesicular Stomatitis Virus." *Journal of Virology* 21 (3): 1149–58.

———. 1977b. "Separate Pathways of Maturation of the Major Structural Proteins of Vesicular Stomatitis Virus." *Journal of Virology* 21 (3): 1128–39.

Knipe, D. M., H. F. Lodish, and D. Baltimore. 1977. "Localization of Two Cellular Forms of the Vesicular Stomatitis Viral Glycoprotein." *Journal of Virology* 21 (3): 1121–27.

Knipe, D. M., J. K. Rose, and H. F. Lodish. 1975. "Translation of Individual Species of Vesicular Stomatitis Viral mRNA." *Journal of Virology* 15 (4): 1004–11.

Kohler, G., and C. Milstein. 1975. "Continuous Cultures of Fused Cells Secreting Antibody of Predefined Specificity." *Nature* 256 (5): 495–97.

Kornfeld, R., and S. Kornfeld. 1976. "Comparative Aspects of Glycoprotein Structure." *Annual Review of Biochemistry* 45 (1): 217–37.

———. 1985. "Assembly of Asparagine-Linked Oligosaccharides." *Annual Review of Biochemistry* 54 (1): 631–64.

Kornfeld, S. 1990. "Lysosomal Enzyme Targeting." *Biochemical Society Transactions* 18 (3): 367–74.

———. 2018. "A Lifetime of Adventures in Glycobiology." *Annual Review of Biochemistry* 87 (1): 1–21.

Koshland, D., and D. Botstein. 1980. "Secretion of Beta-Lactamase Requires the Carboxy End of the Protein." *Cell* 20 (3): 749–60.

Kramer, M. H., J.-C. Farré, K. Mitra, M. K. Yu, K. Ono, B. Demchak, K. Licon, M. Flagg, R. Balakrishnan, J. M. Cherry, S. Subramani, and T. Ideker. 2017. "Active Interaction Mapping Reveals the Hierarchical Organization of Autophagy." *Molecular Cell* 65 (4): 761–74.

Kreibich, G., P. Debey, and D. D. Sabatini. 1973. "Selective Release of Content from Microsomal Membranes without Disassembly: I. Permeability Changes Induced by Low Detergent Concentrations." *Journal of Cell Biology* 58 (2): 436–62.

Kreibich, G., A. L. Hubbard, and D. D. Sabatini. 1974. "On the Spatial Arrangememt of Proteins in Microsomal Membranes from Rat Liver." *Journal of Cell Biology* 60 (3): 616–27.

Kreibich, G., and D. D. Sabatini. 1974. "Selective Release of Content from Microsomal Vesicles without Membrane Disassembly. II. Electrophoretic and Immunological Characterization of Microsomal Subfractions." *Journal of Cell Biology* 61 (3): 789–807.

Krohs, U., and W. Callebaut. 2007. "Data without Models Merging with Models without Data." In *Systems Biology: Philosophical Foundations*, edited by F.C. Boogerd, F. J. Bruggeman, J.-H. S. Hofmeyr and H. V. Westerhoff, 181–213. Amsterdam: Elsevier.

Kuhn, T. [1962] 1970. *The Structure of Scientific Revolutions*. Chicago: University of Chicago Press.

Kurzchalia, T. V., M. Wiedmann, A. S. Girshovich, E. S. Bochkareva, H. Bielka, and T. A. Rapoport. 1986. "The Signal Sequence of Nascent Preprolactin Interacts with the 54k Polypeptide of the Signal Recognition Particle." *Nature* 320 (6063): 634–36.

Kvist, S., F. Bregegere, L. Rask, B. Cami, H. Garoff, F. Daniel, K. Wiman, D. Larhammar, J. P. Abastado, G. Gachelin, P. A. Peterson, B. Dobberstein, and P. Kourilsky. 1981. "cDNA Clone Coding for Part of a Mouse H-2d Major Histocompatibility Antigen." *Proceedings of the National Academy of Sciences* 78 (5): 2772–76.

Kvist, S., K. Wiman, L. Claesson, P. A. Peterson, and B. Dobberstein. 1982. "Membrane Insertion and Oligomeric Assembly of Hla-Dr Histocompatibility Antigens." *Cell* 29 (1): 61–69.

LaBonte, M. L. 2017. "Blobel and Sabatini's 'Beautiful Idea': Visual Representations of the Conception and Refinement of the Signal Hypothesis." *Journal of the History of Biology* 12 (7): 347–37.

Lamb, R. A., and P. W. Choppin. 1977. "The Synthesis of Sendai Virus Polypeptides in Infected Cells. III. Phosphorylation of Polypeptides." *Virology* 81 (2): 382–97.

Lande, M. A. 1975. "Direct Association of Messenger RNA with Microsomal Membranes in Human Diploid Fibroblasts." *Journal of Cell Biology* 65 (3): 513–28.

Landecker, H. 2007. *Culturing Life: How Cells Became Technologies.* Cambridge, MA: Harvard University Press.

Lazarowitz, S. G., and P. W. Choppin. 1975. "Enhancement of the Infectivity of Influenza A and B Viruses by Proteolytic Cleavage of the Hemagglutinin Polypeptide." *Virology* 68 (2): 440–54.

Lenard, J., and J. E. Rothman. 1976. "Transbilayer Distribution and Movement of Cholesterol and Phospholipid in the Membrane of Influenza Virus." *Proceedings of the National Academy of Sciences* 73 (2): 391–95.

Lenard, J., and S. J. Singer. 1966. "Protein Conformation in Cell Membrane Preparations as Studied by Optical Rotatory Dispersion and Circular Dichroism." *Proceedings of the National Academy of Sciences* 56 (6): 1828–35.

Leonelli, S. 2016. *Data-Centric Biology.* Chicago: University of Chicago Press.

Lernmark, A., S. J. Chan, R. Choy, A. Nathans, R. Carroll, H. S. Tager, A. H. Rubenstein, H. H. Swift, and D. F. Steiner. 1976. "Biosynthesis of Insulin and Glucagon: A View of the Current State of the Art." *CIBA Foundation Symposium* 41: 7–30.

Levere, T. H. 2001. *Transforming Matter: A History of Chemistry from Alchemy to the Buckyball.* Baltimore: Johns Hopkins University Press.

Lill, R., K. Cunningham, L. A. Brundage, K. Ito, D. Oliver, and W. Wickner. 1989. "SecA Protein Hydrolyzes ATP and Is an Essential Component of the Protein Translocation ATPase of *Escherichia coli.*" *EMBO Journal* 8 (3): 961–66.

Lillie, R. S. 1924. "Reactivity of the Cell." In *General Cytology: A Textbook of Cellular Structure and Function for Students of Biology and Medicine,* edited by E. V. Cowdry, 165–234. Chicago: University of Chicago Press.

Lingappa, V. R., A. Devillers-Thiery, and G. Blobel. 1977. "Nascent Prehormones Are Intermediates in the Biosynthesis of Authentic Bovine Pituitary Growth Hormone and Prolactin." *Proceedings of the National Academy of Sciences* 74 (6): 2432–36.

Lingappa, V. R., F. N. Katz, H. F. Lodish, and G. Blobel. 1978. "A Signal Sequence for the Insertion of a Transmembrane Glycoprotein: Similarities to the Signals of Secretory Proteins in Primary Structure and Function." *Journal of Biological Chemistry* 253 (24): 8667–70.

Littlefield, J. W., E. B. Keller, J. Gross, and P. C. Zamecnik. 1955. "Studies on Cytoplasmic Ribonucleoprotein Particles from the Liver of the Rat." *Journal of Biological Chemistry* 217 (1): 111–23.

Liu, D. 2016. "Visions of Life and Matter." PhD dissertation, University of Wisconsin.

———. 2017. "The Cell and Protoplasm as Container, Object, and Substance, 1835–1861." *Journal of the History of Biology* 50 (4): 889–925.

———. 2018. "Heads and Tails: Molecular Imagination and the Lipid Bilayer." In *Visions of Cell Biology: Reflections Inspired by Cowdry's "General Cytology,"* edited by K. S. Matlin, J. Maienschein, and M. D. Laubichler, 209–45. Chicago: University of Chicago Press.

———. 2019. "The Artificial Cell, the Semipermeable Membrane, and the Life That Never Was, 1864–1901." *Historical Studies in the Natural Sciences* 49 (5): 504–55.

Lodish, H. F., and J. E. Rothman. 1979. "The Assembly of Cell Membranes." *Scientific American* 240 (1): 48–63.

Loftfield, R. B. 1957. "The Biosynthesis of Protein." *Progress in Biophysics and Biophysical Chemistry* 8 (C): 347–86.

López-Moratalla, N., and M. Cerezo. 2011. "The Self-Construction of Living Organisms." In *Information and Living Systems*, edited by G. Terzis and R. Arp, 177–204. Cambridge, MA: MIT Press.

Maccecchini, M. L., Y. Rudin, G. Blobel, and G. Schatz. 1979. "Import of Proteins into Mitochondria: Precursor Forms of the Extramitochondrially Made F1-ATPase Subunits in Yeast." *Proceedings of the National Academy of Sciences* 76 (1): 343–47.

Maccecchini, M. L., Y. Rudin, and G. Schatz. 1979. "Transport of Proteins across the Mitochondrial Outer Membrane: A Precursor Form of the Cytoplasmically Made Intermembrane Enzyme Cytochrome c Peroxidase." *Journal of Biological Chemistry* 254 (16): 7468–71.

Mach, B., C. Faust, and P. Vassalli. 1973. "Purification of 14S Messenger RNA of Immunoglobulin Light Chain That Codes for a Possible Light-Chain Precursor." *Proceedings of the National Academy of Sciences* 70 (2): 451–55.

Maienschein, J. 2018. "Changing Ideas about Cells as Complex Systems." In *Visions of Cell Biology: Reflections Inspired by Cowdry's "General Cytology,"* edited by K. S. Matlin, J. Maienschein, and M. D. Laubichler, 15–45. Chicago: University of Chicago Press.

Maizel, J. V. 2000. "SDS Polyacrylamide Gel Electrophoresis." *Trends in Biochemical Sciences* 25 (12): 590–92.

Makino, S., J. A. Reynolds, and C. Tanford. 1973. "The Binding of Deoxycholate and Triton X-100 to Proteins." *Journal of Biological Chemistry* 248 (14): 4926–32.

Malkin, L. I., and A. Rich. 1967. "Partial Resistance of Nascent Polypeptide Chains to Proteolytic Digestion due to Ribosomal Shielding." *Journal of Molecular Biology* 26 (2): 329–46.

Manolio, T. A., F. S. Collins, N. J. Cox, D. B. Goldstein, L. A. Hindorff, D. J. Hunter, M. I. McCarthy, E. M. Ramos, L. R. Cardon, A. Chakravarti, J. H. Cho, A. E. Guttmacher, A. Kong, L. Kruglyak, E. Mardis, C. N. Rotimi, M. Slatkin, D. Valle, A. S. Whittemore, M. Boehnke, A. G. Clark, E. E. Eichler, G. Gibson, J. L. Haines, T. F. C. Mackay, S. A. McCarroll, and P. M. Visscher. 2009. "Finding the Missing Heritability of Complex Diseases." *Nature* 461 (7265): 747–53.

Martoglio, B., M. W. Hofmann, J. Brunner, and B. Dobberstein. 1995. "The Protein-Conducting Channel in the Membrane of the Endoplasmic Reticulum Is Open Laterally toward the Lipid Bilayer." *Cell* 81 (2): 207–14.

Marton, L. 1934. "Electron Microscopy of Biological Objects." *Nature* 133: 911.

Matlack, K. E., B. Misselwitz, K. Plath, and T. A. Rapoport. 1999. "Bip Acts as a Molecular Ratchet during Posttranslational Transport of Prepro-Alpha Factor across the ER Membrane." *Cell* 97 (5): 553–64.

Matlack, K. E., W. Mothes, and T. A. Rapoport. 1998. "Protein Translocation: Tunnel Vision." *Cell* 92 (3): 381–90.

Matlin, K. S. 2002. "The Strange Case of the Signal Recognition Particle." *Nature Reviews Molecular Cell Biology* 3 (7): 538–42.

———. 2011. "Spatial Expression of the Genome: The Signal Hypothesis at Forty." *Nature Reviews Molecular Cell Biology* 12 (5): 333–40.

———. 2016. "The Heuristic of Form: Mitochondrial Morphology and the Explanation of Oxidative Phosphorylation." *Journal of the History of Biology* 49 (1): 37–94.

———. 2018. "Pictures and Parts: Representation of Form and the Epistemic Strategy of Cell Biology." In *Visions of Cell Biology: Reflections Inspired by Cowdry's "General Cytology,"* edited by K. S. Matlin, J. Maienschein, and M. D. Laubichler, 246–79. Chicago: University of Chicago Press.

———. 2020. "Microscopes and Moving Molecules: The Discovery of Kinesin at the Marine Biological Laboratory." In *Why Study Biology by the Sea,* edited by K. S. Matlin, J. Maienschein, and R. A. Ankeny, 211–48. Chicago: University of Chicago Press.

Matlin, K. S., D. F. Bainton, M. Pesonen, D. Louvard, N. Genty, and K. Simons. 1983. "Transepithelial Transport of a Viral Membrane Glycoprotein Implanted into the Apical Plasma Membrane of Madin-Darby Canine Kidney Cells. I. Morphological Evidence." *Journal of Cell Biology* 97 (3): 627–37.

Matlin, K. S., and M. J. Caplan. 2017. "The Secretory Pathway at 50: A Golden Anniversary for Some Momentous Grains of Silver." *Molecular Biology of the Cell* 28: 229–32.

Matlin, K. S., J. Maienschein, and M. D. Laubichler, eds. 2018. *Visions of Cell Biology: Reflections Inspired by Cowdry's "General Cytology."* Chicago: University of Chicago Press.

Matlin, K. S., J. Maienschein, and M. D. Laubichler. 2018. Introduction to *Visions of Cell Biology: Reflections Inspired by Cowdry's "General Cytology,"* edited by K. S. Matlin, J. Maienschein, and M. D. Laubichler, 1–14. Chicago: University of Chicago Press.

Matlin, K. S., H. Reggio, A. Helenius, and K. Simons. 1982. "Pathway of Vesicular Stomatitis Virus Entry Leading to Infection." *Journal of Molecular Biology* 156 (3): 609–31.

Matlin, K. S., and K. Simons. 1983. "Reduced Temperature Prevents Transfer of a Membrane Glycoprotein to the Cell Surface but Does Not Prevent Terminal Glycosylation." *Cell* 34: 233–43.

Mechler, B., and P. Vassalii. 1975. "Membrane-Bound Ribosomes of Myeloma Cells. III. The Role of Messenger RNA and the Nascent Polypeptide Chain in the Binding of Ribosomes to Membranes." *Journal of Cell Biology* 67: 25–37.

Meek, G. A. 1976. *Practical Electron Microscopy for Biologists.* 2nd ed. New York: John Wiley & Sons.

Melchior, F., T. Guan, B. Paschal, and L. Gerace. 1995. "Biochemical and Structural Analysis of Nuclear Protein Import." *Cold Spring Harbor Symposia on Quantitative Biology* 60: 707–16.

Mellman, I., and W. J. Nelson. 2008. "Coordinated Protein Sorting, Targeting and Distribution in Polarized Cells." *Nature Reviews Molecular Cell Biology* 9 (11): 833–45.

Mellman, I., and K. Simons. 1992. "The Golgi Complex: In Vitro Veritas?" *Cell* 68 (5): 829–40.

Menetret, J. F., A. Neuhof, D. G. Morgan, K. Plath, M. Radermacher, T. A. Rapoport, and C. W. Akey. 2000. "The Structure of Ribosome-Channel Complexes Engaged in Protein Translocation." *Molecular Cell* 6 (5): 1219–32.

Meyer, D. I., and B. Dobberstein. 1980a. "Identification and Characterization of a Membrane Component Essential for the Translocation of Nascent Proteins across the Membrane of the Endoplasmic Reticulum." *Journal of Cell Biology* 87 (2 Pt. 1): 503–8.

———. 1980b. "A Membrane Component Essential for Vectorial Translocation of Nascent Proteins across the Endoplasmic Reticulum: Requirements for Its Extraction and Reassociation with the Membrane." *Journal of Cell Biology* 87 (2): 498–502.

Meyer, D. I., E. Krause, and B. Dobberstein. 1982. "Secretory Protein Translocation across Membranes: The Role of the 'Docking' Protein." *Nature* 297: 647–50.

Meyer, D. I., D. Louvard, and B. Dobberstein. 1982. "Characterization of Molecules Involved in Protein Translocation Using a Specific Antibody." *Journal of Cell Biology* 92 (2): 579–83.

Michaelis, S., and J. Beckwith. 1982. "Mechanism of Incorporation of Cell Envelope Proteins in *Escherichia coli.*" *Annual Reviews in Microbiology* 36 (1): 435–65.

Mihara, K., and G. Blobel. 1980. "The Four Cytoplasmically Made Subunits of Yeast Mitochondrial Cytochrome c Oxidase Are Synthesized Individually and Not as a Polyprotein." *Proceedings of the National Academy of Sciences* 77 (7): 4160–64.

Milstein, C., G. Brownlee, T. Harrison, and M. Mathews. 1972. "A Possible Precursor of Immunoglobulin Light Chains." *Nature New Biology* 239 (91): 117–20.

Moberg, C. L. 2012. *Entering an Unseen World.* New York: Rockefeller University Press.

Mokranjac, D., and W. Neupert. 2009. "Thirty Years of Protein Translocation into Mitochondria: Unexpectedly Complex and Still Puzzling." *Biochimica et Biophysica Acta (BBA)—Molecular Cell Research* 1793 (1): 33–41.

Moore, W. J. 1972. *Physical Chemistry.* Engelwood Cliffs, NJ: Prentice-Hall.

Morange, M. 1998. *A History of Molecular Biology.* Translated by M. Cobb. Cambridge, MA: Harvard University Press.

———. 2020. *The Black Box of Biology.* Cambridge, MA: Harvard University Press.

Morimoto, T., G. Blobel, and D. D. Sabatini. 1972a. "Ribosome Crystallization in Chicken Embryos. I. Isolation, Characterization, and In Vitro Activity of Ribosome Tetramers." *Journal of Cell Biology* 52 (2): 338–54.

———. 1972b. "Ribosome Crystallization in Chicken Embryos. II. Conditions for the Formation of Ribosome Tetramers In Vitro." *Journal of Cell Biology* 52 (2): 355–66.

Moritz, C. P. 2020. "40 Years Western Blotting: A Scientific Birthday Toast." *Journal of Proteomics* 212: 103575.

Morré, D. J., J. Kartenbeck, and W. W. Franke. 1979. "Membrane Flow and Interconversions among Endomembranes." *Biochimica et Biophysica Acta (BBA)—General Subjects* 559: 71–152.

Morrison, T., M. Stampfer, D. Baltimore, and H. F. Lodish. 1974. "Translation of Vesicular Stomatitis Messenger RNA by Extracts from Mammalian and Plant Cells." *Journal of Virology* 13 (1): 62–72.

Moss, L. 1992. "A Kernel of Truth? On the Reality of the Genetic Program." *PSA: Proceedings of the Biennial Meeting of the Philosophy of Science Association* 1992: 335–48.

———. 2003. *What Genes Can't Do.* Cambridge, MA: MIT Press.

Mostov, K. E., A. de Bruyn Kops, and D. L. Deitcher. 1986. "Deletion of the Cytoplasmic Domain of the Polymeric Immunoglobulin Receptor Prevents Basolateral Localization and Endocytosis." *Cell* 47 (3): 359–64.

Mothes, W., S. Prehn, and T. A. Rapoport. 1994. "Systematic Probing of the Environment of a Translocating Secretory Protein during Translocation through the ER Membrane." *EMBO Journal* 13 (17): 3973–82.

Muller, M., and G. Blobel. 1984. "In Vitro Translocation of Bacterial Proteins across the Plasma Membrane of *Escherichia coli.*" *Proceedings of the National Academy of Sciences* 81 (23): 7421–25.

Munro, E., J. Nance, and J. R. Priess. 2004. "Cortical Flows Powered by Asymmetrical Contraction Transport Par Proteins to Establish and Maintain Anterior-Posterior Polarity in the Early *C. elegans* Embryo." *Developmental Cell* 7 (3): 413–24.

Munro, S., and H. R. Pelham. 1987. "A C-Terminal Signal Prevents Secretion of Luminal ER Proteins." *Cell* 48 (5): 899–907.

Musch, A., M. Wiedmann, and T. A. Rapoport. 1992. "Yeast Sec Proteins Interact with Polypeptides Traversing the Endoplasmic Reticulum Membrane." *Cell* 69 (2): 343–52.

Nakai, K. 2000. "Protein Sorting Signals and Prediction of Subcellular Localization." *Advances in Protein Chemistry* 54: 277–344.

Nakai, K., and P. Horton. 2007. "Computational Prediction of Subcellular Localization." *Methods in Molecular Biology* 390 (1): 429–66.

Nathansohn, A. 1904a. "Ueber die Regulation der Aufnahme anorganischer Salze durch die Knollen von Dahlia." *Jahrbücher für wissenschaftliche Botanik* 39: 607–44.

———. 1904b. "Weitere Mitteilungen über die Regulation der Stoffenaufnahme." *Jahrbücher für wissenschaftliche Botanik* 40: 403–42.

Needham, J., and E. Baldwin, eds. 1949. *Hopkins and Biochemistry: Papers Concerning Sir Frederick Gowland Hopkins, OM, PRS, with a Selection of His Addresses and a Bibliography of His Publications.* Cambridge: W. Heffer and Sons.

Nelson, W. J. 2009. "Remodeling Epithelial Cell Organization: Transitions between Front-Rear and Apical-Basal Polarity." *Cold Spring Harbor Perspectives in Biology* 1 (1): a000513.

Neufeld, E. F., T. W. Lim, and L. J. Shapiro. 1975. "Inherited Disorders of Lysosomal Metabolism." *Annual Review of Biochemistry* 44 (1): 357–76.

Neupert, W., and J. M. Herrmann. 2007. "Translocation of Proteins into Mitochondria." *Annual Review of Biochemistry* 76 (1): 723–49.

Nicchitta, C., and G. Blobel. 1989. "Nascent Secretory Chain Binding and Translocation Are Distinct Processes: Differentiation by Chemical Alkylation." *Journal of Cell Biology* 108 (3): 789–95.

———. 1990. "Assembly of Translocation-Competent Proteoliposomes from Detergent-Solubilized Rough Microsomes." *Cell* 60 (2): 259–69.

———. 1993. "Lumenal Proteins of the Mammalian Endoplasmic Reticulum Are Required to Complete Protein Translocation." *Cell* 73 (5): 989–98.

Nicchitta, C., G. Migliaccio, and G. Blobel. 1991. "Biochemical Fractionation and Assembly of the Membrane Components That Mediate Nascent Chain Targeting and Translocation." *Cell* 65 (4): 587–98.

"The Nobel Prize in Physiology or Medicine 1999." 1999. October 11. https://www.nobelprize.org/prizes/medicine/1999/press-release/.

Novick, P., C. Field, and R. Schekman. 1980. "Identification of 23 Complementation Groups Required for Post-Translational Events in the Yeast Secretory Pathway." *Cell* 21 (1): 205–15.

Nyhart, L. K. 1995. *Biology Takes Form.* Chicago: University of Chicago Press.

Oliver, D. B., and J. Beckwith. 1981. "*E. coli* Mutant Pleiotropically Defective in the Export of Secreted Proteins." *Cell* 25 (3): 765–72.

———. 1982. "Identification of a New Gene (SecA) and Gene Product Involved in the Secretion of Envelope Proteins in *Escherichia coli*." *Journal of Bacteriology* 150 (2): 686–91.

Orci, L., and M. S. Pepper. 2002. "Microscopy: An Art?" *Nature Reviews Molecular Cell Biology* 3: 133–37.

Orci, L., M. Ravazzola, P. Meda, C. Holcomb, H.-P. Moore, L. Hicke, and R. Schekman. 1991. "Mammalian Sec23p Homologue Is Restricted to the Endoplasmic Reticulum Transitional Cytoplasm." *Proceedings of the National Academy of Sciences* 88: 8611–15.

Osswald, M., and E. Morais-de-Sá. 2019. "Dealing with Apical-Basal Polarity and Intercellular Junctions: A Multidimensional Challenge for Epithelial Cell Division." *Current Opinion in Cell Biology* 60: 75–83.

Otis, L. 2007. *Müller's Lab.* Oxford: Oxford University Press.

Overton, E. 1899. "Ueber die allgemeinen osmotischen Eigenschaften der Zelle, ihre Vermutlichen Ursachen und ihre Bedeutung für die Physiologie." *Vierteljahrsschrift der Naturforschenden Gesellschaft in Zürich* 44: 88–135.

Palade, G. E. 1951. "Intracellular Localization of Acid Phosphatase: A Comparative Study of Biochemical and Histochemical Methods." *Journal of Experimental Medicine* 94 (6): 535–48.

———. 1952a. "The Fine Structure of Mitochondria." *Anatomical Record* 114 (3): 427–51.

———. 1952b. "A Study of Fixation for Electron Microscopy." *Journal of Experimental Medicine* 95 (3): 285–98.

———. 1953. "An Electron Microscope Study of the Mitochondrial Structure." *Journal of Histochemistry and Cytochemistry* 1 (4): 188–211.

———. 1955a. "A Small Particulate Component of the Cytoplasm." *Journal of Biophysical and Biochemical Cytology* 1 (1): 59–68.

———. 1955b. "Studies on the Endoplasmic Reticulum. II. Simple Dispositions in Cells in Situ." *Journal of Biophysical and Biochemical Cytology* 1 (6): 567–82.

———. 1956a. "Electron Microscopy of Mitochondria and Other Cytoplasmic Structures." In *Enzymes: Units of Biological Structure and Function*, edited by O. H. Gaebler, 185–215. New York: Academic Press.

———. 1956b. "The Endoplasmic Reticulum." *Journal Biophysical and Biochemical Cytology* 2 (4): 85–98.

———. 1956c. "A Small Particulate Component of the Cytoplasm." In *Proceedings of the Third International Conference on Electron Microscopy*, edited by R. Ross, 417–23. London: Royal Microscopical Society.

———. 1959. "Functional Changes in the Structure of Cell Components." In *Subcellular Particles*, edited by T. Hayashi, 64–83. New York: Ronald Press.

———. 1966. "Structure and Function at the Cellular Level." *Journal of the American Medical Association* 198 (8): 143–53.

———. 1975. "Intracellular Aspects of the Process of Protein Synthesis." *Science* 189, no. 4200: 347–58.

Palade, G. E., and A. Claude. 1949a. "The Nature of the Golgi Apparatus: Identification of the Golgi Apparatus with a Complex of Myelin Figures." *Journal of Morphology* 85 (1): 71–111.

———. 1949b. "The Nature of the Golgi Apparatus: Parallelism between Intercellular Myelin Figures and Golgi Apparatus in Somatic Cells." *Journal of Morphology* 85 (1): 35–69.

Palade, G. E., and K. R. Porter. 1954. "Studies on the Endoplasmic Reticulum. I. Its Identification in Cells in Situ." *Journal of Experimental Medicine* 100 (6): 641–56.

Palade, G. E., and P. Siekevitz. 1956a. "Liver Microsomes: An Integrated Morphological and Biochemical Study." *Journal of Cell Biology* 2 (2): 171–200.

———. 1956b. "Pancreatic Microsomes: An Integrated Morphological and Biochemical Study." *Journal of Cell Biology* 2 (6): 671–90.

Palade, G. E., P. Siekevitz, and L. G. Caro. 1962. "Structure, Chemistry and Function of the Pancreatic Exocrine Cell." In *The Exocrine Pancreas*, edited by A. V. S. De Reuck and M. P. Cameron, 23–55. Chichester, UK: John Wiley and Sons.

Pauling, L., R. B. Corey, and H. R. Branson. 1951. "The Structure of Proteins: Two Hydrogen-Bonded Helical Configurations of the Polypeptide Chain." *Proceedings of the National Academy of Sciences* 37 (4): 205–11.

Pease, D. C. 1964. *Histological Techniques for Electron Microscopy*. 2nd ed. New York: Academic Press.

Pelham, H., and R. J. Jackson. 1976. "An Efficient mRNA-Dependent Translation System from Reticulocyte Lysates." *European Journal of Biochemistry* 67: 247–56.

Peter, I. S., E. Faure, and E. H. Davidson. 2012. "Predictive Computation of Genomic Logic Processing Functions in Embryonic Development." *Proceedings of the National Academy of Sciences*: 109 (41): 16434–42.

Pfeffer, S. R., and J. E. Rothman. 1987. "Biosynthetic Protein Transport and Sorting by the Endoplasmic Reticulum and Golgi." *Annual Review of Biochemistry* 56 (1): 829–52.

Pfeffer, W. 1985. *Osmotic Investigations: Studies on Cell Mechanics.* Translated by G. R. Kepner and E. J. Stadelmann. New York: Van Nostrand Reinhold. Originally published as *Osmotische Untersuchengen, Studien zur Zellmechanik* (Leipzig: Verlag Von Wilhelm Engelmann, 1877).

Pinto da Silva, P., and D. Branton. 1970. "Membrane Splitting in Freeze-Etching: Covalently Bound Ferritin as a Membrane Marker." *Journal of Cell Biology* 45 (3): 598–605.

Ploem, J. S. 1989. "Fluorescence Microscopy." In *Light Microscopy in Biology: A Practical Approach,* edited by A. J. Lacey, 163–86. Oxford: IRL Press.

Plowe, J. Q. 1931. "Membranes in the Plant Cell. I. Morphological Membranes at Protoplasmic Surfaces." *Protoplasma* 12: 196–221.

Poritz, M. A., H. D. Bernstein, K. Strub, D. Zopf, H. Wilhelm, and P. Walter. 1990. "An *E. coli* Ribonucleoprotein Containing 4.5S RNA Resembles Mammalian Signal Recognition Particle." *Science* 250 (4984): 1111–17.

Poritz, M. A., V. Siegel, W. Hansen, and P. Walter. 1988. "Small Ribonucleoproteins in *Schizosaccharomyces pombe* and *Yarrowia lipolytica* Homologous to Signal Recognition Particle." *Proceedings of the National Academy of Sciences* 85 (12): 4315–19.

Porter, K. R. 1953. "Observations on a Submicroscopic Basophilic Component of Cytoplasm." *Journal of Experimental Medicine* 97 (5): 727–50.

———. 1955. "The Submicroscopic Morphology of Protoplasm." *Harvey Lectures* 51: 175–228.

Porter, K. R., and H. S. Bennett. 1981. "Recollections on the Beginnings of the *Journal of Cell Biology.*" *Journal of Cell Biology* 91 (3): ix–xi.

Porter, K. R., and J. Blum. 1953. "A Study in Microtomy for Electron Microscopy." *Anatomical Record* 117 (4): 685–709.

Porter, K. R., A. Claude, and E. F. Fullam. 1945. "A Study of Tissue Culture Cells by Electron Microscopy: Methods and Preliminary Observations." *Journal of Experimental Medicine* 81 (3): 233–46.

Porter, K. R., and F. L. Kallman. 1952. "Significance of Cell Particulates as Seen by Electron Microscopy." *Annals of the New York Academy of Sciences* 54 (6): 882–91.

Porter, K. R., and H. P. Thompson. 1948. "A Particulate Body Associated with Epithelial Cells Cultured from Mammary Carcinomas of Mice of a Milk-Factor Strain." *Journal of Experimental Medicine* 88 (1): 15–24.

Potter, V. R., and C. A. Elvehjem. 1936. "A Modified Method for the Study of Tissue Oxidations." *Journal of Biological Chemistry* 114 (2): 495–504.

Powell, A., and J. Dupré. 2009. "From Molecules to Systems: The Importance of Looking Both Ways." *Studies in the History and Philosophy of Biological and Biomedical Sciences* 40 (1): 54–64.

Racker, E., B. Violand, S. O'Neal, M. Alfonzo, and J. Telford. 1979. "Reconstitution, a Way of Biochemical Research; Some New Approaches to Membrane-Bound Enzymes." *Archives of Biochemistry and Biophysics* 198 (2): 470–77.

Randall, L. L., and S. J. Hardy. 1977. "Synthesis of Exported Proteins by Membrane-Bound Polysomes from *Escherichia coli.*" *European Journal of Biochemistry* 75 (1): 43–53.

Randall, L. L., S. J. Hardy, and L. G. Josefsson. 1978. "Precursors of Three Exported Proteins in *Escherichia coli.*" *Proceedings of the National Academy of Sciences* 75 (3): 1209–12.

Randall, L. L., L. G. Josefsson, and S. J. Hardy. 1978. "Processing In Vitro of Precursor Periplasmic Proteins from *Escherichia coli.*" *European Journal of Biochemistry* 92 (2): 411–15.

Rapoport, T. A. 2007. "Protein Translocation across the Eukaryotic Endoplasmic Reticulum and Bacterial Plasma Membranes." *Nature* 450 (7170): 663–69.

———. 2010. "A Preliminary Report on My Life in Science." *Molecular Biology of the Cell* 21 (22): 3770–72.

Rapoport, T. A., D. Klatt, S. Prehn, V. Hahn, and W. E. Höhne. 1976. "Evidence for the Synthesis of a Precursor of Carp Proinsulin in a Cell-Free Translation System." *FEBS Letters* 69 (1): 32–36.

Rapoport, T. A., L. Li, and E. Park. 2017. "Structural and Mechanistic Insights into Protein Translocation." *Annual Review of Cell and Developmental Biology* 33: 369–90.

Rasmussen, N. 1995. "Mitochondrial Structure and the Practice of Cell Biology in the 1950s." *Journal of the History of Biology* 28 (3): 381–429.

———. 1997. *Picture Control.* Stanford, CA: Stanford University Press.

———. 2014. *Gene Jockeys.* Baltimore: Johns Hopkins University Press.

Redman, C. M., and D. D. Sabatini. 1966. "Vectorial Discharge of Peptides Released by Puromycin from Attached Ribosomes." *Proceedings of the National Academy of Sciences* 56 (2): 608–15.

Redman, C. M., P. Siekevitz, and G. E. Palade. 1966. "Synthesis and Transfer of Amylase in Pigeon Pancreatic Microsomes." *Journal of Biological Chemistry* 241 (5): 1150–58.

Reid, D. W., and C. V. Nicchitta. 2015. "Diversity and Selectivity in mRNA Translation on the Endoplasmic Reticulum." *Nature Publishing Group* 16 (4): 221–31.

Remak, R. 1855. *Untersuchungen über die Entwicklung der Wirbelthiere.* Berlin: Reimer.

Reynolds, A. 2008. "Amoebae as Exemplary Cells: The Protean Nature of an Elementary Organism." *Journal of the History of Biology* 41 (2): 307–37.

———. 2018a. "In Search of Cell Architecture: General Cytology and Early Twentieth Century Conceptions of Cell Organization." In *Visions of Cell Biology: Reflections Inspired by Cowdry's "General Cytology,"* edited by K. S. Matlin, J. Maienschein, and M. D. Laubichler, 46–72. Chicago: University of Chicago Press.

———. 2018b. *The Third Lens: Metaphor and the Creation of Modern Cell Biology.* Chicago: University of Chicago Press.

Reynolds, J. A., and C. Tanford. 1970a. "Binding of Dodecyl Sulfate to Proteins at High Binding Ratios: Possible Implications for the State of Proteins in Biological Membranes." *Proceedings of the National Academy of Sciences* 66 (3): 1002–7.

———. 1970b. "The Gross Conformation of Protein-Sodium Dodecyl Sulfate Complexes." *Journal of Biological Chemistry* 245 (19): 5161–65.

Rheinberger, H.-J. 1993. "Experiment and Orientation: Early Systems of In Vitro Protein Synthesis." *Journal of the History of Biology* 26 (3): 443–71.

———. 1997. *Toward a History of Epistemic Things.* Stanford, CA: Stanford University Press.

———. 2010. *On Historicizing Epistemology.* Stanford, CA: Stanford University Press.

Richards, R. J. 2000. "Kant and Blumenbach on the Bildungstrieb: A Historical Misunderstanding." *Studies in the History and Philosophy of Biological and Biomedical Sciences* 31 (1): 11–32.

———. 2002. *The Romantic Conception of Life: Science and Philosophy in the Age of Goethe.* Chicago: University of Chicago Press.

———. 2018. "The Foundations of Archetype Theory in Evolutionary Biology: Kant, Goethe, and Carus." *Republics of Letters* 6 (1): 1–15.

Roberts, R. B., ed. 1958. *Microsomal Particles and Protein Synthesis: Papers Presented at the First Symposium of the Biophysical Society, at the Massachusetts Institute of Technology, Cambridge, February 5, 6, and 8, 1958.* New York: Pergamon Press.

Robertson, J. D. 1959. "The Ultrastructure of Cell Membranes and Their Derivatives." In *The Structure and Function of Subcellular Components*, edited by E. M. Crook, 3–43. Cambridge: Cambridge University Press.

———. 1981. "Membrane Structure." *Journal of Cell Biology* 91 (3 Pt. 2): 189s–204s.

Rodriguez-Boulan, E., and M. Pendergast. 1980. "Polarized Distribution of Viral Envelope Proteins in the Plasma Membrane of Infected Epithelial Cells." *Cell* 20 (1): 45–54.

Rodriguez-Boulan, E., and D. D. Sabatini. 1978. "Asymmetric Budding of Viruses in Epithelial Monlayers: A Model System for Study of Epithelial Polarity." *Proceedings of the National Academy of Sciences* 75 (10): 5071–75.

Roman, R., J. D. Brooker, S. N. Seal, and A. Marcus. 1976. "Inhibition of the Transition of a 40 S Ribosome-Met-tRNA-I-Met Complex to an 80 S Ribosome-Met-tRNA-I-Met-Complex by 7-Methylguanosine-5'-Phosphate." *Nature* 260 (5549): 359–60.

Romisch, K., J. Webb, J. Herz, S. Prehn, R. Frank, M. Vingron, and B. Dobberstein. 1989. "Homology of 54k Protein of Signal-Recognition Particle, Docking Protein and Two *E. coli* Proteins with Putative GTP-Binding Domains." *Nature* 340 (6233): 478–82.

Rose, S. 2011. "Practicing Biochemistry without a Licence?" *EMBO Reports* 12 (5): 381.

Rothblatt, J. A., and D. I. Meyer. 1986a. "Secretion in Yeast: Reconstitution of the Translocation and Glycosylation of α-Factor and Invertase in a Homologous Cell-Free System." *Cell* 44 (4): 619–28.

———. 1986b. "Secretion in Yeast: Translocation and Glycosylation of Prepro-α-Factor In Vitro Can Occur Via an ATP-Dependent Post-Translational Mechanism." *EMBO Journal* 5 (5): 1031–36.

Rothman, J. E. 1973. "The Molecular Basis of Mesomorphic Phase Transitions in Phospholipid Systems." *Journal of Theoretical Biology* 38 (1): 1–16.

———. 2014. "The Principle of Membrane Fusion in the Cell (Nobel Lecture)." *Angewandte Chemie International Edition* 53 (47): 12676–694.

Rothman, J. E., and D. Engelman. 1972. "Molecular Mechanism for the Interaction of Phospholipid with Cholesterol." *Nature New Biology* 237 (71): 42–44.

Rothman, J. E., and E. Fries. 1981. "Transport of Newly Synthesized Vesicular Stomatitis Viral Glycoprotein to Purified Golgi Membranes." *Journal of Cell Biology* 89 (1): 162–68.

Rothman, J. E., F. N. Katz, and H. F. Lodish. 1978. "Glycosylation of a Membrane Protein Is Restricted to the Growing Polypeptide Chain but Is Not Necessary for Insertion as a Transmembrane Protein." *Cell* 15 (4): 1447–54.

Rothman, J. E., and J. Lenard. 1977. "Membrane Asymmetry." *Science* 195 (4280): 743–53.

Rothman, J. E., and H. F. Lodish. 1977. "Synchronised Transmembrane Insertion and Glycosylation of a Nascent Membrane Protein." *Nature* 269 (5631): 775–80.

Rothman, J. E., D. K. Tsai, E. A. Dawidowicz, and J. Lenard. 1976. "Transbilayer Phospholipid Asymmetry and Its Maintenance in the Membrane of Influenza Virus." *Biochemistry* 15 (11): 2361–70.

Rothman, J. E., and F. Wieland. 1996. "Protein Sorting by Transport Vesicles." *Science* 272 (5259): 227–34.

Rothman, S. S. 1975. "Protein Transport by the Pancreas." *Science* 190 (4216): 747–53.

Ruska, E. 1988. "The Development of the Electron Microscope and Electron Microscopy (Nobel Prize Lecture 1986)." *Bulletin of the Electron Microscopy Society of America* 18 (2): 53–61.

Sabatini, D. D. 2005. "In Awe of Subcellular Complexity: 50 Years of Trespassing Boundaries within the Cell." *Annual Review of Cell and Developmental Biology* 21 (1): 1–33.

Sabatini, D. D., K. Bensch, and R. J. Barrnett. 1964. "Cytochemistry and Electron Microscopy: The Preservation of Cellular Ultrastructure and Enzymatic Activity by Aldehyde Fixation." *Journal of Cell Biology* 17: 19–58.

Sabatini, D. D., and G. Blobel. 1970. "Controlled Proteolysis of Nascent Polypeptides in Rat Liver Cell Fractions. II. Location of the Polypeptides in Rough Microsomes." *Journal of Cell Biology* 45 (1): 146–57.

Sabatini, D. D., G. Blobel, Y. Nonomura, and M. R. Adelman. 1971. "Ribosome-Membrane Interaction: Structural Aspects and Functional Implications." *Advances in Cytopharmacology* 1: 119–29.

Sabatini, D. D., G. Ojakian, M. A. Lande, J. Lewis, W. Mok, M. Adesnik, and G. Kreibich. 1975. "Structural and Functional Aspects of the Protein Synthesizing Apparatus in the Rough Endoplasmic Reticulum." In *Control Mechanisms in Development*, edited by R. H. Meints and E. Davies, 151–80. New York: Plenum Press.

Sabatini, D. D., Y. Tashiro, and G. E. Palade. 1966. "On the Attachment of Ribosomes to Microsomal Membranes." *Journal of Molecular Biology* 19 (2): 503–24.

Sailer, A., A. Anneken, Y. Li, S. Lee, and E. Munro. 2015. "Dynamic Opposition of Clustered Proteins Stabilizes Cortical Polarity in the C. elegans Zygote." *Developmental Cell* 35 (1): 131–42.

Sanders, S. L., K. M. Whitfield, J. P. Vogel, M. D. Rose, and R. W. Schekman. 1992. "Sec61p and Bip Directly Facilitate Polypeptide Translocation into the ER." *Cell* 69 (2): 353–65.

Sanger, F., and E. O. P. Thompson. 1953a. "The Acid Sequence in the Glycyl Chain of Insulin. 2. The Investigation of Peptides from Enzymic Hydrolases." *Biochemical Journal* 53 (3): 366–74.

———. 1953b. "The Amino Acid Sequence of the Glycyl Chain of Insulin. 1. The Identification of Lower Peptides from Partial Hydrolysates." *Biochemical Journal* 53 (3): 353–66.

Sanger, F., and H. Tuppy. 1951a. "The Amino-Acid Sequence in the Phenylalanyl Chain of Insulin. 1. The Identification of Lower Peptides from a Partial Hydrolysate." *Biochemical Journal* 49 (4): 463–81.

———. 1951b. "The Amino-Acid Sequence in the Phenylalanyl Chain of Insulin. 2. The Investigation of Peptides from Enzymatic Hydrolysates." *Biochemical Journal* 49 (4): 481–90.

Sapp, J. 2018. "Epigenetics and Beyond." In *Visions of Cell Biology: Reflections Inspired by Cowdry's "General Cytology,"* edited by K. S. Matlin, J. Maienschein, and M. D. Laubichler, 183–208. Chicago: University of Chicago Press.

Schatz, G. 2000. "Interplanetary Travels—a Scientific Autobiography." In *Comprehensive Biochemistry: Personal Recollections VI*, edited by G. Semenza and R. Jaenicke, 449–530. Amsterdam: Elsevier Science.

Schechter, I. 1973. "Biologically and Chemically Pure mRNA Coding for a Mouse Immunoglobulin L-Chain Prepared with the Aid of Antibodies and Immobilized Oligothymidine." *Proceedings of the National Academy of Sciences* 70 (8): 2256–60.

Schechter, I., R. Guyer, D. J. McKean, and W. Terry. 1975. "Partial Amino Acid Sequence of the Precursor of Immunoglobulin Light Chain Programmed by Messenger RNA In Vitro." *Science* 188 (4184): 160–62.

Scheele, G., B. Dobberstein, and G. Blobel. 1978. "Transfer of Proteins across Membranes: Biosynthesis In Vitro of Pretrypsinogen and Trypsinogen by Cell Fractions of Canine Pancreas." *European Journal of Biochemistry* 82 (2): 593–99.

Schekman, R. 2013. "Nobel Lecture: Genes and Proteins That Control the Secretion Pathway." https://www.nobelprize.org/prizes/medicine/2013/schekman/lecture/.

Schickore, J. 2018. "Methodological Reflections in General Cytology in Historical Perspective." In *Visions of Cell Biology: Reflections Inspired by Cowdry's "General Cytology,"* edited by K. S. Matlin, J. Maienschein, and M. D. Laubichler, 73–99. Chicago: University of Chicago Press.

Schmid, S. L., A. Sorkin, and M. Zerial. 2014. "Endocytosis: Past, Present, and Future." *Cold Spring Harbor Perspectives in Biology* 6 (12): a022509.

Schmidt, G. W., A. Devillers-Thiery, H. Desruisseaux, G. Blobel, and N.-H. Chua. 1979. "NH2-Terminal Amino Acid Sequences of Precursor and Mature Forms of the Ribulose-1,5-Bisphosphate Carboxylase Small Subunit from *Chlamydomonas reinhardtii.*" *Journal of Cell Biology* 83 (3): 615–22.

Schneider, W. C. 1946a. "Intracellular Distribution of Enzymes: The Distribution of Succinic Dehydrogenase, Cytochrome Oxidase, Adenosinetriphosphatase, and Phosphorus Compounds in Normal Rat Liver and in Rat Hepatomas." *Cancer Research* 6 (12): 685–90.

———. 1946b. "Intracellular Distribution of Enzymes: The Distribution of Succinic Dehydrogenase, Cytochrome Oxidase, Adenosinetriphosphatase, and Phosphorus Compounds in Normal Rat Tissues." *Journal of Biological Chemistry* 165 (2): 585–93.

Schor, S., P. Siekevitz, and G. E. Palade. 1970. "Cyclic Changes in Thylakoid Membranes of Synchronized *Chlamydomonas reinhardi.*" *Proceedings of the National Academy of Sciences* 66 (1): 174–80.

Schultze, M. 1860. "Die Gattung Cornuspira unter den Monothalamien und Bemerkungen über die Organisation und Fortpflanzung der Polythalamien." *Archiv für Naturgeschicte* 26: 287–310.

Schwann, T. 1839. *Mikroskopische Untersuchungen ueber die Ueberstimmung in der Struktur und dem Wachstum der Thiers und Planzen.* Berlin: Sanders'chen Buchhandlung.

———. 1847. *Microscopical Researches into the Accordance in the Structure and Growth of Animals and Plants.* Translated by Henry Smith. London: Sydenham Society.

Scott, C. P. G. 1916. "Amblystoma Not Ambystoma." *Science* 44 (1131): 309–11.

Seifriz, W. 1918. "Observations on the Structure of Protoplasm by Aid of Microdissection." *Biological Bulletin* 34 (5): 307–24.

———. 1921. "Observations on Some Physical Properties of Protoplasm by Aid of Microdissection." *Annals of Botany* 35 (138): 269–96.

———. 1936. *Protoplasm.* New York: McGraw-Hill.

Shan, S., and P. Walter. 2004. "Co-Translational Protein Targeting by the Signal Recognition Particle." *FEBS Letters* 579 (4): 921–26.

Shannon, C. E. 1948. "A Mathematical Theory of Communication." *Bell System Technical Journal* 27: 379–423.

Shields, D., and G. Blobel. 1977. "Cell-Free Synthesis of Fish Preproinsulin, and Processing by Heterologous Mammalian Microsomal Membranes." *Proceedings of the National Academy of Sciences* 74 (5): 2059–63.

———. 1978. "Efficient Cleavage and Segregation of Nascent Presecretory Proteins in a Reticulocyte Lysate Supplemented with Microsomal Membranes." *Journal of Biological Chemistry* 253 (11): 3753–56.

Shorter, J., and G. Warren. 2002. "Golgi Architecture and Inheritance." *Annual Review of Cell and Developmental Biology* 18 (1): 379–420.

Siekevitz, P. 1952. "Uptake of Radioactive Alanine In Vitro into the Proteins of Rat Liver Fractions." *Journal of Biological Chemistry* 195 (2): 549–65.

Siekevitz, P., and R. Nagin. 1966. "Chemical Warfare in Vietnam." *Science* 152 (3718): 15.

Siekevitz, P., and G. E. Palade. 1958a. "A Cytochemical Study on the Pancreas of the Guinea Pig. III. In Vivo Incorporation of Leucine-1-C14 into the Proteins of Cell Fractions." *Journal of Biophysical and Biochemical Cytology* 4 (5): 557–66.

———. 1958b. "A Cytochemical Study on the Pancreas of the Guinea Pig. I. Isolation and Enzymatic Activities of Cell Fractions." *Journal of Biophysical and Biochemical Cytology* 4 (2): 203–18.

———. 1958c. "A Cytochemical Study on the Pancreas of the Guinea Pig. II. Functional Variations in the Enzymatic Activity of Microsomes." *Journal of Biophysical and Biochemical Cytology* 4 (3): 309–18.

———. 1960. "A Cytochemical Study on the Pancreas of the Guinea Pig. V. In Vivo Incorporation of Leucine-1-C14 into the Chymotrypsinogen of Various Cell Fractions." *Journal of Biophysical and Biochemical Cytology* 7 (4): 619–30.

———. 1966. "Distribution of Newly Synthesized Amylase in Microsomal Subfractions of Guinea Pig Pancreas." *Journal of Cell Biology* 30 (3): 519–30.

Siekevitz, P., and P. C. Zamecnik. 1981. "Ribosomes and Protein Synthesis." *Journal of Cell Biology* 91 (3): 53s–65s.

Silhavy, T. J., M. J. Casadaban, H. A. Shuman, and J. R. Beckwith. 1976. "Conversion of Beta-Galactosidase to a Membrane-Bound State by Gene Fusion." *Proceedings of the National Academy of Sciences* 73 (10): 3423–27.

Silhavy, T. J., H. A. Shuman, J. Beckwith, and M. Schwartz. 1977. "Use of Gene Fusions to Study Outer Membrane Protein Localization in *Escherichia coli*." *Proceedings of the National Academy of Sciences* 74 (12): 5411–15.

Silverstein, S. C., R. M. Steinman, and Z. A. Cohn. 1977. "Endocytosis." *Annual Review of Biochemistry* 46: 669–722.

Simon, H. A. 1996. *The Sciences of the Artificial*. 3rd ed. Cambridge, MA: MIT Press.

Simon, S. M., and G. Blobel. 1991. "A Protein-Conducting Channel in the Endoplasmic Reticulum." *Cell* 65 (3): 371–80.

———. 1992. "Signal Peptides Open Protein-Conducting Channels in *E. coli*." *Cell* 69 (4): 677–84.

Simon, S. M., G. Blobel, and J. Zimmerberg. 1989. "Large Aqueous Channels in Membrane Vesicles Derived from the Rough Endoplasmic Reticulum of Canine Pancreas or the Plasma Membrane of *Escherichia coli*." *Proceedings of the National Academy of Sciences* 86 (16): 6176–80.

Simon, S. M., C. S. Peskin, and G. F. Oster. 1992. "What Drives the Translocation of Proteins?" *Proceedings of the National Academy of Sciences* 89 (9): 3770–74.

Singer, S. J. 1971. "The Molecular Organization of Biological Membranes." In *Structure and Function of Biological Membranes*, edited by L. I. Rothfield, 145–222. New York: Academic Press.

———. 1975. "Architecture and Topography of Biologic Membranes." In *Cell Membranes: Biochemistry, Cell Biology, and Pathology*, edited by G. Weissman, 35–44. New York: HP Publishing.

Singer, S. J., P. A. Maher, and M. P. Yaffe. 1987a. "On the Transfer of Integral Proteins into Membranes." *Proceedings of the National Academy of Sciences* 84 (7): 1960–64.

———. 1987b. "On the Translocation of Proteins across Membranes." *Proceedings of the National Academy of Sciences* 84 (4): 1015–19.

Singer, S. J., and G. L. Nicolson. 1972. "The Fluid Mosaic Model of the Structure of Cell Membranes." *Science* 175 (4023): 720–31.

Sjöstrand, F. S. 1956. "Recent Advances in the Biological Applications of the Electron Microscope." In *Proceedings of the Third International Conference on Electron Microscopy*, edited by R. Ross, 26–36. London: Royal Microscopical Society.

———. 1962. "Critical Evaluation of Ultrastructural Patterns with Respect to Fixation." In *The Interpretation of Ultrastructure*, edited by R. J. C. Harris, 47–68. New York: Academic Press.

Slack, J. M. W. 1983. *From Egg to Embryo*. Cambridge: Cambridge University Press.

Slot, J. W., and H. J. Geuze. 1983. "Immunoelectron Microscopic Exploration of the Golgi Complex." *Journal of Histochemistry and Cytochemistry* 31 (8): 1049–56.

Stadler, M. 2009. "Assembling Life: Models, the Cell, and the Reformations of Biological Science, 1920–1960." PhD dissertation, Imperial College, London.

Steck, T. L. 1974. "The Organization of Proteins in the Human Red Blood Cell Membrane: A Review." *Journal of Cell Biology* 62 (1): 1–19.

Steck, T. L., G. Fairbanks, and D. F. H. Wallach. 1971. "Disposition of the Major Proteins in the Isolated Erythrocyte Membrane. Proteolytic Dissection." *Biochemistry* 10 (13): 2617–24.

Stein, W. D. 1986. "James Frederic Danielli, 13 November 1911–22 April 1984." *Biographical Memoirs of Fellows of the Royal Society* 32: 115–35.

Steiner, D. F. 2011. "Adventures with Insulin in the Islets of Langerhans." *Journal of Biological Chemistry* 286 (20): 17399–421.

Steiner, D. F., W. Kemmler, H. S. Tager, and J. D. Peterson. 1974. "Proteolytic Processing in the Biosynthesis of Insulin and Other Proteins." *Federation Proceedings* 33 (10): 2105–15.

Steiner, D. F., and P. E. Oyer. 1967. "The Biosynthesis of Insulin and a Probable Precursor of Insulin by a Human Islet Cell Adenoma" *Proceedings of the National Academy of Sciences* 57 (2): 473.

Stent, G. 1968. "That Was the Molecular Biology That Was." *Science* 160 (3826): 390–95.

Stiegler, N., R. E. Dalbey, and A. Kuhn. 2011. "M13 Procoat Protein Insertion into YidC and SecYEG Proteoliposomes and Liposomes." *Journal of Molecular Biology* 406 (3): 362–70.

Stirling, C. J., J. Rothblatt, M. Hosobuchi, R. Deshaies, and R. Schekman. 1992. "Protein Translocation Mutants Defective in the Insertion of Integral Membrane Proteins into the Endoplasmic Reticulum." *Molecular Biology of the Cell* 3 (2): 129–42.

Strathern, J. N., E. W. Jones, and J. R. Broach, eds. 1982. *The Molecular Biology of the Yeast Saccharomyces: Metabolism and Gene Expression*. Cold Spring Harbor, NY: Cold Spring Harbor Laboratory.

Strohman, R. 2002. "Maneuvering in the Complex Path from Genotype to Phenotype." *Science* 296 (5568): 701–3.

Strome, S., and W. B. Wood. 1983. "Generation of Asymmetry and Segregation of Germ-Line Granules in Early *C. elegans* Embryos." *Cell* 35: 15–25.

Sugimoto, K., H. Sugisaki, T. Okamoto, and M. Takanami. 1977. "Studies on Bacteriophage Fd DNA. IV. The Sequence of Messenger RNA for the Major Coat Protein Gene." *Journal of Molecular Biology* 111 (4): 487–507.

Suprynowicz, F. A., and L. Gerace. 1986. "A Fractionated Cell-Free System for Analysis of Prophase Nuclear Disassembly." *Journal of Cell Biology* 103 (6 Pt. 1): 2073–81.

Swan, D., H. Aviv, and P. Leder. 1972. "Purification and Properties of Biologically Active Messenger RNA for a Myeloma Light Chain." *Proceedings of the National Academy of Sciences* 69 (7): 1967–71.

Tanford, C. 1974. *The Hydrophobic Effect: Formation of Micelles and Biological Membranes*. New York: John Wiley & Sons.

Tashiro, Y., and P. Siekevitz. 1965a. "Localization on Hepatic Ribosomes of Protein Newly Synthesized In Vivo." *Journal of Molecular Biology* 11 (2): 166–73.

———. 1965b. "Ultracentrifugal Studies on the Dissociation of Hepatic Ribosomes." *Journal of Molecular Biology* 11 (2): 149–65.

Teich, M. 1973. "From Enchyme to Cytoskeleton: The Development of Ideas on the Chemical Organization of Living Matter." In *Changing Perspectives in the History of Science:*

Essays in Honour of Joseph Needham, edited by M. Teich and R. Young, 439–71. Boston: D. Reidel.

———. 1992. *A Documentary History of Biochemistry 1770–1940.* Rutherford, NJ: Fairleigh Dickinson University Press.

Tonegawa, S., and I. Baldi. 1973. "Electrophoretically Homogeneous Myeloma Light Chain mRNA and Its Translation In Vitro." *Biochemical and Biophysical Research Communications* 51 (1): 81–87.

Traube, M. 1867. "Experimente zur Theorie der Zellenbildung und Endosmose." *Archiv für Anatomie, Physiologie und wissenschaftliche Medicin:* 87–165.

Trosko, J. E., and H. C. Pitot. 2003. "In Memoriam: Professor Emeritus Van Rensselaer Potter II (1911–2001)." *Cancer Research* 63: 1724.

Ullu, E., and M. Melli. 1982. "Cloning and Characterization of cDNA Copies of the 7SL RNA of Hela Cells." *Nucleic Acids Research* 10: 2209–23.

Ullu, E., S. Murphy, and M. Melli. 1982. "7SL RNA Consists of a 140 Nucleotide Middle-Repetitive Sequence Inserted in an Alu Sequence." *Cell* 29: 195–202.

Van Bortle, K., and V. G. Corces. 2012. "Nuclear Organization and Genome Function." *Annual Review of Cell and Developmental Biology* 28 (1): 163–87.

Van den Berg, B., W. M. Clemons, I. Collinson, Y. Modis, E. Hartmann, S. C. Harrison, and T. A. Rapoport. 2004. "X-Ray Structure of a Protein-Conducting Channel." *Nature* 427 (6969): 36–44.

Veigl, S. J., O. Harman, and E. Lamm. 2020. "Friedrich Miescher's Discovery in the Historiography of Genetics: From Contamination to Confusion, from Nuclein to DNA." *Journal of the History of Biology* 53: 451–84.

Venkei, Z. G., and Y. M. Yamashita. 2018. "Emerging Mechanisms of Asymmetric Stem Cell Division." *Journal of Cell Biology* 217 (11): 3785–95.

Verworn, M. 1899. *General Physiology: An Outline of the Science of Life.* Translated by Frederic S. Lee. London: MacMillan and Co.

Voet, D., and J. G. Voet. 2004. *Biochemistry.* 3rd ed. New York: John Wiley & Sons.

von Heijne, G. 2006. "Membrane-Protein Topology." *Nature Reviews Molecular Cell Biology* 7 (12): 909–18.

von Heijne, G., and C. Blomberg. 1979. "Trans-Membrane Translocation of Proteins: The Direct Transfer Model." *European Journal of Biochemistry* 97 (1): 175–81.

von Plato, A., T. Vilimek, P. Filipkowski, and J. Wawrzyniak. 2013. *Opposition als Lebensform das andere Osteuropa.* Berlin: Lit.

Waechter, C. J., and W. J. Lennarz. 1976. "The Role of Polyprenol-Linked Sugars in Glycoprotein Synthesis." *Annual Review of Biochemistry* 45 (1): 95–112.

Wagner, G. P., and M. D. Laubichler. 2000. "Character Identification in Evolutionary Biology: The Role of the Organism." *Theory in Biosciences* 119: 20–40.

Wallach, D. F. H., and P. H. Zahler. 1966. "Protein Conformations in Cellular Membranes." *Proceedings of the National Academy of Sciences* 56: 1552–59.

Walter, P., and G. Blobel. 1980. "Purification of a Membrane-Associated Protein Complex Required for Protein Translocation across the Endoplasmic Reticulum." *Proceedings of the National Academy of Sciences* 77 (12): 7112–16.

———. 1981a. "Translocation of Proteins across the Endoplasmic Reticulum. III. Signal Recognition Protein (SRP) Causes Signal Sequence-Dependent and Site-Specific Arrest of Chain Elongation That Is Released by Microsomal Membranes." *Journal of Cell Biology* 91 (2 Pt. 1): 557–61.

———. 1981b. "Translocation of Proteins across the Endoplasmic Reticulum. II. Signal Recognition Protein (SRP) Mediates the Selective Binding to Microsomal Membranes of In-Vitro-Assembled Polysomes Synthesizing Secretory Protein." *Journal of Cell Biology* 91 (2 Pt. 1): 551–56.

———. 1982. "Signal Recognition Particle Contains a 7S Rna Essential for Protein Translocation across the Endoplasmic Reticulum." *Nature* 299 (5885): 691–98.

Walter, P., R. Gilmore, and G. Blobel. 1984. "Protein Translocation across the Endoplasmic Reticulum." *Cell* 38 (1): 5–8.

Walter, P., I. Ibrahimi, and G. Blobel. 1981. "Translocation of Proteins across the Endoplasmic Reticulum. I. Signal Recognition Protein (SRP) Binds to In-Vitro-Assembled Polysomes Synthesizing Secretory Protein." *Journal of Cell Biology* 91 (2 Pt. 1): 545–50.

Walter, P., R. Jackson, M. Marcus, V. Lingappa, and G. Blobel. 1979. "Tryptic Dissection and Reconstitution of Translocation Activity for Nascent Presecretory Proteins across Microsomal Membranes." *Proceedings of the National Academy of Sciences* 76 (4): 1795–99.

Warren, G. 1993. "Membrane Partitioning during Cell Division." *Annual Review of Biochemistry* 62 (1): 323–48.

Warren, G., and B. Dobberstein. 1978a. "Protein Transfer across Microsomal Membranes Reassembled from Separated Membrane Components." *Nature* 273 (5663): 569–71.

———. 1978b. "Protein Transfer across Microsomal Membranes: Functional Reassembly from Separated Membrane Components." *Hoppe-Seylers Zeitschrift für physiologische Chemie* 359: 335.

Watanabe, M., and G. Blobel. 1993. "SecA Protein Is Required for Translocation of a Model Precursor Protein into Inverted Vesicles of *Escherichia coli* Plasma Membrane." *Proceedings of the National Academy of Sciences* 90 (19): 9011–15.

Watanabe, M., C. Nicchitta, and G. Blobel. 1990. "Reconstitution of Protein Translocation from Detergent-Solubilized *Escherichia coli* Inverted Vesicles: PrlA Protein-Deficient Vesicles Efficiently Translocate Precursor Proteins." *Proceedings of the National Academy of Sciences* 87 (5): 1960–64.

Waters, M., and G. Blobel. 1986. "Secretory Protein Translocation in a Yeast Cell-Free System Can Occur Posttranslationally and Requires ATP Hydrolysis." *Journal of Cell Biology* 102 (5): 1543–50.

Watson, J. D. 1965. *Molecular Biology of the Gene*. New York: W. A. Benjamin.

———. 1992. "A Personal View of the Project." In *The Code of Codes*, edited by D. Kevles and L. Hood, 164–73. Cambridge, MA: Harvard University Press.

———. 2007. "Growing up in the Phage Group." In *Phage and the Origins of Molecular Biology*, edited by J. Cairns, G. Stent and J. D. Watson, 239–45. Cold Spring Harbor, NY: Cold Spring Harbor Laboratory Press.

Watson, J. D., and F. H. C. Crick. 1953a. "Genetical Implications of the Structure of Deoxyribonucleic Acid." *Nature* 171 (4361): 964–67.

———. 1953b. "Molecular Structure of Nucleic Acids; a Structure for Deoxyribose Nucleic Acid." *Nature* 171 (4356): 737–38.

Watson, M. L. 1955. "The Nuclear Envelope: Its Structure and Relation to Cytoplasmic Membranes." *Journal Biophysical and Biochemical Cytology* 1 (3): 257–70.

Weber, M. 2005. *Philosophy of Experimental Biology*. Cambridge: Cambridge University Press.

Weiss, S. B., G. Acs, and F. Lipmann. 1958. "Amino Acid Incorporation in Pigeon Pancreas Fractions." *Proceedings of the National Academy of Sciences* 44: 189–96.

Wickner, W. 1979. "The Assembly of Proteins into Biological Membranes: The Membrane Trigger Hypothesis." *Annual Review of Biochemistry* 48: 23–45.

———. 1988. "Mechanisms of Membrane Assembly: General Lessons from the Study of M13 Coat Protein and *Escherichia coli* Leader Peptidase." *Biochemistry* 27 (4): 1081–86.

Wickner, W., K. Ito, G. Mandel, M. Bates, M. Nokelainen, and C. Zwizinski. 1980. "The Three Lives of M13 Coat Protein: A Virion Capsid, an Integral Membrane Protein, and a Soluble Cytoplasmic Proprotein." *Annals of the New York Academy of Sciences* 343 (1): 384–90.

Wickner, W., G. Mandel, C. Zwizinski, M. Bates, and T. Killick. 1978. "Synthesis of Phage M13 Coat Protein and Its Assembly into Membranes In Vitro." *Proceedings of the National Academy of Sciences* 75 (4): 1754–58.

Wiedmann, M., D. Görlich, E. Hartmann, T. V. Kurzchalia, and T. A. Rapoport. 1989. "Photocrosslinking Demonstrates Proximity of a 34 Kda Membrane Protein to Different Portions of Preprolactin During Translocation through the Endoplasmic Reticulum." *FEBS Letters* 257 (2): 263–68.

Wiedmann, M., T. V. Kurzchalia, E. Hartmann, and T. A. Rapoport. 1987. "A Signal Sequence Receptor in the Endoplasmic Reticulum Membrane." *Nature* 328 (6133): 830–33.

Wieschaus, E. 2016. "Positional Information and Cell Fate Determination in the Early Drosophila Embryo." *Current Topics in Developmental Biology* 117: 567–79.

Wilson, E. B. 1896. *The Cell in Development and Inheritance*. New York: Macmillan,

———. 1899. "The Structure of Protoplasm." *Science* 10 (237): 33–45.

———. [1925] 1928. *The Cell in Development and Heredity*. 3rd ed. with Corrections. New York: Macmillan.

Wimsatt, W. C. 1972. "Complexity and Organization." *PSA: Proceedings of the Biennial Meeting of the Philosophy of Science Association* 1972: 67–86.

———. 1997. "Aggregativity: Reductive Heuristics for Finding Emergence." *Philosophy of Science* 64 (Supplement): S372–S384.

———. 2006. "Reductionism and Its Heuristics: Making Methodological Reductionism Honest." *Synthese* 151 (3): 445–75.

———. 2007. *Re-Engineering Philosophy for Limited Beings*. Cambridge, MA: Harvard University Press.

Wolfe, P. B., M. Rice, and W. Wickner. 1985. "Effects of Two Sec Genes on Protein Assembly into the Plasma Membrane of *Escherichia coli*." *Journal of Biological Chemistry* 260 (3): 1836–41.

Wolpert, L. 2016. "Positional Information and Pattern Formation." *Current Topics in Developmental Biology* 117: 597–608.

Wolpert, L., C. Tickle, and A. Martinez-Arias. 2015. *Principles of Development*. 5th ed. Oxford: Oxford University Press.

Worliczek, H. L. 2020. "Wege zu einer molekularisierten Bildgebung: Eine Geschichte der Immunofluoreszenzmikroskopie als visuellem Erkenntnisinstrument der modernen Zellbiologie 1959–1980." PhD dissertation, University of Vienna.

Wulf, A. 2015. *The Invention of Nature: Alexander Von Humbolt's New World*. New York: Knopf.

Yarmolinsky, M. B., and G. De La Haba. 1959. "Inhibition by Puromycin of Amino Acid Incorporation into Protein." *Proceedings of the National Academy of Sciences* 45: 1721–29.

Yu, Y. H., Y. Y. Zhang, D. D. Sabatini, and G. Kreibich. 1989. "Reconstitution of Translocation-Competent Membrane Vesicles from Detergent-Solubilized Dog Pancreas Rough Microsomes." *Proceedings of the National Academy of Sciences* 86 (24): 9931–35.

Zamecnik, P. C. 1950. "The Use of Labeled Amino Acids in the Study of the Protein Metabolism of Normal and Malignant Tissues: A Review." *Cancer Research* 10 (11): 659–67.

Zimmerman, D. L., and P. Walter. 1990. "Reconstitution of Protein Translocation Activity from Partially Solubilized Microsomal Vesicles." *Journal of Biological Chemistry* 265 (7): 4048–53.

Zwizinski, C., and W. Wickner. 1980. "Purification and Characterization of Leader (Signal) Peptidase from *Escherichia coli*." *Journal of Biological Chemistry* 255 (16): 7973–77.

INDEX

Page numbers in italics refer to figures.

Lightning Source UK Ltd.
Milton Keynes UK
UKHW010059260422
402038UK00002B/47

9 780226 819235